Electronic Waste

Impact on Health, Animals, and the Environment

Edited by
Srijan Goswami
Moharana Choudhury
Sangita Agarwal

CRC Press
Taylor & Francis Group
Boca Raton London New York

CRC Press is an imprint of the
Taylor & Francis Group, an **informa** business

Designed cover image: Shutterstock

First edition published 2025
by CRC Press
2385 NW Executive Center Drive, Suite 320, Boca Raton FL 33431

and by CRC Press
4 Park Square, Milton Park, Abingdon, Oxon, OX14 4RN

CRC Press is an imprint of Taylor & Francis Group, LLC

ISBN: 978-1-032-94905-5 (hbk)
ISBN: 978-1-032-94906-2 (pbk)
SBN: 978-1-003-58231-1 (ebk)

)I: 10.1201/9781003582311

set in Times
Technologies India Pvt Ltd (Straive)

Contents

Foreword

I extend this letter to the consideration of the editorial team to support the development of the book entitled "**Electronic Wastes: Impact on Health, Animals, and the Environment**" whose publication is planned for CRC Press which represents an excellent opportunity to establish challenges and opportunities for understanding the impact of e-waste on ecosystem and individuals. As our society becomes increasingly dependent on electronic devices, the management of e-waste has emerged as a pressing global concern with far-reaching implications. The book, edited by Srijan Goswami, Moharana Choudhury, and Sangita Agarwal, brings together contributions from leading experts in the field to provide a comprehensive overview of the multifaceted challenges posed by e-waste, including its toxic components, the risks it poses to public health, the disruption it causes to biodiversity, and the environmental pollution it generates. This publication's diverse perspectives and insights make it a valuable resource for researchers, academicians, and even a commoner interested in protecting and saving the environment.

The book is divided into sixparts. The first is called **Understanding E-waste, Sources, and Composition**, and it contains four chapters oriented toward judging how e-waste is a global hazard. The second one deals with the **Impact of e-waste on the ecosystem**, which integrates three chapters that encompass the Effect of e-waste on the atmosphere, hydrosphere, and lithosphere, especially soil. The third one addresses the **Impact of e-waste on biodiversity, insects, microorganisms, plants, crops, food safety, and farming practices**, which integrates six chapters oriented toward the toxic chemicals emanating from e-waste that can disrupt the population and diversity. Agrarian crops grown on polluted soil are raising the alarm about food safety. The fourth one highlights the **Effect of e-waste on human health**, which amalgamates four chapters directed towards the health concerns related to e-waste exposure, including maternal and child health, endocrine disruption, and effects on the nervous system. The fifth one deals with **Managing e-waste**, which incorporates eight chapters, including innovative e-waste recycling technologies, circular economy practices, strategies for remediation of e- waste contaminated soil, occupational safety measures, and challenges in e-waste management. The last partdeals with **Case studies and legal implications of e-waste management**. By compiling expertise and insights from multidisciplinary perspectives, this book seeks to contribute to the growing knowledge on e-waste and inspire concrete actions towards sustainable electronic waste management. Given the above, I offer many congratulations to the editors for such a vital project and to all the collaborators, who are undoubtedly crucial to humanity.

IPGP | Université Paris Cité

Eric D. van Hullebusch
Full Professor in Biogeochemistry of
Engineered Ecosystems
Institut de Physique du Globe de Paris /
Université Paris Cité
1, rue Jussieu
75238 Paris cedex 05 France

Preface

The rapid advancement of technology has profoundly transformed our lives, bringing unprecedented convenience and innovations. However, this technological revolution has also introduced complex challenges particularly in the form of e-wastes. "Electronic Waste: Impact on Health, Animals, and Environment" seeks to address the growing concern of e-wastes, an issue that threatens the delicate balance of our ecosystem and the well-being of all living beings. E-waste encompasses discarded electrical and electronic devices ranging from smartphones and computers to household appliances and industrial machinery. The exponential rise in the volume of e-waste is driven by the rapid obsolescence of electronic products and their ever increasing consumer demands for the latest technological advancements. This book delves into multifaceted impact of e-waste, highlighting its hazardous components, the mechanisms by which these pollutants infiltrate our environment, and their consequent effects on human, animal, and environmental health. The chapters in this book are authored by experts in the domain of environmental sciences, toxicology, public health, and waste management. Each chapter provides a comprehensive analysis of specific aspects of WEEE. Additionally, the book explores innovative strategies for managing and mitigating the effects of e-waste, emphasizing the importance of sustainable practices, regulatory frameworks, and technological innovations in recycling and waste reduction. One of the key aims of this book is to raise awareness about the urgent need for responsible e-waste management. The improper disposal and inadequate recycling of electronic devices lead to the release of harmful substances such as lead, mercury, cadmium, and brominated flame retardants. These pollutants pose significant risks, including neurological damage, reproductive harm, and endocrine disruption in human and wildlife. Furthermore, the environmental contamination resulting from e-waste can lead to the degradation of soil, water, and air quality, with far reaching consequences of biodiversity and ecosystem services. By presenting a holistic view of the WEEE crisis, this book seeks to inform and inspire a wide audience, including policymakers, researchers, industry stakeholders, and the general public. It advocates for a collaborative approach to addressing e-waste, stressing the need for global cooperation and integration of scientific research into policy and practice. In conclusion "Electronic Waste: Impact on Health, Animals, and Environment" serves as a clarion call for urgent action. It underscores the necessity of transitioning towards more sustainable and circular economies, where electronic products are designed for longevity, repairability, and recyclability. We hope this book will be a valuable resource in the global effort to mitigate the threats posed by e-waste, safeguarding the health of our planet and all its inhabitants for future generations.

Srijan Goswami
Moharana Choudhury
Sangita Agarwal

The Editors

Author and Editor of several Medical Science Books (PG-Research category), **Dr. Srijan Goswami** is pursuing a Ph.D. in Biotechnology from Department of Biotechnology, University of North Bengal, India. He is a Physician of Biochemic System of Medicine and also holds a specialization in Medical Biotechnology and Clinical Nutrition. He received his professional training on Community Nutrition and Child Care, Medical Microbiology, Medical Biochemistry, and Diabetes Counselling from reputed medical colleges in Kolkata, India. He is experienced in teaching at a diverse range of Paramedical and Lifesciences Departments of reputed colleges and institutions. He is presently associated as Research Coordinator, Biomedical Research and Allied Sciences Division, Voice of Environment, Assam, India.

Dr. Goswami has around 11 years of experience in the field of teaching and research and has published over 4 books (+8 in the pipeline) and ~40 research papers in national and international journals. He has independently and successfully guided around ten undergraduate and postgraduate students in their research work and completion and publications in the past 3 years.

Dr. Goswami has received four Letters of Commendation for Global Community Services and Certificate of Honor for his contribution in the field of Science and Knowledge from the *International Journal of Agricultural Research, Sustainability and Food Sufficiency* (IJARSFS), *International Journal of Advances in Medical Sciences and Biotechnology* (IJAMSB), and *International Journal of Health, Safety and Environment* (IJHSE) in 2019, 2020, 2021, and 2023. The *Current Research in Nutrition and Food Science* (Enviro Research Group) and *Academia Scholarly Journals* have awarded him the Certificate of Excellence as a Reviewer in 2020, 2021, and 2023. He is actively associated with the Editorial Review Board of 15 international journals.

Moharana Choudhury is an Indian researcher and environmentalist. He is a Research Coordinator in the Environmental Research and Management Division, "Voice of Environment" (A Scientific and Environmental Research Organization), Guwahati, Assam, India. He has completed his education as an M Sc, PGD-GIS, PGD-EM, and PG-CGARD from Gurukula Kangri Deemed to be University, Haridwar, Uttrakhand, India, North Eastern Hill University (NEHU), Shillong Meghalaya, India, Tezpur University, Assam, India and National Institute of Rural Development (NIRDPR, Government of India) Hyderabad, India respectively. Choudhury has worked on various scientific projects in various Government departments and has more than 13 years of research and working experience in environmental research and services. He began his career as a Research Intern at the National Institute of Disaster Management in New Delhi, followed by the North Eastern Space Application Center (Department of Space, Government of India/ISRO Unit) as a Junior Research Fellow. Choudhury has authored or edited eight (8) books with internationally reputed publication houses and more than thirty-seven (37) articles in national and internationally reputed journals, proceedings, and more than (30) book chapters. He has written many articles and scientific reports in popular magazines. He is an editor for prestigious international publishing houses such as Elsevier, Royal Society of Chemistry (RSC), Springer, Wiley, CRC Press/Routledge group, Apple Academic Press, Cambridge Scholar Publishing, etc. Choudhury is working on various environmental conservation activities in Assam and the country. He has worked on several initiatives to integrate environmentally sustainable projects for iconic heritage places, wetlands, and ecotourism spots in Assam, India, especially for cleanliness and beating single-use plastic pollution missions. Choudhury contributes to various environmental research, management, social media mass awareness, and research studies. He is also well known in India for his efforts to implement environmental education and safeguard the privileges of environmental degree holders in India.

Dr. Sangita Agarwal works as an Associate Professor, Department of Applied Science, RCC Institute of Information Technology, Kolkata, West Bengal. She received her Bachelor of Science in Chemistry in 1992 from University of Calcutta, West Bengal and Master of Science in Biochemistry in 1994 and then obtained her Ph.D. degree as a UGC-GATE qualified fellow in 2001 from Jadavpur University, West Bengal. She was a project participant in the Indo-Hungarian Exchange program under DST, New Delhi in 1998–99. Dr. Agarwal has 8 years of full-time post-doctoral research experience in various capacities. She also worked as a Visiting Academic in Monash University, Melbourne, Victoria during April-September 2008, under the Endeavour fellowship, sponsored by the Govt. of Australia. She is a recipient of DST Women Scientist fellowship, New Delhi, Govt. of India, during 2009–2012.

Dr. Agarwal has been teaching for more than 14 years and has been involved in research activities for over 3 decades. She has more than 50 scientific publications to her credit, 3 books (+3 in the pipeline), and more than 15 book chapters in the sphere of Health, Toxicity, Pharmacokinetics, Environmental Science, and Environmental Biology. She is a Life Member of the Indian Chemical Society. She is an active member of NSS in her institute and has been involved in a number of events to raise awareness and consciousness of students related to sustainable development. Dr. Agarwal has delivered invited talks at various institutes on topics related to environmental pollution like plastic pollution, e-waste management, green IT, and water conservation.

Contributors

Abdul Rafey
Department of Chemical Engineering
Indian Institute of Technology Delhi
New Delhi, India

Abhay Vasantrao Waware
Jawaharlal Nehru Engineering College
MGM University
Aurangabad, India

Aditya Kumar
National Institute of Mental Health
Klecany, Czech Republic

America Metzdorff
Facultad de Ciencias
Instituto de Ciencias Químicas, Independencia 641
Universidad Austral de Chile
Valdivia, Chile

Ananda Sahithi Taduru
St. Ann's College for Women
Hyderabad, India

Anil Pappachan
P2 Basic Seed Farm, National Silkworm Seed Organization (NSSO), Central Silk Board
Ministry of Textiles, Government of India
Yelagiri Hills, India

Anupa Ruwali
St. Joseph's Health Centre
Toronto, Ontario, Canada

Anwesha Gohain
Department of Botany
Arunachal University of Studies
Namsai, India

Apurba Ratan Ghosh
Department of Environmental Science
The University of Burdwan
Bardhaman, India

Ardra Lekshmi A
Department of Chemistry
School of Physical Sciences
Amrita Vishwa Vidyapeetham, Amritapuri Campus
Kollam, India

Arvind Laxmikant Chel
Department of Mechanical Engineering
Jawaharlal Nehru Engineering College
MGM University
Aurangabad, India

Ashi Ramavat
School of Healthcare
Rishihood University
Sonipat, India

Avinash Sharma
Department of Agricultural Sciences
Arunachal University of Studies
Namsai, India

Avnish Chauhan
Department of Environmental Sciences
Graphic Era Hill University
Dehradun, India

Ayan Chatterjee
School of Health Science and Technology
Medhavi Skills University
Sikkim, India

Banani Ray Chowdhury
Bengal Institute of Technology
Kolkata, India
Baranagar Baghajatin Social Welfare Organization
Baranagar, India

Basem E. Kestha
Department of Chemistry
Faculty of Science
Tanta University
Tanta, Egypt

Bindu Subhadra
Department of Biological and Earth Sciences
Arkansas Tech University
Russellville, Arkansas, United States

Charu Arora
Department of Chemistry
Guru Ghasidas Vishwavidyalaya
Bilaspur, India

Chaudhery Mustansar Hussain
Department of Chemistry and EVSC
New Jersey Institute of Technology
Newark, New Jersey, United States

Deepthi Achuthavarier
Modelling Program Division, Office of Science and Technology Integration
National Weather Service, NOAA
College Park, Maryland, United States

Dharitri Borah
Department of Environmental Science
Arunachal University of Studies
Namsai, India

Dhiraj Kumar
Department of Zoology
Hansraj College
University of Delhi
New Delhi, India

Dibya Das
Department of Pharmaceutical Technology
JIS University, Agarpara Campus,
Kolkata, India

Dipti Bharti
Department of Applied Science & Humanities
Darbhanga College of Engineering
Darbhanga, India

Faisal Zia Siddiqui
Department of Chemical Engineering
Faculty of Engineering & Technology
Aligarh Muslim University
Aligarh, India

Farhana Rahman
Institute for Sustainable Futures
University of Technology Sydney (UTS)
Ultimo, New South Wales, Australia

Geetanjali Kaushik
Department of Civil Engineering
Hi Tech Institute of Technology
Aurangabad, India

Govind Dhage
Hi Tech Institute of Technology
Aurangabad, India

Gourav Samajdar
Department of Pharmaceutical Technology
JIS University, Agarpara Campus
Agarpara, Kolkata

Heeba Fatima
St. Ann's College for Women
Hyderabad, India

Himangshu Sekhar Maji
Department of Pharmaceutical Technology
JIS University, Agarpara Campus
Kolkata, India

Induja Samanta
Bengal Institute of Technology
Kolkata, India

Jagadeesh Chandra Bose K
University Institute of Biotechnology
Chandigarh University
Mohali, India

Jyoti Chowdhry
Maharishi Dayanand University
Rohtak, India

Jyoti Sarwan
University Institute of Biotechnology
Chandigarh University
Mohali, India

Kalyani S. Kondekar
Hi-Tech Institute of Technology
Aurangabad, India

Kamidi Rahul
Central Sericultural Research & Training Institute, Central Silk Board
Ministry of Textiles, Government of India
Berhampore, India
Industrial Insect and Sericulture Division
National Institute of Agricultural Sciences, RDA
Wanju-gun, Republic of Korea

Katsuhiko Ito
Laboratory of Sericultural Science
Department of Science of Biological Production
Graduate School of Agriculture
Tokyo University of Agriculture and Technology
Tokyo, Japan

Khalid Stanikzai
Department of Forestry and Natural Resource
Kabul University
Kabul, Afghanistan

Kusumlata Gayakwad
Department of Biotechnology
Guru Ghasidas Vishwavidyalaya
Bilaspur, India

Lakshika
University Institute of Biotechnology
Chandigarh University
Mohali, India

Lekshmi S
Department of Chemistry
School of Physical Sciences
Amrita Vishwa Vidyapeetham, Amritapuri Campus
Kerala, India

Laxmi Kant Bhardwaj
Amity Institute of Environmental, Toxicology, Safety and Management (AIETSM)
Amity University
Noida, India

Manab Deb Adhikari
Department of Biotechnology
University of North Bengal
Darjeeling, India

Manoj Chandra Garg
Amity Institute of Environmental Science (AIES)
Amity University
Noida, India

Mansi Chawla
The Energy and Resources Institute (TERI)
New Delhi, India

Meenakshi Sharma
Department of Chemistry
IEC University
Baddi, India

Megha Santra
Department of Environmental Science
The University of Burdwan
Burdwan, India

Monjeet Sonowal
Department of Zoology
Arunachal University of Studies
Namsai, India

Monruskin M. Calma
President Calma Expeditions Inc.
Brockton, Massachusetts, United States

Muhammad Umair Chaudhary
Islamia University
Bahawalpur, Pakistan

Mukhtar Iderawumi Abdulraheem
Henan Agricultural University
Zhengzhou, China

Navamallika Gogoi
Faculty of Science
Sadiya College
Tinsukia, India

Nguyen Tran Hai Bang
CUU Long Delta Rice Research Institute
Can Tho, Vietnam

Palas Samanta
Department of Environmental Sciences
Sukanta Mahavidyalaya
University of North Bengal
Dhupguri, India

Pallabi Roy Chowdhury
Baranagar Baghajatin Social Welfare Organization
Baranagar, India

Pankaj Kanti Jodder
Department of Geography and Earth Science
University of North Carolina at Charlotte
Charlotte, North Carolina, United States

Parna Chatterjee
School of Healthcare
Rishihood University
Sonipat, India

Pooja Makwana
Central Sericultural Research & Training Institute, Central Silk Board
Ministry of Textiles, Government of India
Berhampore, India

Poonam Singhal
St. Ann's College for Women
Hyderabad, India

Prangya Rath
Amity Institute of Environmental Sciences (AIES)
Amity University
Noida, India

Pratapa M G
ICAR-Central Institute of Fisheries Education
Mumbai, India

Preeta Bose
Department of Pharmaceutical Technology
JIS University, Agarpara Campus
Agarpara, Kolkat

Raghunath Satpathy
School of Biotechnology
Gangadhar Meher University
Sambalpur, India

Rajat Pratap Singh
Department of Biotechnology
Guru Ghasidas Vishwavidyalaya
Bilaspur, India

Rajesh Dhankhar
Department of Environmental Science
Maharshi Dayanand University
Rohtak, India

Rakesh Choudhary
Department of Civil Engineering
National Institute of Technology Delhi
New Delhi, India

Rakshitha Rani N
Department of Microbiology
JSS Academy of Higher Education and Research
Mysuru, India

Rani Jha
Symbiosis Open Education Society
Pune, India

Richa Chaurasia
University Institute of Biotechnology
Chandigarh University
Mohali, India

Richa Saxena
Department of Biotechnology
Invertis University
Bareilly, India

Richard Baker
School of Chemistry
University of St Andrews
St. Andrews, United Kingdom

Rishikesh S. Aher
Hi-Tech Institute of Technology
Aurangabad, India

Rwitabrata Mallick
Department of Environmental Science
Amity School of Life Science
Amity University
Gwalior, India

Sabina Pradhan
Department of Botany
Banwarilal Bhalotia College
Asansol, India

Sandeep Poddar
Lincoln University College
Selangor, Malaysia

Sanjeev Kumar
Department of Chemistry
Chaman Lal Mahavidyalaya, Landhaura
Haridwar, India

Saptaswa Chakraborty
Bengal Institute of Technology
Kolkata, India

Saufishan T A
Department of Chemical Engineering
Faculty of Engineering & Technology
Aligarh Muslim University
Aligarh, India

Sayari Bhunia
March Friechel's Lab
Pharmakologische Institut
Medizinischen Fakultät
Heidelberg, Germany

heetal Kumari
uity Institute of Environmental
cience (AIES)
University
India

Narwal
t of Environmental Science
ayanand University

borty
earcher

iology
ience College

ogy

Smitha Chandran S
Department of Chemistry
School of Physical Sciences
Amrita Vishwa Vidyapeetham, Amritapuri Campus
Kollam, India

Sobia Usman
Department of Media Sciences
SZABIST University
Karachi, Pakistan

Sougata Karjee
Department of Mathematics
Sukanta Mahavidyalaya
University of North Bengal
West Bengal, India

Soutrik Bhattacharya
Bengal Institute of Technology
Kolkata, India

Sucheta Mandal
Department of Botany
Banwarilal Bhalotia College
Asansol, India

Sudipta Saha
Health Research Board
Dublin, Ireland

Sukhendu Dey
Department of Environmental Science
The University of Burdwan
Burdwan, India

Susmita Dhar
Department of Law
Sister Nivedita University
Kolkata, India

Talluri Rameshwari K R
Department of Microbiology
JSS Academy of Higher Education and Research
Mysuru, India

Tirtha Pratim Mandal
College of Staten Island
City University of New York
Staten Island, New York, United States

Ushmita Gupta Bakshi
Hahnemann Homoeo Pharmaceutical Private Limited
Sodepur, India

Vaishali Soni
PGIMER-Post Graduate Government Institute of Medical Education & Research
Chandigarh, India

Vatsala Soni
PEC-Punjab Engineering College
Chandigarh, India

Vijay Kumar Srivastava
The Maharaja Sayajirao University of Baroda
Vadodara, India

Vishnu Rajput
Head of Soil Health Laboratory
Southern Federal University
Rostov-on-Don, Russia

Zhang Shi
Department of Agricultural Sciences
Senyang Agricultural University
Shenyang, China

other stakeholders have been collaborating to develop global policies that address specific challenges associated with the trade of WEEE and used equipment under the Basel Convention. Furthermore, on March 21, 2018, seven United Nations organizations signed a Letter of Intent during the World Summit on the Information Society (WSIS) Forum. This agreement aims to facilitate increased cooperation in developing a UN E-Waste Coalition to manage electronic waste. The signatories have committed to enhancing collaboration, establishing alliances, and assisting Member States in addressing the global WEEE issue in alignment with their objectives. The International Labour Organization (ILO), International Trade Centre (ITC), International Telecommunications Union (ITU), United Nations Environment Programme (UNEP), United Nations University (UNU), United Nations Human Settlement (UN-Habitat), United Nations Industrial Development Organization (UNIDO), United Nations Institute for Training and Research (UNITAR), World Health Organization (WHO), and the Secretariat of Basel, Rotterdam, and Stockholm Conventions, most of which are headquartered in Geneva, are essential components of the coalition.

The alliance, directed by the Secretariat of the United Nations Environment Management Group (UNEMG) until October 31, 2020, receives support from the World Economic Forum and the World Business Council for Sustainable Development (WBCSD). The provisional secretariat of the coalition is currently located at UNEP. The International Electrotechnical Commission (IEC), founded in 1906, is the foremost global body responsible for advancing and disseminating international standards pertaining to electrical, electronic, and associated technologies, collectively known as "electrotechnology." The IEC platform allows companies, governments, and industries to convene, deliberate, and establish essential international standards. All of the worldwide standards of the IEC are based on consensus and represent the needs of important stakeholders from all participating nations in the IEC's activities (The Growing Environmental Risks of E-Waste, n.d.).

1.3 WEEE AND DEGRADATION OF ENVIRONMENT

Environment is the one and most important aspect of our existence. All living organisms are regulated and nurtured by the environment they are living in. Air, water, food, and other substances are all provided by the environment for every living organisms. The global increase in the amount of E-Wastes is potentially degrading the environment that sustains us. Improper disposal of electronics leads to the discharge of harmful compounds into the air, soil, and water, negatively impacting creatures' health. The process of dismantling, shredding, or melting E-Waste items are releases hazardous chemicals and particles into the environment, leading to air, water, and soil contamination. These actions result in permanent harm to the ecosystems. Improper disposal of e-waste in illicit dumping sites or regular landfills can directly infiltrate heavy metals and flame retardants into the soil (Paul et al., 2019). This has the potential to pollute the underlying groundwater or crops that may be cultivated nearby or in the future. Heavy metals in the soil increase the likelihood of crops absorbing these toxins, leading to bioaccumulation and a decline in farming output. Soil contamination is affected by temperature, soil type, pH level, and soil makeup. If these contaminants persist in the soil for a long time, they might cause damage to plants and soil microbes. Animals and other creatures that rely on nature for survival will consume the impacted plants, leading to health problems. Heavy metals in electronic trash, such as mercury, lithium, lead, and barium, can seep deeper into the ground and contaminate groundwater sources such as ponds, streams, rivers, and lakes. This contamination risks animals, plants, and human consumption, making it dangerous. The detrimental effects can result in the devastation of ecosystems, depletion of biodiversity, and adverse genetic alterations in creatures inhabiting freshwater and marine environments (*What Is e-Waste*, n.d.) (*The Growing Environmental Risks of E-Waste*, n.d.), (*E-Waste & Its Negative Effects on the Environment | Elytus*, n.d.).

1.3.1 Impact on Air, Water, and Soil Quality

E-waste recycling processes, release particulate matter into the atmosphere leading to respiratory problems, cardiovascular diseases, and systemic inflammation. Additionally, it contributes to air pollution and climate change (Suthar et al., 2018). Burning of electronic components releases toxic gases, including dioxins, furans, and polycyclic aromatic hydrocarbons (PAHs), as well as VOCs such as benzene and toluene. Human, animals and plants living in close proximity to pollutants generated from E-wastes possess the risk of acute and long-term health problems like cardiopulmonary complications, neuromuscular abnormalities Communities in close proximity to these pollutants may be at risk for both acute and long-term health problems, such as respiratory irritation, neurological abnormalities, and an elevated risk of cancer (Song et al., 2014). Lead, mercury, cadmium, and chromium are just a few of the heavy elements found in e-waste that can move into water bodies from poorly managed dump sites or from recycling units. Due to their persistence in the environment, these metals have the ability to build up in sediments and aquatic life, which can cause bioaccumulation and biomagnification in the food chain. By contaminating fish and seafood, polluted water sources endanger both human health and aquatic ecosystems (Baldé et al., 2017). Persistent organic pollutants (POPs), such as polychlorinated biphenyls (PCBs), brominated flame retardants (BFRs), and phthalates, present in e-waste can also leach into water bodies, increasing risks to water quality and aquatic life. These chemicals are resistant to degradation and can persist in the environment for long periods, leading to bioaccumulation in aquatic organisms and potential toxicity to humans and wildlife (Liu et al., 2018). Improper disposal of e-waste in landfills or open dumping sites can result in the release of heavy metals into the soil through leaching and runoff. These metals can accumulate in soil, posing risks to plant growth, microbial activity, and ecosystem health. Soil contamination with heavy metals from e-waste can also impact agricultural productivity and food safety, as crops grown in contaminated soil may accumulate toxic substances (Xu et al., 2019). Acidic leachates from e-waste can contribute to soil acidification, leading to changes in soil pH and nutrient availability. Soil degradation resulting from e-waste contamination can impair soil fertility, water retention, and microbial activity, affecting ecosystem functioning and biodiversity. Remediation of e-waste-contaminated soils presents significant challenges due to the persistence and mobility of contaminants (Islam et al., 2020; Mondal et al., 2023). Implementation and enforcement of stringent regulations for e-waste management are essential to minimize environmental contamination and protect air, water, and soil quality. Governments should establish policies for proper disposal, recycling, and treatment of e-waste, including bans on hazardous substances and extended producer responsibility (EPR) schemes. Promoting environmentally sound e-waste recycling practices, such as formal recycling facilities equipped with proper pollution control measures, can help minimize emissions of toxic pollutants into the environment. Encouraging the reuse and refurbishment of electronic devices and components can also reduce the generation of e-waste and alleviate pressure on natural resources.

1.3.2 Impact on Biodiversity and Wildlife

The rapid increase of electronic waste (e-waste) presents a significant threat to global ecosystems and biodiversity. E-waste, comprised of discarded electronic devices and components, contains a complex mixture of hazardous substances that can leach into the environment, posing risks to wildlife and biodiversity. This section assesses the impact of e-waste on wildlife and biodiversity, examining the mechanisms of exposure, ecological consequences, and emerging mitigation strategies. Wildlife species are directly exposed to e-waste contaminants through scavenging, nesting, or inhabiting contaminated areas. Birds, mammals, and reptiles may ingest e-waste particles or use them in nest construction, leading to direct exposure to toxic substances. Additionally, wildlife may suffer injuries or fatalities from entanglement in electronic components or electrocution from exposed wires (Rahman et al., 2022). E-waste pollution indirectly impacts wildlife through the contamination of air, water, and soil. Pollutants released from e-waste recycling sites or leachate from landfills can enter ecosystems, leading to bioaccumulation in food chains and exposure of wildlife to toxic substances. Aquatic organisms, such as fish and amphibians, are particularly vulnerable to e-waste pollution due to their direct reliance on contaminated water sources (Zhang et al., 2021a).

growing concern for farm animals, as they may be exposed to hazardous substances through contaminated air, water, soil, and feed. E-waste contains a variety of toxic chemicals, including heavy metals, flame retardants, and plasticizers, which can accumulate in the environment and enter the food chain. Farm animals may be exposed to e-waste pollutants through contaminated feedstuffs. Feed ingredients sourced from areas with e-waste recycling or disposal activities may contain elevated levels of heavy metals or organic contaminants. Animals consuming contaminated feed may accumulate these pollutants in their tissues, leading to potential health risks (Nnorom et al., 2021). Livestock may also be exposed to e-waste contaminants through contaminated water sources or grazing on polluted pasturelands. Leachates from e-waste disposal sites or runoff from recycling facilities can contaminate surface water and groundwater, as well as soil. Animals consuming or coming into contact with contaminated water or soil may ingest or absorb hazardous substances (Gadde et al., 2020). Exposure to e-waste pollutants can lead to various health problems in farm animals, including toxicity, organ damage, and reproductive disorders. Heavy metals such as lead, cadmium, and mercury can accumulate in animal tissues, affecting the liver, kidneys, and nervous system. Organic contaminants like polychlorinated biphenyls (PCBs) and brominated flame retardants (BFRs) may disrupt hormonal balance and immune function, increasing susceptibility to diseases (Ogunbanwo et al., 2022). Chronic exposure to e-waste pollutants can impair animal growth, reproduction, and milk or egg production. Heavy metal toxicity, for example, may lead to reduced feed intake, growth retardation, and poor reproductive performance in livestock. Contaminants in animal-derived products, such as meat, milk, and eggs, may also affect their quality and marketability, posing economic challenges for farmers (Chen et al., 2021a, 2021b). E-waste pollutants can bioaccumulate in animal tissues and transfer to humans through the consumption of contaminated animal products. Heavy metals and organic contaminants can accumulate in meat, milk, and eggs, posing risks to human health. Long-term exposure to low levels of e-waste pollutants in the diet may lead to adverse health effects, including neurotoxicity, carcinogenicity, and developmental abnormalities (Wu et al., 2021). Ensuring the safety of food products from e-waste-affected areas presents regulatory challenges for food safety authorities.

Monitoring and controlling the levels of e-waste contaminants in animal products require robust surveillance systems, analytical methods, and regulatory standards. Implementing measures to prevent and mitigate e-waste pollution in agricultural areas is essential for safeguarding food safety and public health (Nekhavhambe et al., 2020).

1.5 WEEE AND THREAT TO HUMAN HEALTH

Individuals are exposed to electronic garbage (e-waste) through various complex pathways. Various aspects, such as the nature of exposure, length of exposure, and the possible inhibitory, synergistic, or additive effects of successive exposures, determine the impacts of E-waste on human health. When considered alone, it is challenging to ascertain the impact of being exposed to a specific molecule or ingredient associated with E-waste. Individuals who come into direct touch with polluted soil, dust, air, water, or food sources might be exposed to contaminants through breathing, consuming, or absorbing them through the skin. Exposure to e-waste-related pollutants during pregnancy, infancy, or youth can lead to the development of obesity, asthma, and neurodevelopmental issues. According to Marie-Noel Brune Drisse, the primary author from the WHO, consuming a single chicken egg from Agbogbloshie, a waste site in Ghana, may result in the absorption of chlorinated dioxins that exceed the European Food Safety Authority's daily limit by 220 times. The health risks posed by electronic waste cannot be understated. Toxic metals and other materials get into the soil, water, and air when electronic items are just dumped into landfills or recycled using harmful methods. The examples of some toxic and lethal substances found in E-Wastes include Lead (damages brain and neurological function), Americium (a carcinogen), Mercury (causes neurotoxicity, muscular degeneration, and infertility), Cadmium (causes respiratory complications), Sulfur (known to damage eyes, liver, kidneys and heart), and Chromium (a carcinogen). The toxic and lethal substances from E-Wastes leads to endocrine dysfunction, neuromuscular dysfunction, cognitive and behavioral defects, chronic degeneration of vital organs, increase in carcinogenicity and metastasis, damage to immune system, damage to DNA and other biomolecules, reproductive system dysfunction, weakening of bone structure (*What Is e-Waste*, n.d.) (*The Growing Environmental Risks of E-Waste*, n.d.), (*E-Waste & Its Negative Effects on the Environment | Elytus*, n.d.).

1.5.1 Endocrine Disrupting Chemicals from E-Wastes

Endocrine disrupting chemicals (EDCs) are substances that interfere with the normal functioning of the endocrine system, potentially leading to adverse health effects in humans and wildlife. EDCs can be found in various sources, including electronic wastes (e-wastes), which contain a complex mixture of chemicals. This review aims to examine the presence of EDCs in e-waste and their impact on human health and the environment. BFRs are commonly used as flame retardants in electronics to reduce the risk of fire.

However, these chemicals can leach from e-waste into the environment, posing risks to human health and wildlife. Studies have identified BFRs such as polybrominated diphenyl ethers (PBDEs) in e-waste recycling facilities and surrounding areas, highlighting their potential as EDCs (Zhang et al., 2021b) (Labunska et al., 2014). Phthalates are plasticizers added to electronic devices to improve flexibility and durability. These chemicals have been linked to endocrine disruption, reproductive abnormalities, and developmental disorders. Phthalates can migrate from e-waste into soil and water, contaminating the environment and posing risks to ecosystems and human health (Liu et al., 2018). E-waste contains heavy metals such as lead, mercury, and cadmium, which have known endocrine disrupting properties. Chronic exposure to these metals through e-waste recycling activities or improper disposal can disrupt hormone signaling pathways, leading to reproductive dysfunction, neurodevelopmental disorders, and other health problems (Song et al., 2014). Exposure to EDCs from e-waste can have detrimental effects on human health, particularly during critical developmental stages. Prenatal exposure to EDCs has been associated with adverse outcomes such as low birth weight, neurobehavioral deficits, and impaired reproductive function. Long-term exposure may increase the risk of hormone-related cancers, metabolic disorders, and immune dysfunction (Kumar et al., 2017). E-waste recycling processes can release EDCs into the environment, contaminating soil, water, and air. These chemicals can bioaccumulate in the food chain, posing risks to wildlife and ecosystems. EDCs have been linked to reproductive abnormalities in aquatic organisms, population declines in wildlife species, and disruptions to ecosystem functioning (Lv et al., 2024).

1.5.2 Effects of E-Wastes on Human Nervous System

Electronic waste (e-waste) contains a complex mixture of hazardous substances, including heavy metals, flame retardants, and organic pollutants, which have been associated with neurotoxic effects. Exposure to e-waste pollutants poses risks to the nervous system, affecting cognitive function, behavior, and neurological development. Heavy metals such as lead, mercury, and cadmium are common contaminants in e-waste and have well-established neurotoxic effects. These metals can cross the blood-brain barrier and accumulate in the brain, disrupting neurotransmitter function, oxidative balance, and synaptic plasticity. Chronic exposure to heavy metals has been linked to cognitive impairments, neurodevelopmental disorders, and neurodegenerative diseases (Ji et al., 2021). Brominated flame retardants (BFRs), polychlorinated biphenyls (PCBs), and phthalates are organic pollutants found in e-waste that exhibit neurotoxic properties. These compounds can interfere with neuronal signaling, disrupt endocrine function, and induce oxidative stress in the brain. Prenatal and early-life exposure to organic pollutants has been associated with neurodevelopmental disorders, including autism spectrum disorder (ASD) and attention deficit hyperactivity disorder

(ADHD) (Cicchetti et al., 2020). Epidemiological studies have reported associations between e-waste exposure and adverse neurodevelopmental outcomes in children living in e-waste recycling communities.

Elevated levels of heavy metals and organic pollutants in maternal blood, cord blood, and breast milk have been linked to deficits in cognitive function, attention, and behavior in exposed children. Longitudinal studies have demonstrated persistent neurodevelopmental effects of e-waste exposure throughout childhood and adolescence (Zheng et al., 2021a; 2021b). Emerging evidence suggests a potential link between e-waste exposure and neurodegenerative diseases, such as Alzheimer's disease (AD) and Parkinson's disease (PD). Epidemiological studies have observed higher levels of heavy metals in the brains of individuals with neurodegenerative disorders, implicating environmental exposures as risk factors. Chronic exposure to e-waste pollutants may contribute to neuroinflammation, oxidative stress, and protein aggregation, leading to neurodegeneration (Bai et al., 2022). The neurotoxic effects of e-waste pollutants represent a significant public health burden, particularly in low- and middle-income countries where informal recycling practices are prevalent. Vulnerable populations, including pregnant women, infants, and children, are at higher risk of neurodevelopmental effects due to their heightened susceptibility and exposure levels. Addressing the neurotoxicity of e-waste requires comprehensive strategies for pollution prevention, environmental remediation, and public health interventions (Chen et al., 2021a; 2021b). Strengthening regulatory measures and enforcement mechanisms is essential for reducing e-waste pollution and protecting neurological health. Implementing e-waste management policies, recycling standards, and occupational safety guidelines can help minimize exposure to neurotoxic substances and promote safer recycling practices. Multisectoral collaboration among government agencies, industry stakeholders, and public health experts is needed to address the complex challenges of e-waste neurotoxicity (Wu et al., 2022).

1.5.3 Effects of E-Wastes on Human Maternal and Child Health

Electronic waste (e-waste) contamination has emerged as a global public health concern, with potential implications for maternal and child health. Maternal exposure to e-waste pollutants during pregnancy, as well as childhood exposure through various pathways, can lead to adverse health outcomes. Maternal exposure to e-waste pollutants, such as heavy metals, brominated flame retardants, and phthalates, can disrupt reproductive health. These chemicals may interfere with hormone regulation, leading to menstrual irregularities, infertility, and pregnancy complications. Exposure during pregnancy can also increase the risk of adverse birth outcomes, including preterm birth, low birth weight, and congenital anomalies (Kumar et al., 2022b). E-waste pollutants have been associated with neurodevelopmental disorders in children, but maternal exposure during pregnancy may also impact fetal brain development. Prenatal exposure to heavy metals like lead and mercury has been linked to cognitive impairments, behavioral problems, and neurodevelopmental disorders, such as autism spectrum disorder and attention deficit hyperactivity disorder (ADHD) (Gao et al., 2021). Children are particularly vulnerable to the health effects of e-waste pollutants due to their rapid growth and development.

Exposure to heavy metals, persistent organic pollutants (POPs), and other toxic chemicals during critical periods of development can disrupt organ systems, impair cognitive function, and increase the risk of chronic diseases later in life. Early-life exposure to e-waste pollutants may have long-lasting health consequences for children (Hu et al., 2021). E-waste recycling activities, such as burning of electronic components or open-air dismantling, can release airborne pollutants that pose respiratory risks to children living in nearby communities. Inhalation of particulate matter, heavy metals, and volatile organic compounds (VOCs) can exacerbate respiratory conditions, such as asthma, bronchitis, and pneumonia, leading to respiratory symptoms and impaired lung function (Song et al., 2020). Maternal exposure to e-waste pollutants can occur through inhalation, ingestion, and dermal contact during e-waste recycling, informal dismantling, or living in proximity to e-waste disposal sites. These pollutants can cross the placental barrier and accumulate in fetal tissues, affecting fetal development and increasing the risk of adverse pregnancy outcomes (Zhang et al., 2021b).

Children may be exposed to e-waste pollutants through multiple pathways, including inhalation of indoor and outdoor air pollution, ingestion of contaminated dust and soil, and consumption of contaminated food and water. Childhood exposure to e-waste pollutants can occur in e-waste recycling communities, informal settlements, and areas with improper e-waste disposal practices (Tong et al., 2022). Raising awareness among pregnant women, caregivers, and communities about the health risks associated with e-waste exposure is critical. Providing education on proper e-waste disposal, hygiene practices, and dietary choices can empower individuals to reduce exposure and protect maternal and child health. Community-based interventions and health promotion programs can promote safer e-waste management practices (Yang et al., 2021).

1.5.4 E-Wastes and Occupational Hazards and Safety Measures

The management and recycling of electronic waste (e-waste) pose significant occupational health risks to workers involved in various stages of the e-waste recycling process. E-waste recycling facilities are often characterized by poor working conditions, inadequate safety measures, and exposure to hazardous substances. Workers in e-waste recycling facilities are exposed to a wide range of hazardous chemicals, including heavy metals (e.g., lead, mercury, cadmium), flame retardants, and organic pollutants. Inhalation of fumes, dust, and vapors during dismantling, shredding, and incineration processes can lead to respiratory problems, neurological disorders, and cancer (Li et al., 2021). E-waste recycling operations involve manual handling of bulky and heavy electronic equipment, which increases the risk of musculoskeletal injuries, strains, and fractures. Workers may also be exposed to mechanical hazards, such as sharp edges, broken glass, and moving machinery, leading to cuts, lacerations, and crush injuries (Yang et al., 2020). The informal nature of e-waste recycling, often characterized by low wages, long working hours, and lack of job security, contributes to psychosocial stress among workers.

Inadequate training, limited access to healthcare, and social stigma associated with informal recycling activities further exacerbate mental health issues, including anxiety, depression, and burnout (Dordevic et al., 2022). Providing workers with appropriate PPE, including respirators, gloves, goggles, and protective clothing, is essential for minimizing exposure to hazardous substances and physical hazards. PPE should be properly fitted, regularly inspected, and replaced as needed to ensure effectiveness (Zhou et al., 2021). Implementing engineering controls, such as local exhaust ventilation systems, enclosed workstations, and automated processes, can help reduce exposure to airborne contaminants and mitigate physical hazards. Engineering controls should be designed to minimize worker contact with e-waste pollutants and improve workplace safety (Lee et al., 2020). Providing comprehensive training and education programs for e-waste recycling workers is essential for raising awareness of occupational hazards and promoting safe work practices. Training should cover proper handling and disposal of e-waste, use of PPE, emergency procedures, and health surveillance (Liu et al., 2021). Governments and regulatory agencies play a crucial role in ensuring the enforcement of occupational health and safety regulations in the e-waste recycling industry. Establishing and enforcing standards for workplace ventilation, exposure limits, waste management, and worker protection can help reduce occupational hazards and improve safety outcomes (Wang et al., 2022).

1.6 WEEE AND ONE-HEALTH APPROACH – ADDRESSING E-WASTE CRISIS

From the previous sections, we came to know the impact of E-wastes on the environment and ecosystems, animal and plant species, and human population. In this section we try to understand the connection between them. For this let us consider a scenario. A person buys electrical and electronic equipment (EEE) from a showroom, and after using it for a few years that person buys an upgraded version of the EEE device and discards the old one. The discarded EEE device finds its way to the dump yard, where it gets dismantled, shredded, and burned. As a consequence, the toxic and lethal chemicals from the discarded EEE get leached into the surrounding environment and degrade the quality of air, water, and soil. The food crops and plants growing in the chemically contaminated soil takes up these toxic and lethal substances inside them. The animals (both wild and farm-grown) consume those toxic and diseased plants, and also, the animals consuming these plant sources are consumed by predatory animals and human beings as a part of natural dietary processes. This contamination through the food chain leads to bioaccumulation and biomagnification of toxic and lethal substances from E-Wastes into the human system. As learned in the previous section that these chemicals are potential neuro-endocrine disruptors, carcinogens, and immune system disruptors, and their accumulation in the body through the food chain, through exposure to contaminated soil and water bodies, and breathing in the air mixed with particulate E-Wastes leads to life-threatening chronic illnesses requiring serious medical interventions. But to overcome this crisis, only treating the patients with sophisticated medical technologies and advanced pharmacological substances is not going to solve the issue as this problem is not unidimensional (or not just limited to the physical body of the patient).

Integrated multisectoral and multidimensional approach is required to contain the anthropogenic crisis of electronic Wastes from Electrical and Electronic Equipment (WEEE). Recognizing that animal health, human health and environmental health are interrelated and inseparable are fundamental to planning strategies for overcoming the global E-Waste crisis. This is where the "One Health Approach" becomes relevant (Figure 1.1).

Electronic waste (e-waste) pollution poses complex challenges to environmental, human, and animal health, necessitating a holistic and interdisciplinary approach for effective mitigation. The One Health framework emphasizes the interconnectedness of human, animal, and environmental health and provides a comprehensive strategy for addressing e-waste pollution. Adopting sustainable e-waste management practices, such as reduce, reuse, and recycle (3Rs), is fundamental to preventing environmental contamination. Implementing extended producer responsibility (EPR) policies, eco-design standards, and product stewardship programs can incentivize manufacturers to produce more durable and recyclable electronic products, reducing e-waste generation (Schroeder et al., 2020). Promoting circular economy principles, such as resource recovery and materials recycling, can minimize the extraction of raw materials and reduce environmental impacts associated with e-waste disposal. Investing in innovative recycling technologies, e-waste recycling facilities, and informal sector integration can enhance resource efficiency and reduce environmental pollution

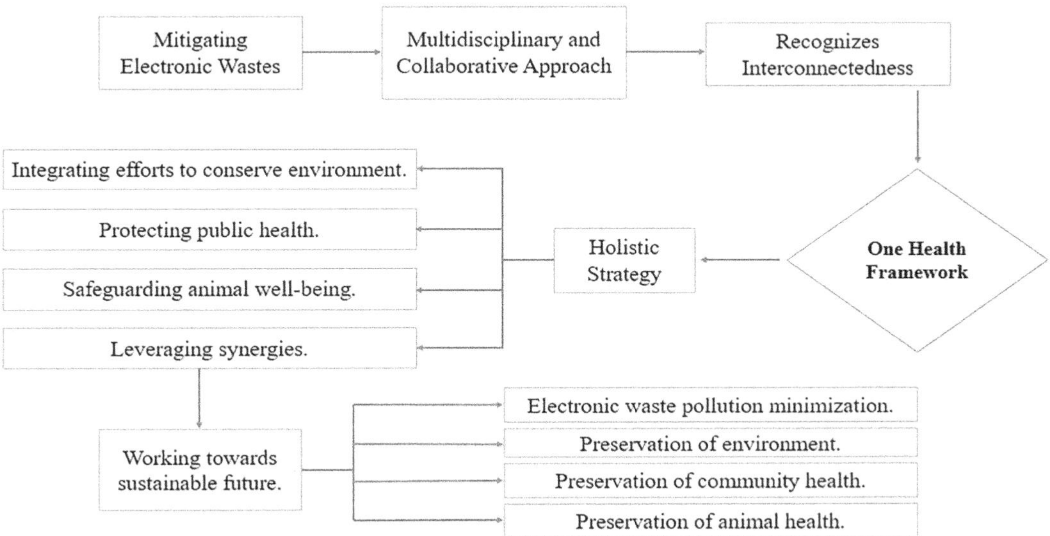

FIGURE 1.1 WEEE and One Health Approach – addressing e-waste crisis.

(Geng et al., 2021). Ensuring the health and safety of e-waste recycling workers is essential for preventing occupational exposure to hazardous substances. Implementing occupational health and safety regulations, providing training and personal protective equipment (PPE), and promoting informal sector formalization can mitigate health risks and improve working conditions (Baldé et al., 2020).

Protecting the health of communities living near e-waste recycling sites requires addressing environmental contamination and reducing exposure pathways. Conducting health risk assessments, providing healthcare access, and implementing community-based interventions, such as environmental education and health promotion, can empower communities to mitigate health risks and advocate for environmental justice (Awasthi et al., 2021). Assessing the impact of e-waste pollution on animal health is essential for understanding ecosystem health and identifying potential risks to human health. Integrating veterinary surveillance systems, wildlife health monitoring programs, and biomonitoring studies can facilitate early detection of emerging health threats and inform ecosystem management strategies (van der Meer et al., 2020). Fostering interdisciplinary collaboration among environmental scientists, public health professionals, veterinarians, and policymakers is critical for implementing One Health interventions to mitigate e-waste pollution. Establishing One Health task forces, research consortia, and policy platforms can promote knowledge sharing, capacity building, and coordinated action to address e-waste challenges (Schnitzer et al., 2021). Mitigating electronic waste pollution requires a multidisciplinary and collaborative approach that recognizes the interconnectedness of environmental, human, and animal health. The One Health framework offers a holistic strategy for addressing e-waste challenges by integrating efforts to conserve the environment, protect public health, and safeguard animal well-being. By leveraging synergies between disciplines and stakeholders, we can work toward a sustainable future where electronic waste pollution is minimized, and the health of ecosystems, communities, and future generations is preserved. Understanding and accepting the inseparable interconnectedness between terrestrial and aquatic ecosystems, plant and animal health, food and feed production and environmental conditions with the state of public health safety is core to the concept of One-Health. One-Health concept emphasizes on the fact that Environmental Health, Animal Health and Human Health are interrelated with each other and cannot be (or should not be) studied in isolated manner. For maintaining sound human health it is very essential to safeguard the animal health and positively restore the environmental health, then only we can think about maintaining public health safety. Thus in order to overcome the issue of global E-Waste Crisis, a holistic, and multidisciplinary and multisectoral approach need to be implemented. Because the only effective way to address the global issues like E-Waste Crisis is through open, inter-sector collaborations, including experts and authorities from animal health, environmental sciences, food and nutrition sciences and public health – i.e. One-Health (*What Is e-Waste*, n.d.) (*The Growing Environmental Risks of E-Waste*, n.d.), (*E-Waste & Its Negative Effects on the Environment | Elytus*, n.d.).

REFERENCES

Awasthi, A. K., Li, J., & Wei, Z. (2021). Informal e-waste recycling in developing countries: Review of metal(loid) exposure, health risk, pathways, and public health implications. *Environmental Science and Pollution Research*, 28(15), 18700–18716.

Bai, S., Jiang, X., Man, Y. B., Yang, Y., Luo, J., Zhang, D., … & Wong, M. H. (2022). Heavy metals in brain tissue and its relationship with neurodegenerative diseases: A systematic review. *Environmental Research*, 206, 112494.

Baldé, C. P., Forti, V., Gray, V., Kuehr, R., & Stegmann, P. (2017). The global e-waste monitor–2017: Quantities, flows, and resources. United Nations University, International Telecommunication Union, and International Solid Waste Association.

Baldé, C. P., Wang, F., Kuehr, R., Huisman, J., & Carroll, M. (2020). The Global E-waste Monitor 2020: Quantities, flows and the circular economy potential. United Nations University (UNU)/United Nations Institute for Training and Research (UNITAR) – co-hosted SCYCLE Programme, International Telecommunication Union (ITU) & International Solid Waste Association (ISWA).

Chen, C., Jin, T., Chen, Z., & Jin, Y. (2021b). Environmental heavy metal exposure and neurodevelopmental outcomes in children: A review. *Journal of Hazardous Materials*, 402, 123926.

Chen, J., Wu, S., Dong, Q., & Wei, C. (2021a). Effects of heavy metals on the reproductive systems of birds and mammals. In *Handbook of Environmental Chemistry* (pp. 1–24). Springer, Cham.

Cicchetti, D., Hui, L. M., Ackerman, S., & Izard, C. (2020). Exposures to environmental toxicants and attention deficit hyperactivity disorder in US children. *Environmental Research*, 188, 109723.

Clark, J. (2022). What is E-waste? Definition and Why It's Important. Great Lakes Electronics. https://www.ewaste1.com/what-is-e-waste/

Dordevic, D., Rosic, J., Dordevic, M., & Ristic, D. (2022). Occupational Health Risks in Electronic Waste Recycling and the Potential Impact on Workers' Quality of Life: A Review. *International Journal of Environmental Research and Public Health*, 19(1), 124.

E-Waste & its Negative Effects on the Environment | Elytus. (n.d.). https://elytus.com/blog/e-waste-and-its-negative-effects-on-the-environment.html

Gadde, R., Yadav, A., Srivastava, A. K., & Mishra, V. K. (2020). Heavy metals toxicity in livestock: Biomarkers and oxidative stress. In *Veterinary Science: Current Aspects in Biology, Animal Pathology, Clinic and Food Hygiene* (pp. 175–188). Springer, Singapore.

Gao, Z., Li, J., Li, L., Chen, T., Zhang, Z., Gao, Y., ... & Luo, C. (2021). Maternal exposure to polychlorinated biphenyls and their hydroxylated metabolites in relation to fetal neurobehavioral development: A review. *Science of The Total Environment*, 796, 148918.

Geng, Y., Dong, H., & Xue, B. (2021). Policy instruments for sustainable e-waste management: A review. *Journal of Cleaner Production*, 281, 124428.

Gouin, T., El Banna, S., Nikoo, M., Shishehbore, M. R., Chen, Z., Shih, Y. H., & Xu, X. (2021). Assessment of hazardous substances in e-waste and its environmental impacts: A critical review. *Journal of Cleaner Production*, 311, 127588.

Goulson, D., Nicholls, E., Botías, C., & Rotheray, E. L. (2015). Bee declines driven by combined stress from parasites, pesticides, and lack of flowers. *Science*, 347(6229), 1255957.

Gupta, N., Yadav, K. K., Kumar, V., Shah, A. V., Li, X., & Kumar, S. (2020). A review on current status of municipal solid waste management in India: Challenges and opportunities. *Journal of Environmental Management*, 250, 109500.

Halden, R. U. (2014). E-waste-induced disruption of soil microbiomes and its potential contribution to soil fertility decline and food insecurity. In *E-waste in transition* (pp. 115–130). Routledge.

Horton, A. A., Walton, A., Spurgeon, D. J., Lahive, E., & Svendsen, C. (2019). Microplastics in freshwater and terrestrial environments: Evaluating the current understanding to identify the knowledge gaps and future research priorities. *Science of the Total Environment*, 586, 127–141.

Hu, X., Kang, Y., Zheng, Q., Sun, Y., Zou, H., & Li, J. (2021). Effects of heavy metals on children's health: A review. *Environmental Pollution*, 292, 118404.

Huo, L., Jiang, X., Zhang, H., Zou, Z., Guo, H., Zhang, X., ... & Jiang, F. (2021). Heavy metal contamination in agricultural soils surrounding an e-waste recycling site in China and the associated health risks. *Environmental Pollution*, 275, 116624.

Islam, M. S., Ahmed, M. K., Raknuzzaman, M., Al-Mamun, M. H., Islam, M. K., & Tokumura, M. (2020). Potential ecological and human health risks of heavy metals in surface soils affected by anthropogenic activities in Bangladesh. *Human and Ecological Risk Assessment: An International Journal*, 26(8), 1976–1998.

Ji, Y., Zhang, X., Gao, M., Xu, W., Zhou, Y., Wang, X., ... & Li, W. (2021). Environmental exposure to heavy metals and neurodevelopmental disorders in children: A systematic review and meta-analysis. *Environmental Pollution*, 290, 117994.

Kumar, A., Kumar, V., & Maiti, S. K. (2022b). Heavy metal contamination in food chain and maternal exposure: Impact on fetal health and pregnancy outcomes. *Environmental Pollution*, 292, 118415.

Kumar, A. & Holuszko, M. & Espinosa, D. C. R., 2017. "E-waste: An overview on generation, collection, legislation and recycling practices," Resources, Conservation & Recycling, Elsevier, vol. 122(C), 32–42.

Kumar, V., Mishra, S., & Singh, S. K. (2022a). Toxicity of heavy metals in plants and remediation strategies. In *Plants and Heavy Metals* (pp. 131–153). Springer, Singapore.

Labunska, I., Harrad, S., Wang, M., Santillo, D., & Johnston, P. (2014). Human dietary exposure to PBDEs around E-waste recycling sites in Eastern China. *Environmental Science & Technology*, 48(10), 5555–5564. https://doi.org/10.1021/es500241m

Lee, S., Kim, S., Jung, Y., Moon, C., Son, J. H., Kim, D., ... & Lee, J. (2020). Review on occupational hazards and their prevention measures in electronic waste recycling. *Journal of Korean Society of Occupational and Environmental Hygiene*, 30(1), 37–49.

Li, J., Chen, Y., Liu, J., Zhang, Z., Zhang, Y., & Dong, X. (2021). Health risk assessment of heavy metals in e-waste recycling environments in China. *Environmental Pollution*, 277, 116780.

Liu, S., Xu, Z., Xue, W., Guo, L., & Pang, X. (2018). The analysis of environmental pollution caused by E-waste recycling in China and India. *Journal of Cleaner Production*, 180, 297–307.

Liu, S., Xu, Z., Xue, W., Guo, L., & Pang, X. (2022a). The analysis of environmental pollution caused by E-waste recycling in China and India. *Journal of Cleaner Production*, 180, 297–307.

Liu, X., Han, Z., Cheng, S., Liu, Y., Wang, Z., & Li, Y. (2021). Occupational health and safety risks of e-waste recycling workers: a review. *Environmental Science and Pollution Research*, 28(41), 57749–57765.

Lv, Y.-Z., Luo, X.-J., Qi, X.-M., Guan, K.-L., Zeng, Y.-H., Mai, B.-X., (2024). A comprehensive assessment of external exposure to persistent halogenated organic pollutants for residents in an e-waste recycling site, South China, Environmental Pollution, 343, 123120 ISSN 0269–7491, https://doi.org/10.1016/j.envpol.2023.123120.

Mishra, S., Kumar, P., Raj, A., & Chauhan, P. S. (2021). Electronic waste (e-waste) and its impact on microbial communities: A review. *Environmental Pollution*, 287, 117356.

Mondal, T., Choudhury, M., Kundu, D., Dutta, D., & Samanta, P. (2023). Landfill: An eclectic review on structure, reactions and remediation approach. *Waste Management*, 164, 127–142. https://doi.org/10.1016/j.wasman.2023.03.034

Naidoo, R., Gerkey, D., Hole, D., Pfaff, A., Ellis, A. M., Golden, C. D., & Herrera, D. (2020). Evaluating the impacts of environmental certifications on ecosystem services: A global analysis of the Forest Stewardship Council. *PLoS Biology*, 17(2), e3000216.

Nekhavhambe, T. J., Muchuweti, M., & Muredzi, P. (2020). Heavy metal residues in water, soil, and vegetables from farms around an electronic waste recycling site in Bulawayo, Zimbabwe. *Environmental Monitoring and Assessment*, 192(11), 698.

Nnorom, I. C., Osibanjo, O., & Ogwuegbu, M. (2021). Heavy metal contamination of animal feed and fodder from an electronic waste dumpsite in Lagos State, Nigeria. *Journal of Environmental Management*, 285, 112103.

Ogunbanwo, S. T., Onalaja, A. O., & Ajidagba, B. O. (2022). E-waste contamination in farm animals: Potential health implications. *Environmental Pollution*, 293, 118566.

Patel, M., Shah, K., & Upadhyay, A. (2021). Environmental Impacts of Electronic Waste on Human Health and Biodiversity: A Review. *International Journal of Advanced Science and Technology*, 30(12), 2612–2624.

Paul, S., Choudhury, M., Deb, U., Pegu, R., Das, S., & Bhattacharya, S. S. (2019). Assessing the ecological impacts of ageing on hazard potential of solid waste landfills: A green approach through vermitechnology. *Journal of Cleaner Production*, 236, 117643. https://doi.org/10.1016/j.jclepro.2019.117643

Rahman, M. M., Liu, Y., Chan, Y. S., Adedoyin, F. F., Al Mamun, A., & Xu, Y. (2022). E-waste-induced metal pollution in soils and its impact on human health and the environment: A review. *Science of the Total Environment*, 804, 150087.

Reed, M., Thomas, G. O., Goudie, A. S., & Horwell, C. J. (2020). The environmental and human health challenges of e-waste. In *E-Waste Management* (pp. 1–20). Springer, Cham.

Schnitzer, T., & Basu, N. (2021). Heavy metal contamination of e-waste: Implications for environmental and human health. *Environmental Science and Pollution Research*, 28(41), 57139–57140.

Schroeder, P., Padel, S., & Wynne, B. (2020). The political economy of the circular economy: A comparative analysis of e-waste recycling in the UK and Brazil. *Environmental Innovation and Societal Transitions*, 34, 135–147.

Song, Q., Jiang, L., & Liu, H. (2020). E-waste: An overview on generation, collection, legislation and recycling practices. *Science of The Total Environment*, 733, 139356.

Song, Q., Li, J., & Hu, Y. (2014). The potential risks of e-waste pollution on human health: a critical review. *Science of the Total Environment*, 468–469, 391–406.

Suthar, S., Singh, S., & Purohit, P. (2018). Impact of E-waste on air and soil pollution: A review. *Journal of Environmental Science, Toxicology and Food Technology*, 12(2), 1–9.

The Growing Environmental Risks of E-Waste. (n.d.). https://www.genevaenvironmentnetwork.org/resources/updates/the-growing-environmental-risks-of-e-waste/

Tong, T., Hu, Y., Guo, Z., Liu, L., Lu, S., & Zhou, W. (2022). Current status of e-waste pollution and its human health impact: A review. *Environmental Pollution*, 293, 118603.

van der Meer, M., Panka, D., & Bodnar, K. (2020). Electronic waste: A growing global challenge for animal health and welfare. *Veterinary Record*, 187(6), 219–221.

Wang, C., Xu, M., Wang, X., Cao, Z., Xu, P., Chen, Y., … & Sun, Y. (2021). Electronic waste (e-waste) exposure affects soil microbial community structure and metabolic potential in agricultural soil. *Environmental Pollution*, 289, 117957.

Wang, Y., Xu, Z., Lu, D., Chen, L., Zhang, X., & Ma, C. (2022). A review on occupational health hazards and safety measures in e-waste recycling. *Journal of Cleaner Production*, 343, 130240.

What is e-Waste. (n.d.). E-Waste. https://globalewaste.org/what-is-e-waste/

Wu, J., Lu, X., Liu, Y., Wang, Y., Sun, Y., & Han, Y. (2022). Neurotoxicity of electronic waste: A review of epidemiological evidence and potential mechanisms. *Environmental Research*, 205, 112473.

Wu, Y., Lu, D., Chen, L., Cheng, S., Zhang, X., Lin, Y., … & Liu, J. (2021). Human health risks associated with heavy metal concentrations in soil, dust, and animal-based foodstuffs from e-waste recycling areas in South China. *Environmental Pollution*, 278, 116838.

Xiao, R., Ma, Z., Li, X., Yu, J., & Liu, Y. (2021). Pollution characteristics and health risk assessment of heavy metals in soil around e-waste recycling site. *Journal of Environmental Management*, 279, 111807.

Xu, P., Zhao, Z., Li, Z., Wang, C., Han, Y., Chen, Y., … & Sun, Y. (2020). Heavy metals in e-waste affects soil microbial community structure and metabolic potential. *Environmental Pollution*, 262, 114276.

Xu, X., Lv, B., Sun, Y., Zhang, Y., Zhao, X., Wu, J., & Wu, J. (2019). Heavy metal pollution in agricultural soils around a typical E-waste recycling area in southeast China and associated health risks. *Environmental Science and Pollution Research*, 26(12), 12394–12404.

Yang, Q., Liu, L., Gong, M., Yang, L., Zhou, W., & Lu, Y. (2021). A review on human health risks and measures to reduce the impacts of e-waste. *Environmental Chemistry Letters*, 19(4), 2763–2784.

Yang, X., Wang, C., Wu, Q., Ma, Z., Tang, X., & Wang, Y. (2020). Occupational hazards and safety measures in e-waste recycling: A review. *Journal of Cleaner Production*, 254, 120050.

Zhang, Y., Huang, C. H., Zhong, Y., Peng, L., Wang, X., & Sun, J. (2021a). E-waste-related environmental pollution and ecological risks: A review. *Environmental Pollution*, 269, 116192.

Zhang, Z., Zhang, Z., Yang, L., & Yang, D. (2021b). Health risk assessment of heavy metals and polycyclic aromatic hydrocarbons (PAHs) exposure via dietary intake of vegetables from e-waste recycling areas in China. *Environmental Pollution*, 272, 115979.

Zheng, S., Chen, L., Huang, L., Yan, X., Yang, J., & Luo, K. (2021a). Health risk assessment of heavy metals and PAHs in agricultural soils near e-waste recycling site in South China: Implication for child health. *Environmental Pollution*, 287, 117646.

Zheng, Y., Xu, Y., Guo, X., Liu, S., & Sun, G. (2021b). Prenatal and postnatal exposure to e-waste leads to neurobehavioral deficits in rats: Role of autophagy dysregulation. *Environment International*, 156, 106761.

Zhou, H., Liu, L., Su, S., Yang, Y., Liu, T., Liu, B., … & Xue, Y. (2021). Occupational hazards and safety measures for e-waste recycling: A systematic review. *Environmental Research*, 195, 110817.

2 Electronic Wastes
The Manmade Crisis

Navamallika Gogoi
Faculty of Science, Sadiya College, Tinsukia, India

Dharitri Borah
Department of Environmental Science, Arunachal University of Studies, Namsai, India

Moharana Choudhury
Environmental Research and Management Division, Voice of Environment (VoE), Guwahati, India

2.1 ELECTRONIC WASTES – THE MANMADE CRISIS

Electronic waste refers to all the discarded electrical equipment that is of no value to their consumers anymore. It is defined as one of the most speedy developing solid waste classes in the ecosystem (Rajpal et al., 2024; Perkins et al., 2014). E-waste is a global environmental and health issue because of its release of huge volume pollutants like polychlorinated biphenyls (PCBs), tetrabromobisphenolA (TBBPA), polybrominated biphenyl (PBB), hazardous metals, etc., along with other hazardous components like volatile organic compounds, heavy metals, etc.

The predicted worldwide e-waste generation estimated 53.6 (Mt) in the year 2019 and it is projected that by 2030 the global total is likely to reach 74.7 Mt (Forti et al., 2020). According to the previous report, 20–50 Mt e-waste is generated every year (Wang et al., 2016). Categorization of e-waste is difficult due to its diverse composition.It contains more than 54 categories, and these 54 types can be grouped into 6 major types and contribution in the year 2019 (million Mt) (Figure 2.1) (Rajesh et al., 2022).

According to a previous study, there are 5.15 billon cell phones and 4.57 billion internet users worldwide (Kemp, 2022). The quick development and availability of e-product worldwide, poor infrastructure, lack of consciousness, poor recycling technology, etc., negatively impact the atmosphere, increasing the e-waste pollution (Rajesh et al., 2022).

E-waste consists of a diverse range of components like plastic, glass, metals (hazardous and valuable) as shown in Figure 2.1 (Ilankoon et al., 2018). E-waste has been defined as one of the most complex categories of waste to handle because of its nature of feature to a constant specificities. Previous study reports that only 20% of the e-waste is partially used, remaining 80% turned out in landfills, gradually polluting the land and water around it. Each year, the world generates almost 50 metric tons (Mt) of electronic garbage, with industrialized regions like Europe and the United States contributing more than their fair share. As of now, Asia generates the most e-waste (24.9 Mt), after that Americas (13.1 Mt), Europe (12 Mt), Africa (2.9 Mt), and Oceania (0.7 Mt) (Forti et al., 2020). Europe produces 16.2 kg/capita, the United States produces 13.3 kg/capita, and Asia produces 5.6 kg/capita for consumption (Forti et al., 2020).

DOI: 10.1201/9781003582311-2

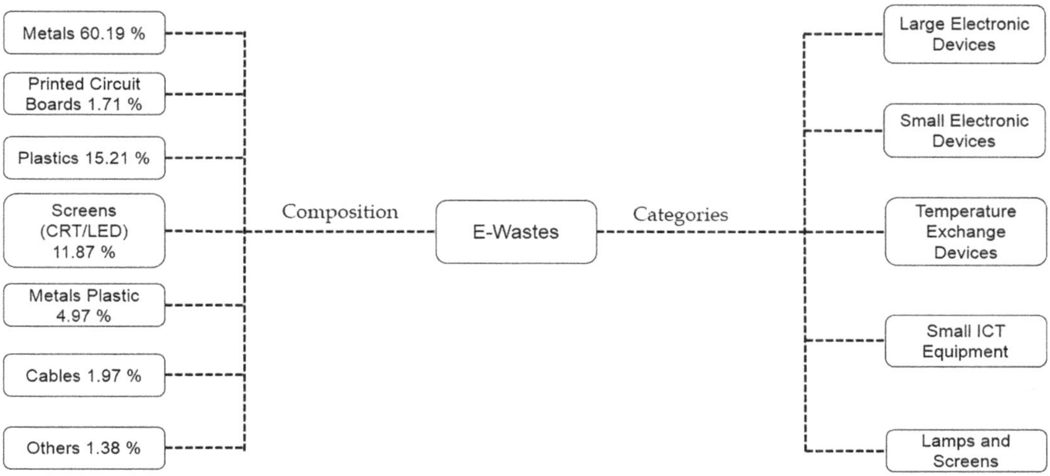

FIGURE 2.1 Composition and categories of electronic wastes.

2.2 LEACHING OF E-WASTES-ASSOCIATED TOXICANTS INTO THE ENVIRONMENT

Most leachates are composed of a variety of toxic organic pollutants (Heavy metal, NH_3/N compounds, aromatic organic compounds) (Youcai, 2018). The process of dissolving contaminants in e-waste makes leachate's composition more complicated, converting it to more poisonous when combined with industrial and municipal solid waste in landfills (Paul et al. 2019). According to the results of a recent study, leachate produced at an open e-waste dumpsite are toxic to living beings and may affect cell division, breaks chromosomes, damage genetic information, mutations (Purchase et al., 2020). The basic causes of landfill leachate are rain falling on the landfill, flow of water above ground into the subsurface, flow of surface water into the ground, breakdown of organic matter (Youcai, 2018). Leachate has different characteristics depending on the type of garbage, the amount of moisture present, where the garbage was produced, and the climate. As a result of their high concentrations of organic and inorganic pollutantas, leachates pose serious health risks to humans and other organisms (Mondal et al., 2023). In addition, a laboratory study showed that leachates formed from broken e-waste contain much higher amounts of metals than the leachates generated from other garbage (Kiddee et al., 2013).

2.3 VALUABLE AND REUSABLE MATERIALS FROM E-WASTES

Many significant materials can be recycled and used again in e-waste, including ferrous/nonferrous metal, Al, Au, Pt, Ag, Zn, Cu gold, platinum, silver, zinc, copper, plastic, palladium, and relatively minor amounts of indium, gallium, and rare earth metals, such as neodymium. A range of valuable materials found in e-waste and can be reused to potentially create new raw materials. Precious metals are found in electronic trash in every industry. Because they include copper, gold, silver, palladium, and other precious metals, recycling them becomes increasingly crucial. Recovery of these valuable metals from garbage requires less work and energy than mining and is more profitable. The various recoverable precious material against different electronic equipment are shown in Table 2.1. Of course, it is impossible to assume that electronic garbage from such disparate locations will contain an equal quantity of valuable metal content. The quantity/quality of valuable metals differs depending on the type of material and even the brand, although the exact amount is unknown. Because they include copper, gold, silver, palladium, and other precious metals, recycling them becomes increasingly crucial.

TABLE 2.1
Rate of Recoverable Material from Various E-products (Arya et al., 2021)

Material	Rate
Gold	559,5000/kg
Silver	67,100/kg
Ferrous	28,000/tone
Copper	425,000/tone
Plastic	425,000/tone

2.4 CURRENT APPROACHES IN WASTE MANAGEMENT AND RECYCLING PROCESS

The term "recycling" refers to the practice of returning used materials to the material cycle for processing in order to minimize environmental impact and maximize resources salvaged from them. It's a crucial step in getting rid of trash and getting something useful out of it. Generally E-waste exposures treat in two ways, informal (open burning, melting, heating of e-waste, etc.) and formal way. Formal recycling process is comparatively safe and tolerable for environment. However in developing nations E-waste management worked on improve environmental condition by reusing e-waste and heal the precious and rare metals. Energy costs can be drastically reduced by using recycled materials instead of new ones. The goal of waste management is to decrease the pollution of e-waste on human healthiness and the atmosphere by collecting, transporting, and disposing of or processing and recycling (Figure 2.2A).

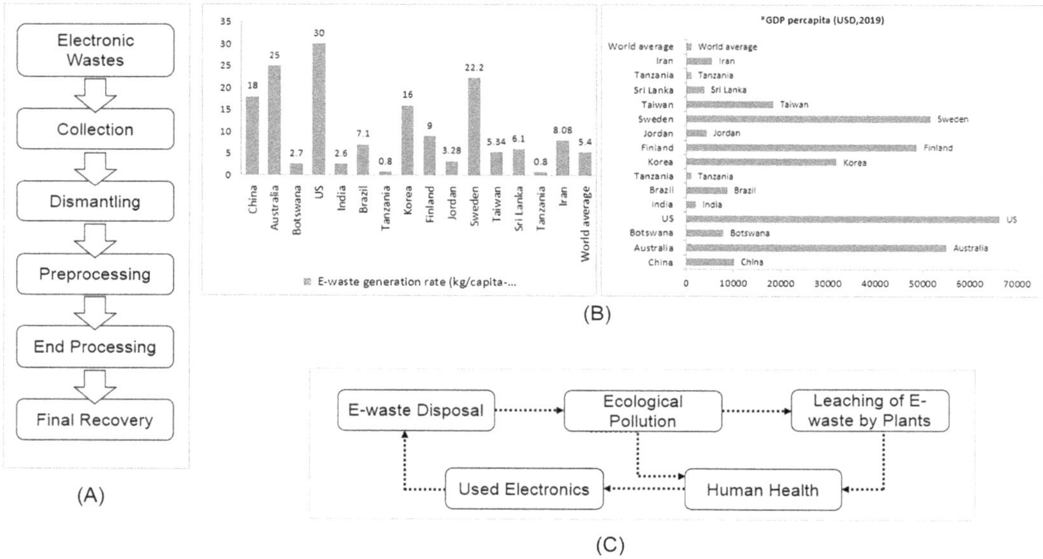

FIGURE 2.2 (A) E-waste recycling process (Abdelbasir et al., 2018); (B) Rate of E-waste generation (kg/capita/year) and GDP of various nations (Fadaei 2022); (C) E-waste cycle on human health.

Three primary fuel sources of e-waste management are "the collection network and infrastructure," "the recycling industry," and "supporting activities," which include things like "awareness raising" and "monitoring and reporting systems" (Tanskanen, 2013). The reprocessing of e-waste materials is just one of the many difficulties that may be tackled by the various e-waste management systems (Ongondo et al. 2011; Agarwal et al. 2023). All parties involved should receive the proper education, guidelines, and is a significant volume of e-waste being dumped in unregulated dump sites across a number of developing nations, particularly those with low or middle incomes. Similar to other waste streams, electronic trash recycling is highly prevalent in the underground economy (Jalili et al., 2021). Popular public treatment options include chemical, physical, and biological processes. Both pyro-metallurgy and hydro-metallurgy fall under the umbrella term "chemical technology," which is often considered to be the best method. The physical methodology is a mechanical process that involves sorting out the e-waste's components according to their unique physical qualities. Disassembly, pulverization, and magnetic separation are the primary techniques. Extraction of heavy metals by microbial leakage is the goal of the biological method. This method is still developing and has limited potential (Islam and Mahmud, 2018). Increased use of electronic items can be attributed to a number of factors, including societal shifts, technological advancements, and the widespread availability of low-cost electronic tools. More effort should be put into community management of e-waste because it poses a growing threat to the environment. Furthermore, not substantial research has been done in this area, as seen by the scholarly writings. Most countries in transition and development still use unreliable e-waste management practices. E-waste management creates a serious risk to governments because of the dangerous metals it contains, but it also has the potential to be economically lucrative because of the abundance of valuable metals within it. In today's world, around 70% of these dangerous compounds found in the environment are produced by discarded electronic devices (Abalansa et al., 2021).

2.5 ENVIRONMENTAL IMPLICATIONS OF ELECTRONIC WASTE RECYCLING

The e-waste recycling areas are heavily polluted with HMs like Pb, Zn, Ag, Cd, Cu, Hg, polycyclic aromatic hydrocarbons (PAHs), polychlorinated biphenyls (PCBs), etc. Interestingly, about 50–80% of e-wastes particularly in Asia is dumped in 'developing countries'from developed nations. However, presently, e-waste production in developing/underdeveloped countries has also increased by 200–500% (2017) in the countries like India, China, and South Africa (Narayanasamy et al., 2021). The maximum e-waste pollutants from recycling have been estimated in China. Globally, a very less quantity (15–20%) of e-wastes is recycled and reused. In addition to domestic wastes, the recycling sites (air/land) are contaminated pertaining to the lack of appropriate facilities for separation storage, transportation, treatment, and disposal. Recycling of e-wastes involves the collection, sorting/disassembly for mechanical/ chemical processing followed by final recovery. Even in a modern factory, disassembly involves releasing heavy dust with noise pollution (Awasthi et al., 2022). The unrestrained burning brings about pollution to air/water. Also, primitive recycling techniques involve men/women, and children leading to unprotected exposure to toxic emissions The unauthorized and illegal import, backyard recycling about material recovery from e-wastes from cyanide leaching, open sky incineration, and smelting operation yield fewer metals with high pollutants to the environment Incineration of e-wastes results in toxic gases/fumes, which requires a suitable process technology to reduce the environmental consequence. Dumping of waste into the soil without any pre-treatment is aided by the weak environmental laws/regulations/economic conditions of a developing country. It impacts the environment including the biota. Interestingly, there are only a few authorized recycling facilities. Countries like India are still dependent on unorganized sectors for recycling, often with untrained workers. These informal sectors are needed to be controlled by appropriate policies. Heavy metals are from improper recycling in soil cause water pollution through run-off. Toxicants are released, deep in the soil and make water contaminated. Recycling of a few e-waste compositions like printed circuit board (PCB) is difficult due to the heterogeneity in the product (organic materials/metals/glass fibers). Recycling of e-wastes also releases 'persistent organic pollutants' (POPs) brominated flame retardants (BFRs), polychlorinated biphenyls, hexabromocyclododecane, polybrominated biphenyls, di-brominated di-phenyl ethers, poly-chlorinated or poly-brominated dioxins, and di-benzofuransdioxins, and per-fluoroalkyl)] with long half-lives. Organic pollutants like PAHs, PCBs, BFRs and polychlorinated dibenzo-p-dioxin/furans (PCDD/Fs) are released during thermal recycling processes. The POPs that originated from dismantling/smelting are polychlorinated dibenzofurans, polychlorinated biphenyls, and dioxins like polychlorinated dibenzodioxins. The usual methods of recycling involve high energy input. The gases released during the recycling are dependent upon the composition of the e-wastes. The air is intoxicated with polybrominateddiphenyl ethers (PBDEs), PAHs, PCDD/Fs, As, Cd, Cr, Cu, Mn, Ni, Pb, Zn, etc. (Kiddee et al., 2013). Pyrolysis of computers-PCB release gases like H_2 (4.6%), CO (27%), CO_2 (51%), CH_4 (10.3%), C_2H_4 (0.6%), C_2H_6 (1.9%), C_3H_6 (2%), C_2H_8 (1%), C_4H_8 (0.8%), C_4H_{10} (0.5%), etc. (Gupta et al., 2014). Soil is acidified by the acid melting of computer chips. Also, 'illegal e-waste recycling' creates water deficiency pertaining to water contamination resulting from the dumping of wastes in rivers from recycling facilities (Rao, 2014). Breaking/removal of the copper yoke from 'cathode ray tubes' causes the release of Pb, Ba, and other heavy metals/toxic phosphor in the groundwater. Hydrocarbons, heavy metals, and brominated substances are released into the environment while processing electronic chips/gold-plated compounds (Pinto, 2008). The by-products resulting from primitive recycling enter the 'food chain' and bioaccumulate. Also, the PBDE in indoor air was found to be more (46–350 pg/m^3) than in the outdoor conditions of a waste recycling facility (Kiddee et al., 2013). The residual wastes in recycling present higher metallic contents with serious environmental issues. Also, it is emphasized that it requires electronic manufacturing with an eco-friendly design (Awasthi et al., 2022). Often the metal concentration in the environmental samples is more than the 'international limits'. Pollution due to contamination by e-waste requires more research for better comparison amongst developed/developing countries (Houessionon et al., 2021). Also, manual handling of the reusable valued high metal/nonmetal is preferable. Globally 280 tons of PCBs (14mg/Kg) are released from condensers/transformers. Fire retardants/plastics release 34,000 tons antimony (1700mg/Kg) globally. About 41,000 mg/Kg of Cu is present in cabling, which is responsible for releasing 8, 20,000 tons of global emissions (Gaidajis et al., 2010). Also, Ni (10,300mg/Kg) is emitted from rechargeable Ni-Cd/Ni MH batteries/electron gun in CRT is responsible for 2, 06,000 tons of emission globally. Although recycling has a lesser impact on the environment than dumping/burning, it should be combined with appropriate technology (Gaidajis et al., 2010). Also, recycling is less effective if the recyclable materials are needed to be carried over a long distance consuming more energy. Recycling needs high initial investments for advanced technological investments with low pollutant emissions. While recycling, a reduction in the use of hazardous materials in product formulation would be expedient to reduce toxic emissions. Notably, recycling becomes more difficult when the e-wastes are composed of different materials. Efficient recycling with less environmental impact is only possible with one material type. The environmental impact is also dependent upon the recycling efficiency of materials. For a desktop computer, the recyclable efficiency of Ag (0.02%) is 98%. However, Pb (6.3%) is having a recyclable efficiency of 5% (Man et al., 2013). Mobile phones are devised of plastic/metal covers; printed wiring board; battery; low amount (0.01%) of precious metal/rare earth materials. In the UK, 88% of 'large household appliances' are recycled, while only 27% is recycled for 'IT/computers' (Tanskanen, 2013). It is required to optimize the 'value chain' for the entire process of product development for an economical and sustainable recycling process (Figure 2.2B).

2.6 SOCIO-ECONOMIC IMPLICATIONS OF E-WASTE RECYCLING

As economic prospect e-waste recycling is a major practice for enhance economic growth of an organization. Pandey and Govind (2014) revels that in the 21st century IT and electronic industries are the key elements of global economy. The raw material cost associated with the 53.6 Mt of e-waste expected to be produced in 2019 and predicted to be $57 billion USD (Forti et al., 2020). Thus, the developing economies that import Cu, Co, and Ni can limit their excessive extraction and over-reliant on imports through e-waste recovery and endure stability against price rises, controlling manners, and tactical nature by traders. One more economic factor in the e-waste management is the possibility of job growth. Each region of the world has a diverse influence on the global volume of e-waste and the influences of this. Quantities and patterns of e-waste generation, as well as in-house e-waste regulation rules and treatment of e-waste, including activities by promoters, vary vastly between developed and developing parts of the world. In general, advanced economies have significant per capita e-waste generation, e-waste regulation, nearly 100 percent e-waste management by the organized and official sectors, and many schemes like take-back schemes by market participants. As of 2019, Asia produced 24.9 million tons (Mt) of e-waste, followed by the United States (13.1 Mt), Europe (12 Mt), Africa (2.9 Mt), and Oceania (0.7 Mt) (Forti et al., 2020). E-waste already accounts for 1–2% of total solid trash in developing and transitioning nations, and this percentage is only expected to rise in the coming years (Sadeghi et al., 2020). Collection, sorting, and pre-processing (i.e., arranging, disassembling, treatment using mechanical techniques, and end-processing, i.e., treatment and disposal) are the three main steps of the e-waste recycling chain (Méndez-Fajardo et al., 2020). Extending the useful life of electronics and recycling their electrical components has a greater monetary payoff than it might first appear, since fully functional gadgets are worth much more than the sum of their component parts. A circular system in which resources are not mined, used, and discarded preserves more value in the electronics industry and creates more good jobs.

2.7 PUBLIC HEALTH IMPLICATIONS OF E-WASTE RECYCLING

Toxic chemicals produced and carried via various natural pathways have been the primary means by which the health consequences of e-waste vulnerability on humans has been measured. The human body's exposure to e-waste causes a wide range of ailments, including neurological issues, immunological problem, reproductive system, genetic disorder, DNA damage, haematological, respiratory, cardiovascular changes (Parvez et al., 2021).

Li and Achal, 2020, have presented a complete assessment of the health risks of human being contact to numerous hazardous substances related with e-waste recycling (Li and Achal, 2020). In Figure 2.2C, we see a summary of how discarded electrical equipment might harm people. According to reports, people who live in close proximity to e-waste processing facilities are highly vulnerable to the toxic chemicals that are released and move via various ecological channels (Awasthi et al., 2018). Particularly vulnerable were children, who were exposed to contaminants in a variety of ways, including through their mothers' breast milk (Li et al., 2017). Exposure to high levels of lead ($B = 0.196$, p 0.001) and cadmium ($B = 0.148$, p 0.05) in the blood of children was found to be related with an increased threat of asthma and respiratory symptoms, as well as an elevated risk of bronchitis (OR, 2.37; 95% CI, 1.30–4.32; p 0.01) in people living near e-waste sites (Zeng et al., 2016). Apart from these leachates of e-waste can change DNA structure, DNA breakage and induce mutation (Youcai, 2018). Numerous studies have shown that dioxins (found in hair, human milk), PCB, PBDE, and PBB (found in hair) from e-waste disassembly facilities entering the human body all the way through foodstuff, water, soil and air can cause serious health problems (Zhao et al., 2008). In Table 2.2, we can see the consequences of e-waste toxicants on public health.

TABLE 2.2
Different Type of E-waste Pollutants, Sources, and Their Impacts on Human Health

E-waste Material	Component of Electrical and Electronic	Health Issues
Nickel (Ni)	Cathode Ray Tube (CRT), Semiconductors, pigments	Carcinogenic, Skin allergy
Barium (Ba)	Fluorescent lamp	Muscle weakness, Brain swelling
Cadmium (Cd)	Chip resister and semiconductor	Damage Kidneys and Respiratory System
Lead (Pb)	Batteries, glass panels lasers, LEDs, circuit boards	Damage CNS (Central Nervous System), reproductive system, blood system and kidney.
Chromium (Cr)	Dyes/pigments	Bronchitis
Mercury (Hg)	Copy machines, LCDs, printer circuit board	Damage CNS and Immune system.
Lithium (Li)	Li-ion batteries, cell phones photographic equipment.	Harmful for nursing baby, damage to cardiomyocytes.
Arsenic (As)	Semiconductors, diodes, microwaves, LEDS	Gastrointestinal problems, cardiovascular disease'
Polychlorinated biphenyls (PCBs)	Coolants in generator, Florescent lamp, glue, plastic	Damage CNS and immune system.
Chlorofluorocarbons (CFCs)	Fridge and coolers	Bronchial constriction, damage immune, CNS system.
Polyvinyl chloride (PVC)	Cabling and computer body	Hormone disruption and damage immune system.
Polybrominated diphenyl ethers (PBDEs)	Flame retardants for electronic devices.	Genetic damage, carcinogenic

REFERENCES

Abalansa, S., Mahrad, B. E., Icely, J., & Newton, A. (2021). Electronic waste, an environmental problem exported to developing countries: The GOOD, the BAD and the UGLY. *Sustainability*. https://doi.org/10.3390/su13095302

Abdelbasir, S., El-Sheltawy, C. T., & Abdo, D. M. (2018). Green processes for electronic waste recycling: A review. *Journal of Sustainable Metallurgy*, *4*(2), 295–311. https://doi.org/10.1007/s40831-018-0175-3

Agarwal, S., Darbar, S., Saha, S., Choudhury, M., & Singh, R. P. (2023). E-waste management using different cost-effective, eco-friendly biological techniques: An overview. In *Elsevier eBooks* (pp. 205–235). Elsevier BV. https://doi.org/10.1016/b978-0-323-90463-6.00026-9

Arya, S., Gupta, A., & Bhardwaj, A. (2021). Economic analysis of electronics waste materials: A region based study. *Materials Today: Proceedings*, *46*, 10239–10243. https://doi.org/10.1016/j.matpr.2020.11.750

Awasthi, A. K., Awasthi, M. K., Mishra, S., Sarsaiya, S., & Pandey, A. (2022). Evaluation of E-waste materials linked potential consequences to environment in India. *Environmental Technology and Innovation*, *28*, 102477. https://doi.org/10.1016/j.eti.2022.102477

Awasthi, A. K., Wang, M., Awasthi, M. K., Wang, Z., & Li, J. (2018). Environmental pollution and human body burden from improper recycling of e-waste in China: A short-review. *Environmental Pollution*, *243*, 1310–1316. https://doi.org/10.1016/j.envpol.2018.08.037

Fadaei, A.(2022). Proceedings of the International Academy of Ecology and Environmental Sciences, *12*(4), 281–293.

Forti, V., Balde, C. P., Kuehr, R., & Bel, G. (2020). The Global E-waste Monitor 2020: Quantities, flows and the circular economy potential. *United Nations University/United Nations Institute for Training and Research, International Telecommunication Union, and International Solid Waste Association*. https://collections.unu.edu/view/UNU:7737

Gaidajis, G., Angelakoglou, K., & Aktsoglou, D. (2010). E-waste: Environmental Problems and Current Management. *Journal of Engineering Science and Technology Review*, *3*(1), 193–199. https://doi.org/10.25103/jestr.031.32

Gupta, S., Modi, G., Saini, R., & Agarwala, V. (2014). A review on various electronic waste recycling techniques and hazards due to its improper handling. *ResearchGate*. https://www.researchgate.net/publication/299290452_A_review_on_various_electronic_waste_recycling_techniques_and_hazards_due_to_its_improper_handling

Houessionon, M. G. K., Ouendo, E., Bouland, C., Takyi, S. A., Kedote, N. M., Fayomi, B., Fobil, J. N., & Basu, N. (2021). Environmental heavy metal contamination from electronic waste (E-waste) recycling activities worldwide: A systematic review from 2005 to 2017. *International Journal of Environmental Research and Public Health*, *18*(7), 3517. https://doi.org/10.3390/ijerph18073517

Ilankoon, I., Ghorbani, Y., Chong, M. N., Herath, G., Moyo, T., & Petersen, J. (2018). E-waste in the international context – A review of trade flows, regulations, hazards, waste management strategies and technologies for value recovery. *Waste Management*, *82*, 258–275. https://doi.org/10.1016/j.wasman.2018.10.018

Islam, T., & Mahmud, M. a. P. (2018). Application of Material Flow Analysis (MFA) in Electronic Waste (E-Waste) management: A review. In *Proceedings*. MDPI. https://doi.org/10.3390/proceedings2231457

Jalili, D., Dehghani, M., Fadaei, A., & Alimohammadi, M. (2021). Assessment of Airborne Bacterial and Fungal Communities in Shahrekord Hospitals. *Journal of Environmental and Public Health*, *2021*, 1–7. https://doi.org/10.1155/2021/8864051

Kemp, S. (2022, December 5). *Digital use around the world in July 2020 - We Are Social USA*. We Are Social USA. https://wearesocial.com/us/blog/2020/07/digital-use-around-the-world-in-july-2020/

Kiddee, P., Naidu, R., & Wong, M. H. (2013). Electronic waste management approaches: An overview. *Waste Management*, *33*(5), 1237–1250. https://doi.org/10.1016/j.wasman.2013.01.006

Li, W., & Achal, V. (2020). Environmental and health impacts due to e-waste disposal in China – A review. *Science of the Total Environment*, *737*, 139745. https://doi.org/10.1016/j.scitotenv.2020.139745

Li, X., Tian, Y., Zhang, Y., Ben, Y., & Lv, Q. (2017). Accumulation of polybrominated diphenyl ethers in breast milk of women from an e-waste recycling center in China. *Journal of Environmental Sciences-china*, *52*, 305–313. https://doi.org/10.1016/j.jes.2016.10.008

Man, M., Naidu, R., & Wong, M. H. (2013). Persistent toxic substances released from uncontrolled e-waste recycling and actions for the future. *Science of the Total Environment*, *463–464*, 1133–1137. https://doi.org/10.1016/j.scitotenv.2012.07.017

Méndez-Fajardo, S., Böni, H., Vanegas, P., & Sucozhañay, D. (2020). Improving sustainability of E-waste management through the systemic design of solutions: the cases of Colombia and Ecuador. In *Elsevier eBooks* (pp. 443–478). Elsevier BV. https://doi.org/10.1016/b978-0-12-817030-4.00012-7

Mondal, T., Choudhury, M., Kundu, D., Dutta, D., & Samanta, P. (2023). Landfill: An eclectic review on structure, reactions and remediation approach. Waste Management, *164*, 127–142. https://doi.org/10.1016/j.wasman.2023.03.034

Narayanasamy, M., Dhanasekaran, D., & Thajuddin, N. (2021). Bioremediation of noxious metals from e-waste printed circuit boards by Frankia. *Microbiological Research*, *245*, 126707. https://doi.org/10.1016/j.micres.2021.126707

Ongondo, F., Williams, I. H., & Cherrett, T. (2011). How are WEEE doing? A global review of the management of electrical and electronic wastes. *Waste Management*, *31*(4), 714–730. https://doi.org/10.1016/j.wasman.2010.10.023

Pandey, P., & Govind, M. (2014). Social repercussions of e-waste management in India: a study of three informal recycling sites in Delhi. *International Journal of Environmental Studies*, *71*(3), 241–260. https://doi.org/10.1080/00207233.2014.926160

Parvez, S. M., Jahan, F., Brune, M. N., Gorman, J. F., Rahman, M. J., Carpenter, D., Islam, Z., Rahman, M., Aich, N., Knibbs, L. D., & Sly, P. D. (2021). Health consequences of exposure to e-waste: an updated systematic review. *The Lancet Planetary Health*, *5*(12), e905–e920. https://doi.org/10.1016/S2542-5196(21)00263-1

Paul, S., Choudhury, M., Deb, U., Pegu, R., Das, S., & Bhattacharya, S. S. (2019). Assessing the ecological impacts of ageing on hazard potential of solid waste landfills: A green approach through vermitechnology. *Journal of Cleaner Production*, *236*, 117643. https://doi.org/10.1016/j.jclepro.2019.117643

Perkins, D., Drisse, M. B., Nxele, T. C., & Sly, P. D. (2014). E-Waste: A global hazard. *Annals of Global Health*, *80*(4), 286. https://doi.org/10.1016/j.aogh.2014.10.001

Pinto, V. N. (2008). E-waste hazard: The impending challenge. *Indian Journal of Industrial Medicine*, *12*(2), 65. https://doi.org/10.4103/0019-5278.43263

Purchase, D., Abbasi, G., Bisschop, L., Chatterjee, D., Ekberg, C., Ermolin, M. S., Fedotov, P. S., Garelick, H., Isimekhai, K., Kandile, N. G., Reuter, M. A., Matharu, A. S., Miller, B. S., Pineda, A. M., Popoola, O. E., Retegan, T., Ruedel, H., Serpe, A., Sheva, Y., … Wong, M. H. (2020). Global occurrence, chemical properties, and ecological impacts of e-wastes (IUPAC Technical Report). *Pure and Applied Chemistry*, *92*(11), 1733–1767. https://doi.org/10.1515/pac-2019-0502

Rajesh, R., Kanakadhurga, D., & Prabaharan, N. (2022). Electronic waste: A critical assessment on the unimaginable growing pollutant, legislations and environmental impacts. *Environmental Challenges*, *7*, 100507. https://doi.org/10.1016/j.envc.2022.100507

Rajpal, A., Choudhury, M., Goswami, S., Chakravorty, A., & Raghavan, V. (Eds.). (2024). *Waste Management and Treatment: Advances and Innovations* (1st ed.) CRC Press. https://doi.org/10.1201/9781003258377

Rao, L.N., 2014. Environmental impact of uncontrolled disposal of e-wastes. *International Journal of ChemTech Research*, *6* (2), 1343–1353 (CODEN (USA): 0974-4290).

Sadeghi, M., Fadaei, A., & Ataee, M. (2020). Assessment of hospitals medical waste management in Chaharmahal and Bakhtiari Province in Iran. *Archives of Agriculture and Environmental Science*, *5*(2), 157–163. https://doi.org/10.26832/24566632.2020.0502011

Tanskanen, P. (2013). Management and recycling of electronic waste. *Acta Materialia*, *61*(3), 1001–1011. https://doi.org/10.1016/j.actamat.2012.11.005

Wang, Z., Zhang, B., & Guan, D. (2016). Take responsibility for electronic-waste disposal. *Nature*, *536*(7614), 23–25. https://doi.org/10.1038/536023a

Zeng, X., Xu, X., Zheng, X., Reponen, T., Chen, A., & Huo, X. (2016). Heavy metals in PM2.5 and in blood, and children's respiratory symptoms and asthma from an e-waste recycling area. *Environmental pollution (Barking, Essex: 1987)*, *210*, 346–353. https://doi.org/10.1016/j.envpol.2016.01.025

Zhao Youcai, (ed.) (2018) Leachate Generation and Characteristics, *Pollution Control Technology for Leachate from Municipal Solid Waste*, 1–30, Butterworth-Heinemann, ISBN 9780128158135, https://doi.org/10.1016/B978-0-12-815813-5.00001-2.

Zhao, G., Wang, Z., Dong, M. H., Rao, K., Luo, J., Wang, D., Zha, J., Huang, S., Xu, Y., & Ma, M. (2008). PBBs, PBDEs, and PCBs levels in hair of residents around e-waste disassembly sites in Zhejiang Province, China, and their potential sources. *Science of the Total Environment*, *397*(1–3), 46–57. https://doi.org/10.1016/j.scitotenv.2008.03.010

3 Electronic Wastes, Its Sources, Types, Composition, and Toxicity

Kusumlata Gayakwad and Rajat Pratap Singh
Department of Biotechnology, Guru Ghasidas Vishwavidyalaya, Bilaspur, India

3.1 INTRODUCTION TO ELECTRONIC WASTES

Electrical and Electronic Equipments (EEE) have typically made life easier and more convenient because of their effectiveness. The largest, most inventive, and fastest-growing sector of the world economy over the past century has been the electronic industry, which has fundamentally changed the peoples' way of life (Verma and Prakash, 2020). EEE wastes increase globally along with the exponential rise in the usage of electronic items (Ankit et al., 2021). Furthermore, as the fastest-growing markets for EEE, developing countries are also generating considerable volumes of waste electrical and electronic equipment (WEEE) (Sepúlveda et al., 2010). The term WEEE refers to outdated electrical and electronic devices that contain electromagnetic fields and electric circuits. Electronic wastes (also known as e-wastes) are various forms of abandoned electrical and electronic devices that are not in working condition. E-wastes are the fastest-growing waste stream globally as per the European Directive on Waste Electrical and Electronics Equipment (Rajarao et al., 2014). E-waste is increasing at an annual rate of 3–5% as compare to the typical municipal solid waste. More than 50% of households have access to the internet, televisions, refrigeration and mobile phones, all of which significantly increase the amount of electronic trash globally (Aboughaly and Gabbar, 2020). Over a thousand distinct chemicals such as heavy metals, polyhalogenated organics, and other toxic compounds are found in the e-wastes. Many of these chemicals are toxic in nature and release in the environment from e-wastes (Alam and Bahauddin, 2015; Pathak et al., 2019). Polyhalogenated organics, such as polychlorinated biphenyls (PCBs) and polybrominateddiphenyl ethers (PBDEs), can also be discharged from e-wastes. These toxic components of e-wastes are harmful for all life forms and the environment (Kiddee et al., 2013). WEEE is increasing by around 40 million tonnes annually on a global scale which have a significant detrimental effect on the environment (Pathak et al., 2019; Koli and Hussain, 2019).

3.2 SOURCES OF ELECTRONIC WASTES

Worldwide, EEE is widely utilized in the commercial, industrial, and home appliances sectors as a result of revolutionary advancement in technology (Vats and Singh, 2014). The ways in which we interact, travel, entertain, educate, and take care of ourselves are all impacted by the rise in EEE (Pathak et al., 2019). The e-waste refers to a wide range of used and outmoded electronic equipments and products, including discarded cell phones, TVs, computer mother boards, keyboards, cameras, digital video discs (DVDs), radios, batteries, etc., and industrial and household electronic

DOI: 10.1201/9781003582311-3

machinery like refrigerators, ovens, washing machines, fans, air conditioners, etc. (Verma and Prakash, 2020; Rawat et al., 2020). Different home appliances, information technology and computer equipments (ITC), consumer electronics, toys, sporting goods, medical equipments, automatic dispensers, etc. are the primary sources of e-wastes (Srichandan et al., 2019). These discarded electronic products are the sources of e-wastes which releases a wide variety of hazardous substances (Table 3.1).

TABLE 3.1
Sources of E-wastes and Their Toxic Constituent

S. No.	Sources of Electronic Wastes	Toxic Constituent	Reference
1.	Motherboards, batteries, thermoelectric elements, LEDs, optical fibers, LCD screens, mobile phones, computer, laptop, etc.	Lead	Verma and Prakash, 2020; Sharma et al., 2012
2.	Chip resistors and semi-conductors, CRT screen, toners of printers, motherboards, NiCd-batteries, mobile phones, laptop.	Cadmium	Verma and Prakash, 2020; Kiddee et al., 2013; Pathak et al., 2019
3.	Relays and switches, printed circuit boards, computer alkaline batteries, CRT, LCD, batteries, backlight bulbs or lamps, flat panel displays, thermostats	Mercury	Monika and Kishore, 2010; Verma and Prakash 2020; Sharma et al., 2012; Kiddee et al., 2013
4.	Non-stick cookware, electronics	Perfluoroocta-noic acid (PFOA)	Awuchi et al., 2020
5.	Older photocopy machines, fax machines, photoelectric cells,	Selenium (Se)	Kiddee et al., 2013; Pathak et al., 2019
6.	Semiconductor/PWB Light-emitting diode (LED)	Gallium	Mundada et al., 2004
7.	Red phosphor emitter/CRT	Vanadium	
8.	Glass, solid state devices/CRT, PWB	Silica	
9.	Motherboards, relays and finger clips, beam line components.	Beryllium (Be)	Verma and Prakash, 2020; Kiddee et al., 2013
10.	Sparkplugs, fluorescent lamps and CRT	Barium (Ba)	
11.	Copper wires, circuits, cables, ribbons, coils, pigments.	Copper	Monika and Kishore, 2010
12.	Computer housing and Cabling	Plastics and polyvinyl chloride (PVC)	
13.	Electronic equipment, circuit boards, Plastic housing of electronics	Brominated flame	
14.	Cooling unit, insulation foam	CFC (Chlorofluorocarbon)	Verma and Prakash, 2020
15.	Light emitting diodes, solar cells, LEDs, semiconductors, ovens	Arsenic (As)	
16.	Flame-retardant materials, batteries and cable sheathing	Antimony (An)	
17.	Cell phone batteries, photography and videography, Li-batteries	Lithium	
18.	Rechargeable NiCd-batteries, CRT	Nickel	
19.	Corrosion protection of untreated and galvanized steel plates.	Chromium	Kiddee et al., 2013
20.	Computer batteries, Condensers, Transformers	Polychlorinated biphenyls (PCBs)	Sharma et al., 2012
21.	Cable insulation/coating Plastic housing	Bromine	

3.3 CLASSIFICATION OF ELECTRONIC WASTES

Electronic products are frequently divided into three fundamental categories (Koli and Hussain, 2019).

1. White goods, which include home appliances
2. Brown goods, which comprise televisions and cameras
3. Grey goods, which include computers, printers, and scanners

Electronic waste, or e-waste, is EEE that has been disposed of as waste (Munialo and Odoyo, 2020). WEEE can be classified into the following groups: 1. Climate control devices, 2. Appliances for washing, cleaning, and cooking, 3. TVs and computer monitors, 4. Home appliances, computer hardware, and telephone apparatus, 5. Lighting apparatus (Aboughaly and Gabbar, 2020). According to European Union (EU) directive, electronic wastes or e-waste are divided into ten main categories: (i) large household appliances, (ii) small household appliances, (iii) IT and telecommunications equipment, (iv) consumer equipment, (v) lighting equipment, (vi) toys, leisure, and sports equipment, (vii) medical devices, (viii) monitoring and control instruments, (ix) automatic dispensers, and (x) electrical and electronic tools (Mallawarachchi and Karunasena, 2012; Vats and Singh, 2014; Ganguly, 2016)

3.4 COMPOSITION OF ELECTRONIC WASTES

More than 60% of e-waste is composed of different metals like mercury, cadmium, lead, iron, nickel, gold, aluminum, copper, antimony, arsenic, aluminum, beryllium, and other metals (Monika and Kishore, 2010). Other constituents of e-wastes are plastic (approximately 30%) and other hazardous components (about 2.80%) such as glass, ceramics, steel, silicon, BFRs, chlorinated substances including polyvinyl chlorides (PVCs), polychlorinated biphenyls (PCBs), and other organic compounds (Ganguly, 2016; Rawat et al., 2020). Toxic materials such as heavy metals and flame retardants like pentabromophenol, polybrominateddiphenyl ethers (PBDEs), and tetrabromobisphenol A are found in the components of e-wastes, which include motherboards, capacitors, relays, batteries, cathode ray tubes (CRTs), printers, photocopying machines, and liquid crystal displays (LCDs). These hazardous components are present in generally all EEE (Sharma et al., 2012). Moreover, OctaBDE, DecaBDE, polyvinyl chloride (PVC), polychlorinated biphenyls and Polychlorinated Dibenzo-p-dioxins (PCDDs) are very common in WEEE (Sepúlveda et al., 2010; Khan et al., 2021).

3.5 TOXICITY OF ELECTRONIC WASTES

Electronic wastes are significantly more hazardous as compare to other wastes due to the presence of many hazardous components including heavy metals and toxic chemicals (Kamuhanda, 2017). These hazardous components of e-waste can have a direct or indirect negative impact on human health and the environment (Sharma et al., 2012; Chin and Yong, 2019). Examples of direct impacts include cancer and other abnormalities caused by heavy metals and other toxic chemicals. One illustration of an indirect consequence is the biomagnification of heavy metals (Verma and Prakash, 2020). Prolonged exposure of these toxic components due to routine handling of these materials without protective equipment can damage diverse biological systems and organs (Parvez et al., 2021). Leachates and other harmful emissions from WEEE enhance the risk to damage the natural resources like soils, crops and drinking water (Sepúlveda et al., 2010). Numerous studies have revealed the presence of highly harmful organic contaminants in e-wastes landfills and other disposal sites (Paul et al., 2019). These toxic contaminants including polycyclic aromatic hydrocarbons (PAHs), dibenzofurans, polychlorinated biphenyls, heavy metals and other organic contaminants may negatively affect the liver, kidney, reproductive and nervous system of living organisms as well as pollute the air, soil, and water (Table 3.2). The majority of solders, glass panels, and monitor

TABLE 3.2
Toxic Components of E-waste and Their Harmful Effects

S.No.	Toxic Components of E-waste	Harmful Effects	Reference
1.	Lead	Affect the nervous systems, circulatory system and kidney	Pathak et al., 2019; Monika and Kishore, 2010
2.	Mercury	Affect the nervous systems, kidneys, causes respiratory and skin diseases, bioaccumulation in aquatic organisms	
3.	Cadmium	Acute toxicity due to accumulation in kidney and liver, bioaccumulation in aquatic organisms	Monika and Kishore, 2010
4.	Beryllium (Be)	Carcinogen, warts, inhalation of fumes cause beryllicosis.	
5.	Brominated flame	Affect the endocrine system	
6.	Plastics including PVC	Dioxin produced by burning of plastics responsible for reproductive and developmental problems, interfere with immune system	
7.	Barium	Affect heart, liver and spleen and muscles	
8.	Polychlorinated biphenyls (PCBs)	Damage liver and kidney, affects the reproductive, immune, nervous, and endocrine systems, cause tumors and cancer, bioaccumulation.	Sharma et al., 2012; Kiddee et al., 2013
9.	Perfluorooctanoic acid (PFOA)	Hormonal imbalance, developmental toxicity, immunotoxicity and carcinogen in mice	Awuchi et al., 2020
10.	Selenium (Se)	High concentrations cause selenosis.	Pathak et al., 2019
11.	Hexavalent chromium	Asthmatic bronchitis and DNA damage.	Kiddee et al., 2013
12.	CFC (Chlorofluorocarbon)	Skin cancer.	Kamuhanda, 2017
13.	Arsenic (As)	Cause skin disease and lung cancer.	
14.	Bromine	Affect immune system.	Sharma et al., 2012
15.	Bismuth	Causes skin problems.	Masud et al., 2019
16.	Gallium	Breathing problems, pain on chest.	

gaskets contain lead. It is among the most toxic components that cause neurological and other abnormalities in living organisms. Even at very low exposure levels, mercury has been shown to be a neurotoxic agent that can induce developmental abnormalities and IQ deficits in children (Rawat et al., 2020). Compared to adults, children are more vulnerable to lead poisoning (Monika and Kishore, 2010). E-waste contains nickel (Ni), which is a skin irritant, lung irritant, asthma trigger, and carcinogen. Skin irritation, hair loss, heart and lung damage, and problems with fertility are all caused by antimony (Sb). Brominated Flame Retardants are utilised in circuit boards, cables, and computer casing which disrupt the immune and reproductive systems and responsible for hormonal imbalances (Rawat et al., 2020). Circuit boards also contain hexavalent chromium, which at high concentration has been shown to cause lung cancers (Mundada et al., 2004). Long-term human exposure to arsenic can result in lung and skin cancer. Barium exposure for a brief period of time can cause damage to the heart, liver, and spleen in addition to brain enlargement and muscular weakening (Koli and Hussain, 2019; Agarwal et al., 2023). Chlorinated compounds including polychlorinated biphenyls (PCBs) can bioaccumulate in aquatic organisms. PCBs may have detrimental effects on the immune system, endocrine system, neurological system, reproductive system, and cancer, among other serious health consequences (Khan et al., 2021). Due to their high flammability, PCB and other plastic components of electronic gadgets generate acute toxic dioxins and furans after burning which affect the biological systems and essential components of ecosystem

(Pathak et al., 2019). Dioxins and furans have been linked to cancer, neurological, and reproductive issues. Polyvinyl chlorides (PVC) have adverse effects on the liver and kidneys (Awasthi and Li, 2017).

3.6 CONCLUSION

Due to the development of the latest technologies, the production and usage of EEE have also increased all over the world, resulting in an increase in e-waste. The discarded EEE of commercial, industrial, and home appliances is the source of e-waste. The quantity of e-waste in the form of WEEE is increasing daily. The e-waste comprises various toxic components, including heavy metals (cadmium, lead, mercury, antimony, etc.), chlorinated substances (PCBs and PVCs), and other organic compounds that harm living systems and the environment. These toxic components of e-waste can affect the living systems by damaging the kidneys, liver, nervous systems, circulatory systems, and reproductive system and cause cancer. These hazardous components can contaminate the land, water, and air, which can have an impact on the ecosystem. The release of e-waste into the environment needs to be strictly regulated to protect it from the many hazardous substances found in it.

REFERENCES

Aboughaly, M. H., & Gabbar, H. A. (2020). E-waste Recycling and Management. In Khan, A., Inamuddin, & Asiri, A. (eds), *E-waste Recycling and Management. Environmental Chemistry for a Sustainable World* (pp. 63–80). Springer, Cham. https://doi.org/10.1007/978-3-030-14184-4_4

Agarwal, S., Darbar, S., Saha, S., Choudhury, M., & Singh, R. P. (2023). E-waste management using different cost-effective, eco-friendly biological techniques: an overview. In Singh, P., Verma, P., Singh, R., Ahamad, A., & Batalhão, A. C. S. (eds), *Waste Management and Resource Recycling in the Developing World* (pp. 205–235). Elsevier. https://doi.org/10.1016/B978-0-323-90463-6.00026-9

Alam, M. M., & Bahauddin, K. M. (2015). Electronic Waste in Bangladesh: Evaluating the Situation, Legislation and Policy and Way Forward With Strategy and Approach. *Present Environment and Sustainable Development*, *9*(1), 81–101. https://doi.org/10.1515/pesd-2015-0005

Ankit, Saha, L., Kumar, V., Tiwari, J., Sweta, Rawat, S., Singh, J., & Bauddh, K. (2021). Electronic waste and their leachates impact on human health and environment: Global ecological threat and management. *Environmental Technology & Innovation*, *24*, 102049. https://doi.org/10.1016/j.eti.2021.102049.

Awasthi, A. K., & Li, J. (2017). Management of electrical and electronic waste: A comparative evaluation of China and India. *Renewable & Sustainable Energy Reviews*, *76*, 434–447. https://doi.org/10.1016/j.rser.2017.02.067

Awuchi, C. G., Twinomuhwezi, H., Awuchi, C. G., & Ikechukwu, A. O. (2020). Industrial Waste Management, Treatment, and Health Issues: Wastewater, Solid, and Electronic Wastes. *European Academic Research* *8*(2), 1081–1119.

Chin, Y. E., & Yong, H. N. A. (2019). Barriers and critical success factors towards sustainable hazardous waste management in electronic industries – A review. *IOP Conference Series: Materials Science and Engineering*, *669*(1), 012029. https://doi.org/10.1088/1757-899x/669/1/012029

Ganguly, R. (2016). E-waste management in India – An overview. *International Journal of Earth Sciences and Engineering*, *9*(2), 574–588.

Kamuhanda, J. K. (2017). Assessment of local community awareness on health effects of improper electronic waste disposal in rwanda. *Imperial Journal of Interdisciplinary Research*, *3*(2), ISSN: 2454-1362.

Khan, J., Kumar, A., Giri, A., Pal, D. B., Tripathi, A., & Giri, D. D. (2021). Impact of electronic waste pollutants on underground water. *Groundwater Geochemistry*, 265–281. https://doi.org/10.1002/9781119709732.ch13

Kiddee, P., Naidu, R., & Wong, M. H. (2013). Electronic waste management approaches: An overview. *Waste Management*, *33*(5), 1237–1250. https://doi.org/10.1016/j.wasman.2013.01.006

Koli, S. K. & Hussain, A. (2019). Status of electronic waste management in India: A review. In Hussain, A., & Ahmed, S. (eds.), *Advanced Treatment Techniques for Industrial Wastewater* (pp. 238–250). IGI Global. https://doi.org/10.4018/978-1-5225-5754-8.ch014

Mallawarachchi, H., & Karunasena, G. (2012). Electronic and electrical waste management in Sri Lanka: Suggestions for national policy enhancements. *Resources Conservation and Recycling*, *68*, 44–53. https://doi.org/10.1016/j.resconrec.2012.08.003

Masud, M. H., Akram, W., Ahmed, A., Ananno, A. A., Mourshed, M., Hasan, M., & Joardder, M. U. H. (2019). Towards the effective E-waste management in Bangladesh: a review. *Environmental Science and Pollution Research, 26*(2), 1250–1276. https://doi.org/10.1007/s11356-018-3626-2

Monika, & Kishore, J. (2010). E-Waste management: As a challenge to public health in India. *Indian Journal of Community Medicine, 35*(3), 382. https://doi.org/10.4103/0970-0218.69251

Mundada, M. N., Kumar, S., & Shekdar, A. V. (2004). E-waste: a new challenge for waste management in India. *International Journal of Environmental Studies, 61*(3), 265–279. https://doi.org/10.1080/00207230 42000176060

Munialo, S. W., & Odoyo, C. O. (2020). Impacts of electronic wastes on environment and human health. *International Journal of Scientific & Engineering Research, 11*(5), 1174–1179.

Parvez, S. M., Jahan, F., Brune, M. N., Gorman, J. F., Rahman, M. J., Carpenter, D., Islam, Z., Rahman, M., Aich, N., Knibbs, L. D., & Sly, P. D. (2021). Health consequences of exposure to e-waste: an updated systematic review. *Lancet Planet Health, 5(12)*, e905–e920. https://doi.org/10.1016/S2542-5196(21)00263-1

Pathak, P., Rajiv Ranjan Srivastava, R. R., & Ojasvi. (2019). Environmental Management of E-waste. In Prasad, M. N. V., & Vithanage, M. (eds), *Electronic Waste Management and Treatment Technology* (pp. 103–132). Butterworth-Heinemann. https://doi.org/10.1016/B978-0-12-816190-6.00005-4.

Paul, S., Choudhury, M., Deb, U., Pegu, R., Das, S., & Bhattacharya, S. S. (2019). Assessing the ecological impacts of ageing on hazard potential of solid waste landfills: A green approach through vermitechnology. *Journal of Cleaner Production*, 236, 117643. https://doi.org/10.1016/j.jclepro.2019.117643

Rajarao, R., Sahajwalla, V., Cayumil, R., Park, M., & Khanna, R. (2014). Novel approach for processing hazardous electronic waste. *Procedia Environmental Sciences, 21*, 33–41. https://doi.org/10.1016/j.proenv.2014.09.005

Rawat, S., Verma, L., & Singh, J. A. (2020). Environmental hazards and management of E-waste. In Shukla, V., & Kumar, N. (eds), *Environmental Concerns and Sustainable Development* (pp. 381–398). Springer, Singapore. https://doi.org/10.1007/978-981-13-6358-0_16

Sepúlveda, A., Schluep, M., Renaud, F. G., Streicher, M., Kuehr, R., Hagelüken, C., & Gerecke, A. C. (2010). A review of the environmental fate and effects of hazardous substances released from electrical and electronic equipments during recycling: Examples from China and India. *Environmental Impact Assessment Review, 30*(1), 28–41. https://doi.org/10.1016/j.eiar.2009.04.001

Sharma, P., Fulekar, M. H., & Pathak, B. (2012). E-Waste- A Challenge for Tomorrow. *Research Journal of Recent Sciences, 1*(3), 86–93. http://www.isca.in/rjrs/archive/v1/i3/14.ISCA-RJRS-2012-041_Done.php

Srichandan, H., Mohapatra, R. K., Parhi, P. K., & Mishra, S. (2019). Bioleaching: A Bioremediation Process to Treat Hazardous Wastes. *Soil Microenvironment for Bioremediation and Polymer Production*, 115–129. https://doi.org/10.1002/9781119592129.ch7

Vats, M. C., & Singh, S. K. (2014). E-Waste characteristic and its disposal. *International Journal of Ecological Science and Environmental Engineering, 1*(2), 49–61.

Verma, A., & Prakash, S. (2020). E-wastes and their impact on environment and public health. *International Journal of Applied Research, 6*(9), 164–168. https://doi.org/10.22271/allresearch.2020.v6.i9c.7111

4 Understanding the Global Status of Electronic Wastes

Avnish Chauhan
Graphic Era Hill University, Dehradun, India

Vishnu Rajput
Southern Federal University, Rostov-on-Don, Russia

Khalid Stanikzai
Kabul University, Kabul, Afghanistan

Sanjeev Kumar
Department of Chemistry, Chaman Lal Mahavidyalaya, Haridwar, India

4.1 E-WASTE CRISIS – GLOBAL SCENARIO

The global concern regarding e-wastes is increasing as it is reported that in 2019, there were 53.6 million tonnes produced worldwide; by 2030, that number may reach 74.7 million tonnes. Higher consumption rates, smaller life cycles, and fewer maintenance alternatives for WEEE (Waste from Electrical and Electronic Equipment) are the primary causes of this rapid growth. Unregulated handling of e-waste might have negative effects on the human and surrounding environment (Van Yken et al., 2021). The Brownian economic model has recently caused serious ecosystem loss, environmental damage, and resource depletion. As a result, a new economic paradigm known as the green economy has arisen. Reduce, reuse, and recycle, or the "3Rs," is key component of the circular economy and the green economy (Azapagic and Perdan, 2000; Barbier, 1987). Based on his 3R method, built a polarization-based phase-shifting interferometer system utilizing optical components salvaged from electronic waste. E-waste and outdated electronic gadgets have emerged as a result of rapid advancements in technology (Kiddee et al., 2013). For instance, the life expectancy of a computer devices dropped from 4.60 years in 1992 to 2.30 years in 2015, according to Yazici and Deveci (2013), and Shammim et al. (2015). In 2016, 44.70 million tonnes of electronic garbage were created globally.

This comes to 6.10 kilograms for each person (Ilankoon et al., 2018). Production of e-waste is rising by 3% to 5% yearly (Ilankoon et al., 2018). Both the formal and informal sectors recycle e-waste to a large extent (Agarwal et al., 2023). In industrialized countries, the formal sector handles e-waste, but in developing countries like those in Africa, Nijia, Malaysia, China, and India, informal recycling is more prevalent (Pandey and Govind, 2020).

4.2 RAW MATERIALS PRESENT IN E-WASTES – TOXIC AND NON-TOXIC

The e-products are made up of several electronic components, each of which is created from a different raw material and has unique electro-physical characteristics ranging from conductivity to insulation. Electronics materials and parts may contain more than 60 periodic table elements (United

DOI: 10.1201/9781003582311-4

Nations University, 2017). By weight, the majority of materials used in electronics comprise metals as well as polymers, which are further classified into four main groups.

4.2.1 Metals

Electronics goods are made of a variety of metals. The most significant components, making up around 50% of the product's weight, are steel and iron. Aluminum and copper are two more popular and well-known materials that are employed because of their excellent electrical conductivity and formability. Other metals are utilized in resistors, capacitors, and converters, including nickel, chromium, lead, silver, gold, and tin. The majority of these other metals are utilized in far lesser quantities (Mahore et al., 2023).

4.2.2 Rare-Earth Elements

There are 17 different elements that make up the group of lanthanides known as rare earth elements (REEs) (Atwood, 2013). Despite being used in extremely small amounts most of the time, rare earth elements (REEs) are necessary for many high-tech applications such as phosphors, batteries, lasers, and permanent magnet circuits (Reddy, 2024).

4.2.3 Plastics and Other Petroleum-Based Materials

It is known that plastics are the second-largest group of materials used in electronic equipment, accounting for 20% of the weight of e-waste. Mostly used in electronics, polymers have insulating and heat-resistant qualities (da Silva Müller Teixeira et al., 2021). The manufacture of electric and electronic equipment in the EU uses about 2.6 Mt of plastics yearly, or 5.6% of the overall demand for plastics worldwide (Melo et al., 2024).

4.2.4 Minerals and Non-Metallic Materials

Silicon is one of the fundamental materials used in e-products. The use of specific metalloid (or semimetal) materials, like silicon, enables many of the fundamental technological aspects of e-products. The primary substrate material used in the manufacture of semiconductors and microchips is silicon and its variants. Antimony, bismuth, cobalt, garnet, fluorite, magnesium, talc, and bismuth are additional non-metal or semimetal components. For its insulating qualities, ceramics and other materials are frequently used (Andrews et al. 2011).

4.3 TOTAL ANNUAL GLOBAL GENERATION OF E-WASTES

The total annual global production of e-waste is a noteworthy environmental distress. As of latest assessments, around 53.60 million metric tons (Mt) of e-waste were produced worldwide in 2019 (Nithya et al., 2021). This amount is expected to rise sharply, reaching approximately 74.70 million metric tons by 2030 if present trends continue (Forti et al., 2020). The rise in e-waste generation is driven by rapid technological advancements, risen use of electronic devices, and shorter product life cycles (Rene et al., 2021).

Despite the potential for material recovery, merely about 17.40% of e-waste was recorded as being gathered and appropriately recycled in 2019 (Van Yken et al., 2021). The remaining e-waste often ends up in landfills or is informally processed in developing countries (Mondal et al., 2023), leading to environmental pollution and health hazards for local communities (Ackah, 2017). E-waste production totals 50 million tonnes annually over the world. That is growing by roughly 5% annually and is the same as 6,000 Eiffel Towers. With 18.2 million tonnes created in 2016, Asia generated the most e-waste, pursued by the United States region (11.3 million tonnes), Africa region (2.20 million tonnes), and Oceania region (0.70 million tonnes).

TABLE 4.1

Rank	Country	E-Waste Produced (Kt.)	Recycling Rate
First	China	10129.00	16.00%
Second	USA	6918.00	15.00%
Third	India	3230.00	1.00%
Fourth	Japan	2569.00	22.00%
Fifth	Brazil	2143.00	0%
Sixth	Russia	1631.00	6%
Seventh	Indonesia	1618.00	n/a
Eight	Germany	1607.00	52.00%
Ninth	UK	1598.00	57.00%
Tenth	France	1362.00	56.00%

(*Source:* https://theroundup.org/global-e-waste-statistics/#Key_Facts) the total volume produced (shown in KiloTons) (Data access on 13 June 13, 2024)

Asia generated 16.0 Mt of electronic waste in 2014. While India, China, South Korea, Japan, Hong Kong, Vietnam, Bhutan, Cyprus, and Turkey all have national e-waste laws. The Philippines and Jordan have not yet given their approval. The three main Asian countries with the biggest absolute amounts of e-waste are China (6 million tonnes), Japan (2.2 million tonnes), and India (1.7 million tons). In just 16 years, the production of e-waste is expected to nearly triple to 74.70 Mt by year 2030, a 9.20 Mt increase from 2014. The increasing size of e-waste is caused by the increased rates of consumption pattern, shorter life cycles, and fewer choices for repair of EEE. Asia generated the most amount of e-waste (24.90 Mt) in 2019, with the Americas producing 13.1 Mt and Europe producing 12.0 Mt. While Africa and Oceania produced the least amount of e-waste (2.90 Mt and 0.70 Mt, respectively). Europe creates more e-waste than any other region in the world, with 16.20 kg produced per person. Oceania produced about 16.10 kg per person and America produced 13.30 kg per person, while Asia and Africa produced only 5.60 kg and 2.50 kg per person, respectively. The raw materials used to make the world's e-waste in 2019 were valued at over US$57 billion. The three main elements that go into this value are gold, copper, and iron. With a currently reported recovery and recycling rate of 17.40%, 4 Mt of raw materials will be made available for recycling, and 10 billion dollars' worth of resources will be responsibly recovered from global e-waste. There is a chance. By 2021, it is projected that 57.4 million tons (one million tonnes) of e-waste will be created worldwide. Europe produces the most electronic waste per person (16.2 kg), with Oceania coming in second (16th).

The chart above lists the top 10 countries that yield the maximum e-waste, sorted by total quantity produced (given in KiloTons). The nation that creates the most electronic waste worldwide is China. The United States is the leader in e-waste output, with India coming in third (Table 4.1).

Six broad categories that closely match the features of their waste management are used to classify the 54 EEE product types.

- **Temperature exchange equipment**: Equipment for freezing and cooling temperatures is another name for this type of equipment. Temperature control equipment, such as air conditioners, refrigerators, and freezers, significantly contributes to e-waste generation. These equipment contain dangerous materials. Their growing usage and frequent upgrades drive substantial volumes of e-waste globally.
- **Screens and monitors**: Common equipment includes things like Computer, TV, laptops, notebooks, tablets etc.
- **Lamps**: Common lighting fixtures include LED (bulbs, panelight, concealed), high intensity discharge lamps as well as fluorescent lights etc.

- **Large equipment**: Solar panels, huge printers, dishwashers, clothes dryers, electric ovens, and dryers are examples of commonplace appliances.
- **Miniature devices**: Some examples of common small gadgets are vacuum cleaners, oven, geysers, mobile charger, toasters, electric kettles, microwaves, video cameras, calculators, music system, radio sets, small medical devices as well as small monitoring instruments.
- **Miniature IT and telecommunications equipment**: Mobiles, GPS units, calculators, desktop computers, routers, printers, and phones are examples of common equipment.

4.4 E-WASTE CRISIS – INDIAN SCENARIO

According to a UN estimate, 48.50 million tonnes of electronic waste were generated globally in 2018. In addition to highlighting some alarming data—just 20% of total garbage gets recycled—this figure emphasizes the rising significance of recycling. If we keep going in this path, the UN estimates that by 2050, we will have produced 120 million tonnes of electronic garbage. The region of South Asia is starting to understand how important proper e-waste management is. While several other nations are looking into enacting legislations are along similar lines. Since 2011, India has had e-waste management legislation in place, which limits the collection of e-waste to authorized dismantlers and recyclers. There are 312 authorized recyclers in India, with the ability to process about 800 kt yearly. Formal recycling facilities have been established mostly due to Indian regulation. Official recycling capacity is still underutilized, nevertheless, as most of the waste is still managed by the unofficial sector. There are 31 authorized PROs offering compliance services, such as the management of awareness campaigns and the collecting and channeling of e-waste to authorized recycling facilities (Figure 4.1).

Global e-waste production is predicted to have reached 59.4 Mt (million metric tonnes) in 2022 and 74.7 Mt in 2030. There has been an annual growth in e-waste statistics since the start of data gathering in 2014.

The 53.6 Mt is composed of the following parts: Oven, vacuum cleaners, ceiling and wall fans, kettles, toasters, shavers, hair dryers, radios, music systems and toys make up the 17.4 million-ton category of small equipment. Large appliances, are including ovens, cookers, dishwashers, and washing machines, weighing 13.1 metric tons. 10.8 metric tons of temperature-exchange equipment are comprising freezers, refrigerators, and AC etc. Tablets, laptops, monitors, and televisions are totaling 6.7 metric tons of screens and displays. Weighing 4.7 metric tonnes, the small IT and telecom equipment consisted of GPS units, wireless routers, cell phone covers, and pocket calculators. 0.9 Million LEDs, bulbs, and lights.

While this varies by region, globally, less than 20.00% of e-waste is reprocessed; of this total, 17.4% is known to have been properly collected and recycled. Over the past five years, this amount has declined as a percentage of all garbage produced. Europe leads the world in collection and recycling rate (42.5%). Asia trailed behind with a meagre 11.7%. Africa has the lowest rate, at 0.9%.

India generates less than 2.0% of the world's e-waste every year through recycling, placing it behind the US, China, Japan, and Germany as the world's top five producers of e-waste. Since 2018, India has become a major producer of e-waste, producing over two million tonnes yearly, and importing a large amount of e-waste from other countries. Computer and its related equipment accounts for nearly 70.0% of all e-waste in India, according to a study on the subject conducted by ASSOCHAM) and KPMG. The remaining e-waste is sourced from domestic sources such as cell phones (12.0%), electrical equipment (8.0%), and medical equipment (7.0%).

When everything is taken into account, the annually growth rate of electronic trash is amid 5% and 10%. 146,000 tonnes of e-waste were created in India each year (Borthakur and Sinha, 2013). Nevertheless, these quantities only include e-waste generated domestically; they do not take into consideration the importation of trash, both legal and illicit. The reason is the massive influx of wet waste electrical and electronic equipment (WEEE) from other countries into India. Over 95.0% of

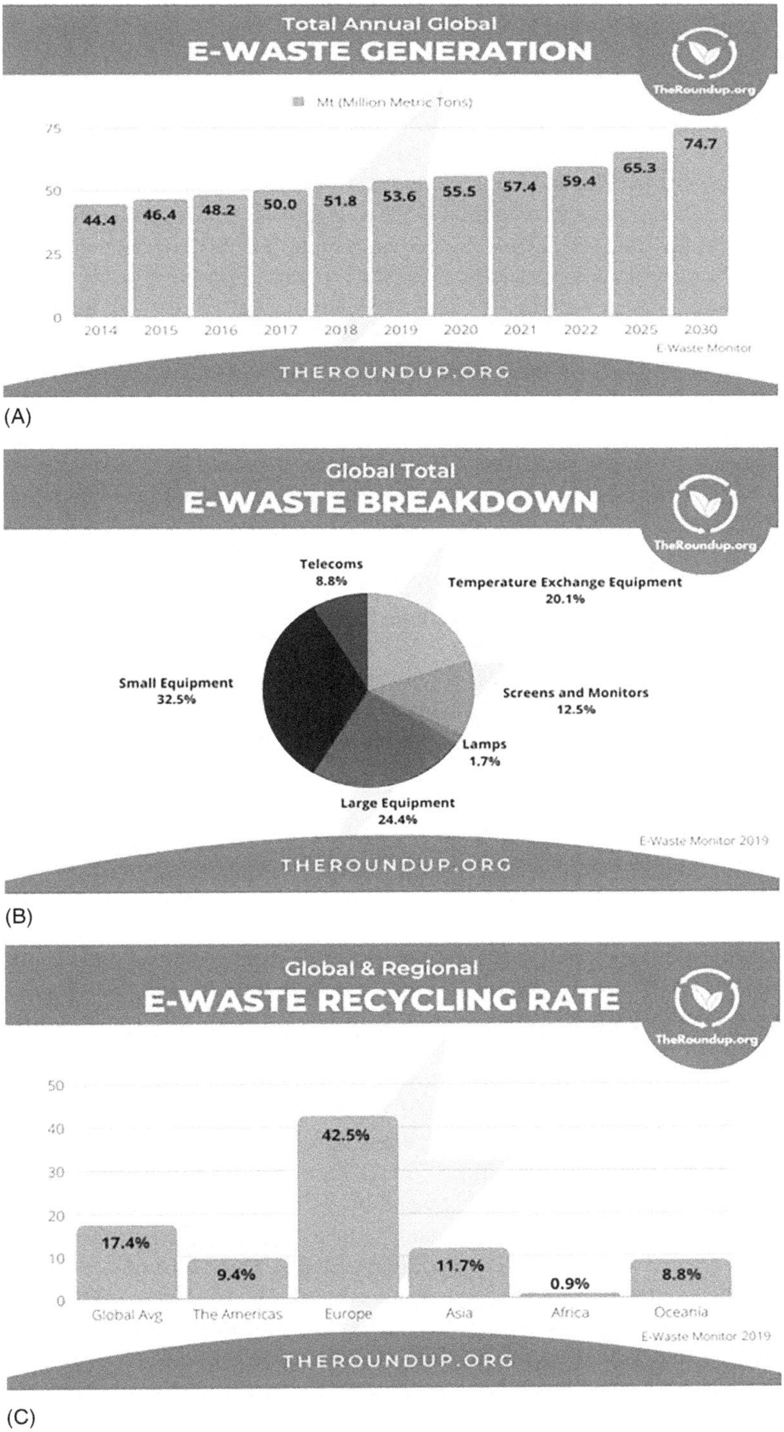

FIGURE 4.1 (A) Total annual global e-waste generation (Source: https://theroundup.org/global-e-waste-statistics/#Key_Facts) the total volume produced (shown in KiloTons); (B) Global total e-waste generation (Source: https://theroundup.org/global-e-waste-statistics/#Key_Facts); (C) Global and regional e-waste recycling rate (Source: https://theroundup.org/global-e-waste-statistics/#Key_Facts).

all electronic waste in India is recycled by the unorganized sector (Kaur, 2018). According to Saha et al., (2021), electronic trash in India is starting to pose a serious threat to human health and the environment.

India is the "Third largest producer of electronic trash in the world," generating about 2.0 million tonnes of e-waste yearly and importing an unknown amount from other countries. Computer devices comprise about 70% of all electronic waste created annually; telecom equipment accounts for 12.0%, health equipment for 8%, and electric equipment for 7.0%. The government as well as public sector organization account for approximately 75% of electronic waste, with individual homes contributing only 16% of the total (Pti, 2016). E-waste is a complicated collection of materials that may contain up to 1,000 hazardous substances. Only 20% of the 50 million tonnes of electronic and electrical waste manufactured worldwide every year gets formally recycled, according to a recent UN assessment.

In developing nations, most of the leftover material is either thrown away or recycled sporadically (UNEP, 2019). Global e-waste production in 2014 was close to 41.8% mmt (Baldé et al., 2015). Of that, the BRICS nations produced 25.0% of the e-waste. With about 1.6 mmt produced in 2014, India is the second-largest producer of e-waste, only surpassed by China (Ghosh et al., 2016). In 2018–19, 33 lakh tonnes of plastic waste were produced. There would be 34 lakh tonnes of plastic trash produced year in 2019–20, an increase of one lakh tonnes. The Ministry of Environment and Forests reports that the amount of plastic waste produced increases by 3% every year. There are 468 licensed recyclers and dismantlers operating in 22 states, and their combined processing capacity is 13.85 lakh tonnes of e-waste (Athrady, 2022). India handled almost 3.4 lakh tons of electronic waste in 2020–21, up from 69,414 tonnes in 2017–18. The ability of the country to collect and treat e-waste has increased fourfold in the last four years. But only 11% of the country's garbage is recycled. The CPCB, has reported an increase in e-waste creation in recent years. India came in third place behind China and the United States in terms of e-waste production in 2019–2020 with 1,014,961.2 tons. The disposal of electronic waste is becoming a major environmental and public health issue. In 22 states, the CPCB has 2,808 collection locations and over 468 authorized recyclers. With 468 recyclers, the total capacity for the amount of waste generated in the country is more than 13 lakh tonnes. India would have produced 10 lakh tonnes of e-waste by the end of 2020, up from more than 7 lakh tonnes in 2017–18 (Rana, 2021). India is currently the nation that produces the most e-waste, behind the US and China. India produced 1,014,961.2 tonnes of e-waste for 21 distinct forms of EEE during the 2019–2020 fiscal year (CPCB). Many rules governing the management and disposal of electronic waste exist in India, although they are not always adhered to. These laws include the Hazardous Wastes (Management and Handling) Amendment Rules from 2003, the E-waste (Management and Handling) Rules from 2011, and the Guidelines for Environmentally Sound Management of E-waste from 2008. India only collected 10% of the projected volume of electronic garbage (e-waste) that was produced there in 2018–19 and 3.5% of that amount in 2017–18 (Central Pollution Control Board) (Figure 4.2).

According to the research, India produced 771,215 tonne of e-waste in the next fiscal year and 708,445 tonne in 2017–18. The amount increased by 32% to 1,014,961 tonnes in 2019–20. The report printed on December 18, 2020 states that, in accordance with the regulations, the collection objectives for 2017–18 and 2018–19 were, respectively, 35,422 tonnes and 1, 54,242 tonnes. However, in both years, the actual collection was lower, at 25,325 tonnes in 2017–18 and 78,281 tonnes in 2018–19One of the 28 nations with a high risk of e-waste congestion identified by a new mathematical model that maps the global flow of hazardous rubbish over the past 20 years is India. This might provide challenges for the efficient treatment of the material in the years to come. Between 7 and 10 billion tonnes of e-waste are produced worldwide each year, of which 300 to 500 million tonnes are detrimental—toxic, flammable, and exceedingly harmful for human health. A significant portion of these waste items are exported as used commodities between countries, including obsolete electronics, whose volume has risen by 50% in the last three decades.

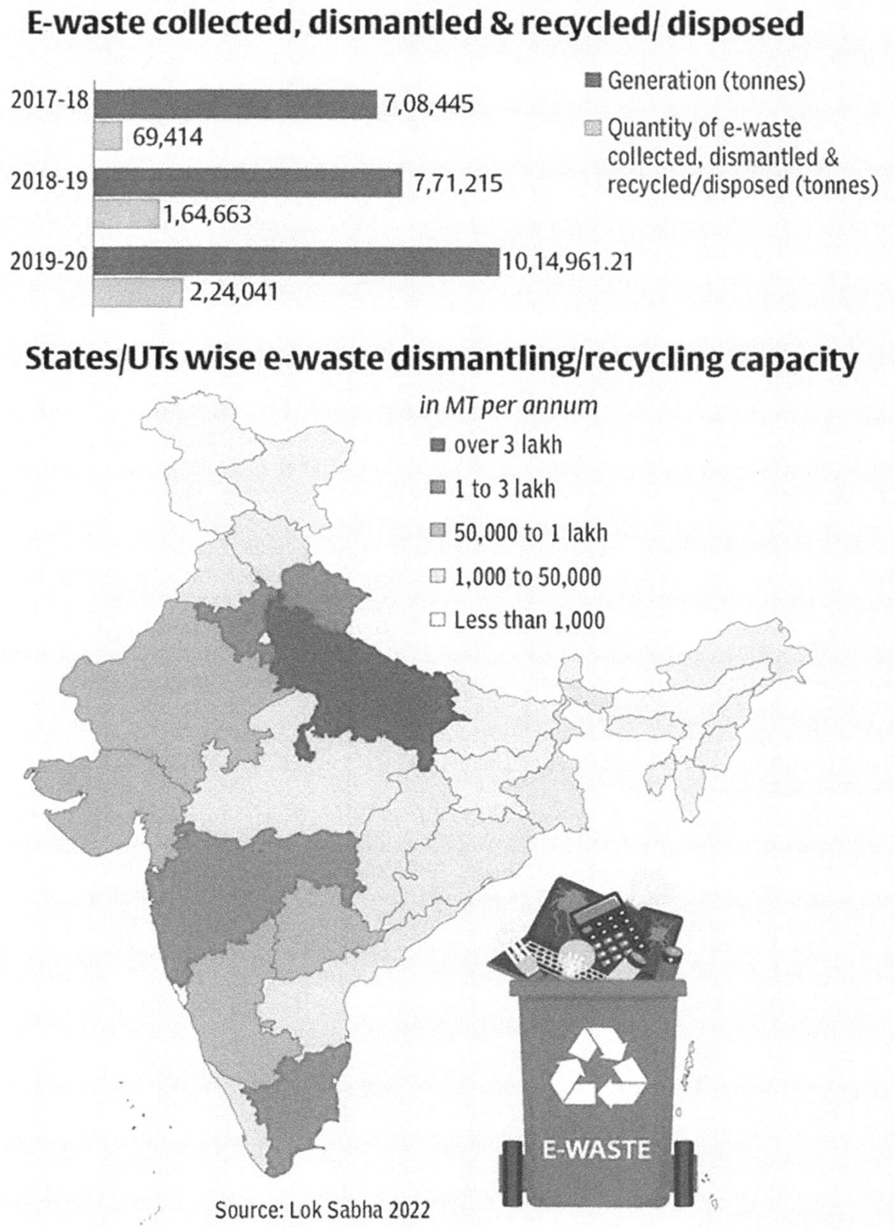

FIGURE 4.2 E-waste Indian scenario.

Source: (https://www.thehindubusinessline.com/data-stories/data-focus/around-78-of-indias-e-waste-is-not-being-collected-or-disposed-by-the-government/article65406820.ece).

In India, 10,14,961 tonnes of electronic garbage were produced in 2019–20; only 22,7% of that amount was collected, disassembled, recycled, or discarded. In India, between 2018 and 2019, 21.35 percent of all generated e-waste was managed, assembled, dismantled, and recycled, up from 9.79 percent between 2017 and 2018.

REFERENCES

Ackah, M. (2017). Informal E-waste recycling in developing countries: review of metal (loid) s pollution, environmental impacts and transport pathways. *Environmental Science and Pollution Research*, 24(31), 24092–24101.

Agarwal, S., Darbar, S., Saha, S., Choudhury, M., & Singh, R. P. (2023). E-waste management using different cost-effective, eco-friendly biological techniques: an overview. In *Elsevier eBooks* (pp. 205–235). Elsevier BV. https://doi.org/10.1016/b978-0-323-90463-6.00026-9

Andrews, R., Shah, K. M., Wilkinson, J. S., & Gartland, A. (2011). Effects of cobalt and chromium ions at clinically equivalent concentrations after metal-on-metal hip replacement on human osteoblasts and osteoclasts: Implications for skeletal health. *Bone*, *49*(4), 717–723. https://doi.org/10.1016/j.bone.2011.06.007

Athrady, A. (2022, February 14). Generation of e-waste in India sees 31% annual growth. *Deccan Herald*. https://www.deccanherald.com/national/generation-of-e-waste-in-india-sees-31-annual-growth-1081505.html

Atwood, D. A. (Ed.). (2013). *The rare earth elements: fundamentals and applications*. John Wiley & Sons.

Azapagic, A., & Perdan, S. (2000). Indicators of Sustainable Development for Industry. *Chemical Engineering Research & Design*, *78*(4), 243–261. https://doi.org/10.1205/095758200530763

Baldé, K., Wang, F., Kuehr, R., & Huisman, J. (2015). The Global E-waste Monitor – 2014. *ResearchGate*. https://www.researchgate.net/publication/275152363_The_Global_E-waste_Monitor_-_2014

Barbier, E. B. (1987). The Concept of Sustainable Economic Development. *Environmental Conservation*, *14*(2), 101–110. https://doi.org/10.1017/s0376892900011449

Borthakur, A., & Sinha, K. (2013). Generation of electronic waste in India: Current scenario, dilemmas and stakeholders. *African Journal of Environmental Science and Technology*, *7*(9), 899–910. https://doi.org/10.5897/ajest2013.1505

da Silva Müller Teixeira, F., de Carvalho Peres, A. C., Gomes, T. S., Visconte, L. L. Y., & Pacheco, E. B. A. V. (2021). A review on the applicability of life cycle assessment to evaluate the technical and environmental properties of waste electrical and electronic equipment. *Journal of Polymers and the Environment*, 29, 1333–1349.

Forti, V., Baldé, C. P., Kuehr, R., & Bel, G. (2020). The global e-waste monitor 2020. United Nations University (UNU), International Telecommunication Union (ITU) & International Solid Waste Association (ISWA), Bonn/Geneva/Rotterdam, 120.

Ghosh, S. K., Debnath, B., Baidya, R., De, D., Li, J., Ghosh, S. K., Zheng, L., Awasthi, A. K., Liubarskaia, M. A., Ogola, J., & Tavares, A. R. (2016). Waste electrical and electronic equipment management and Basel Convention compliance in Brazil, Russia, India, China and South Africa (BRICS) nations. *Waste Management & Research*, 34(8), 693–707. https://doi.org/10.1177/0734242x16652956

Ilankoon, I., Ghorbani, Y., Chong, M. N., Herath, G., Moyo, T., & Petersen, J. (2018). E-waste in the international context – A review of trade flows, regulations, hazards, waste management strategies and technologies for value recovery. *Waste Management*, 82, 258–275. https://doi.org/10.1016/j.wasman.2018.10.018

Kaur, B. (2018, June 28). *Can India manage its toxic e-waste?* DownToEarth. Retrieved April 27, 2023, from https://www.downtoearth.org.in/news/waste/can-india-manage-its-toxic-e-waste-60891

Kiddee, P., Naidu, R., & Wong, M. H. (2013). Electronic waste management approaches: An overview. *Waste Management*, 33(5), 1237–1250. https://doi.org/10.1016/j.wasman.2013.01.006

Mahore, S., Mahore, R. Y., & Dhoble, S. J. (2023). Economics of electronic waste (e-waste): a critical review of negative externalities and the circular economy. *360-Degree Waste Management*, 1, 215–237.

Melo, H. D. B., de Souza-Araujo, J., Garzón, L. G. B., Macedo, J. C., Cardoso, R., Mancini, S. D., … & Rosa, A. H. (2024). Concentrations and legislative aspects of PBDEs in plastic of waste electrical and electronic equipment in Brazil. *Science of The Total Environment*, 906, 167349.

Mondal, T., Choudhury, M., Kundu, D., Dutta, D., & Samanta, P. (2023). Landfill: An eclectic review on structure, reactions and remediation approach. *Waste Management*, 164, 127–142. https://doi.org/10.1016/j.wasman.2023.03.034

Nithya, R., Sivasankari, C., & Thirunavukkarasu, A. (2021). Electronic waste generation, regulation and metal recovery: a review. *Environmental Chemistry Letters*, 19, 1347–1368.

Pandey, P., & Govind, M. (2020). Socio-technological challenges in formalization of E-waste recycling in India. In *Elsevier eBooks* (pp. 243–262). Elsevier BV. https://doi.org/10.1016/b978-0-12-817030-4.00016-4

Pti. (2016, October 18). *India fifth largest producer of e-waste: study*. https://www.thehindu.com/sci-tech/energy-and-environment/India-fifth-largest-producer-of-e-waste-study/article14340415.ece

Rana, R. (2021). The Logical Indian. *The Logical Indian*. https://thelogicalindian.com/trending/electronic-waste-generation-32756

Reddy, R. R. (2024). Prospective Potential Applications and Emerging Tendencies in Rare Earth Materials. *Rare Earth: A tribute to the late Mr. Rare Earth, Professor Karl Gschneidner*, 164, 177–210.

Rene, E. R., Sethurajan, M., Ponnusamy, V. K., Kumar, G., Dung, T. N. B., Brindhadevi, K., & Pugazhendhi, A. (2021). Electronic waste generation, recycling and resource recovery: Technological perspectives and trends. *Journal of Hazardous Materials*, 416, 125664.

Saha, L., Kumar, V., Tiwari, J., Rawat, S., Singh, J., & Bauddh, K. (2021). Electronic waste and their leachates impact on human health and environment: Global ecological threat and management. *Environmental Technology & Innovation*, *24*, 102049.

Shamim, A., Mursheda, A.K. & Rafiq, I. (2015). E-Waste Trading Impact on Public Health and Ecosystem Services in Developing Countries. *International Journal of Waste Resources*, *05*(04). https://doi.org/10.4172/2252-5211.1000188

United Nations Environment Programme (UNEP) Press Release. (2019). UN Report: Time to seize opportunity, tackle challenge of e-waste.

United Nations University, International Telecommunication Union, and International Solid Waste Association. (2017). *The Global E-waste Monitor 2017: Quantities, Flows and Resources*. UNU Collections. http://collections.unu.edu/view/UNU:6341

Van Yken, J., Boxall, N. J., Cheng, K. W. E., Nikoloski, A. N., Moheimani, N. R., & Kaksonen, A. H. (2021). E-Waste Recycling and Resource Recovery: A Review on Technologies, Barriers and Enablers with a Focus on Oceania. *Metals*, *11*(8), 1313. https://doi.org/10.3390/met11081313

Yazici, E. Y., & Deveci, H. (2013). Extraction of metals from waste printed circuit boards (WPCBs) in H2SO4–CuSO4–NaCl solutions. *Hydrometallurgy*, *139*, 30–38. https://doi.org/10.1016/j.hydromet.2013.06.018

5 Electronic Wastes and Contamination of Air

Sukhendu Dey, Apurba Ratan Ghosh, and Megha Santra
The University of Burdwan, Bardhaman, India

Sougata Karjee and Palas Samanta
Sukanta Mahavidyalaya, University of North Bengal, Dhupguri, India

5.1 INTRODUCTION

"E-waste," is sometimes misunderstood as computers that are no longer in use or IT equipment. It's been described in various ways. The European Union describes e-waste as "waste electrical and electronic equipment, including all components, subassemblies, and consumables that are part of the product when it is discarded." According to Directive 75/442/EEC Article I (a), the definition of waste is "any substance or object that the owner must dispose of in accordance with national laws." E-waste is defined by the Basel Action Network (BAN) as an ever-expanding category of electronic devices, including computers that their owners have thrown away, cell phones, stereo systems, and other disposable electronic equipment in addition to large household appliances like refrigerators and air conditioners. E-waste is defined by the OECD as "any household appliance that consumes electricity and has reached the end of its life cycle."

The rate and volume of electronic waste production are rising significantly due to the ongoing need for sophisticated equipment. Between 20 to 50 million tonnes (Mt) of waste are produced annually from electronic waste, which significantly adds to the world's waste problem (Alabi et al., 2021). The quantity of e-waste generated worldwide in 2019 was estimated to be 53.6 million metric tonnes or 7.3 kilograms per person. E-waste is predicted to create more than 74 million metric tonnes by 2030, making it the trash source with the fastest pace of growth. The primary causes include:

- Shorter lifespan.
- Fewer alternatives for maintenance.
- A rise in the use of electrical and electronic equipment (EEE).

Concerning the production of e-waste in 2019, Asia accounted for the largest share, producing 24.9 million metric tonnes. According to Forti et al. (2020), North America accounted for 13.1 million metric tonnes, followed by European countries (12 million metric tonnes), African countries (2.9 million metric tonnes), and Oceania (0.7 million metric tonnes).

Many countries either burn or dump their electronic waste in landfills, which has been found to have detrimental effects on the environment (Alabi et al., 2021). However, a significant portion of generated e-waste, especially from wealthy nations, has allegedly been trafficked to developing countries to bridge the digital divide (Bridgen et al., 2005; Alabi et al., 2021). Because of this, most of these developing countries currently engage in sporadic reprocessing and uncontrolled dumping in municipal and unstructured dumping sites, streets, rivers, and

DOI: 10.1201/9781003582311-5

residential areas. E-waste poses a risk to the environment and public health during disposal and combustion because of the poisonous, hazardous, and complex substances it contains (Ansari et al., 2019). Numerous authors have reported that e-waste has contaminated the environment. This chapter provides the most recent information on the global e-waste contamination issue, focusing on air contamination.

5.2 TYPES AND COMPOSITION OF ELECTRONIC WASTES

Various forms of E-waste are classified based on European Union's WEEE Directive (Table 5.1). Printers, laptops, radios, stereos, mobile phones, television sets are just a few examples of the many electronic gadgets that are a good source of E-waste.

E-waste is made up of a variety of metals and non-metals, which gives it a distinctive and varied makeup. E-waste typically contains 30% organics (such as flame retardants, polymers, and glass fiber), 30% ceramics (like mica, silica, and alumina), and 40% inorganics (*e.g.*, ferrous and non-ferrous metals). Base metals (such as copper, iron, aluminum, and tin), noble metals (such as silver, palladium and gold), heavy metals (such as nickel, beryllium, zinc, chromium, cadmium, lead and mercury), and rare earth metals (such as tantalum, gallium, and platinum groups) are examples of inorganic compounds found in e-waste (Kaya, 2016). A typical constituent of E-waste is presented in the Figure 5.1A.

TABLE 5.1
Categorization of E-wastes

Category	Category Name	Category Label	Examples
Category 1	Larger household appliances	LHA	Ovens, microwave, ceiling fans, refrigeration, washers and dryers, and air conditioners.
Category 2	Smaller household appliances	SHA	Coffee machines, grinders, haircutting appliances, vacuum cleaners, toasters, tooth brushing, drying and shaving.
Category 3	IT and telecommunications	ICT	Cellular phones, minicomputers, laptop, tablet, mainframe computers, pads, scanners
Category 4	Consumer equipment	CE	Acoustic amplifiers, binaural recorders, stereoscopic screens, transmitters, and camcorders
Category 5	Lighting equipment	Lighting	Discharge lights, LED, fluorescence bulbs, solar panels.
Category 6	Electrical and electronic tools	E&E tools	Device like twisting, sawing, crushing, boring, cutting holes, bending, bending, Drills, saws, sewing machinery, soldering equipment, or related operations of metal/wood.
Category 7	Toys, leisure and sports equipment	Toys	Dolls, recreational products, and athletic items, e-trains or rally car parts, computer games, and electric-powered sports gear
Category 8	Medical devices	Medical devices	Oxygen tanks, cranial examination tools, nuclear detectors, radiation tools, and dialysis tools.
Category 9	Monitoring & control devices	M&C	Thermometers, warming controls, and smoke alarms
Category 10	Automated dispensers	Dispensers	Hot or cold containers, solid objects, food and beverage-related electrical devices.

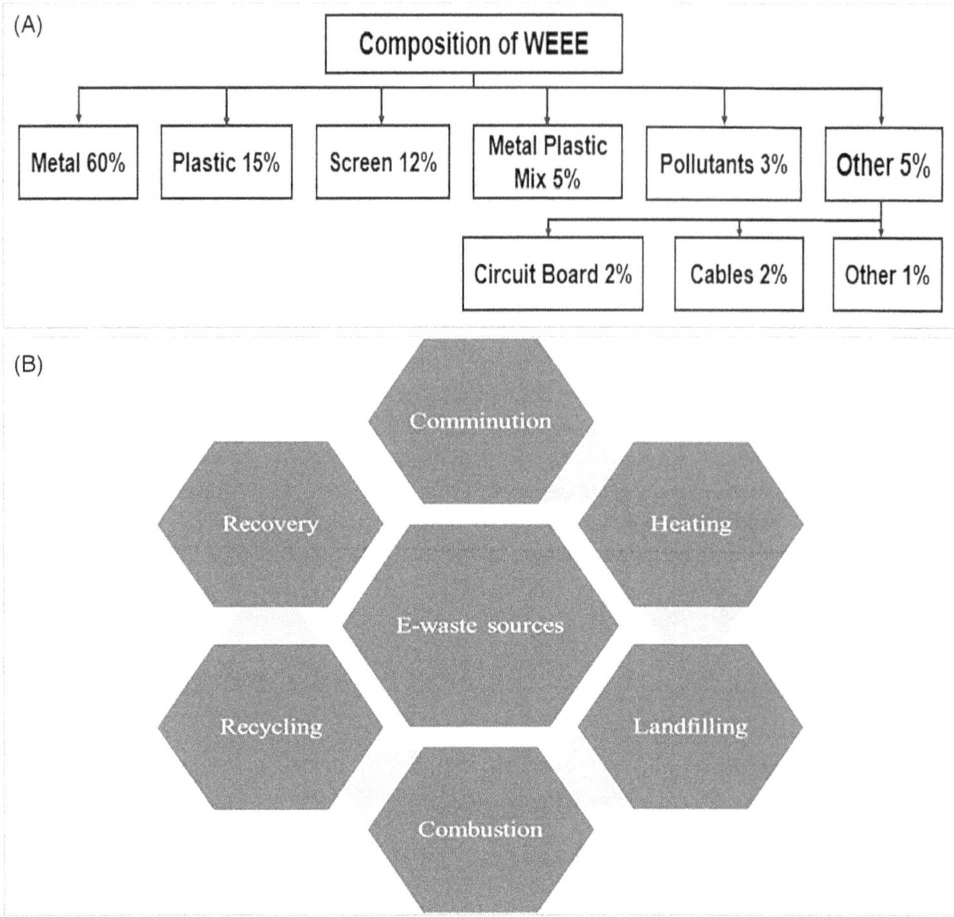

FIGURE 5.1 (A) Distribution of e-wastes ingredients (Andeobu et al., 2021); (B) Sources of e-wastes hazardous ingredients.

5.3 E-WASTE-DERIVED CONTAMINANTS IN THE ATMOSPHERE

The hazardous ingredients of E-waste are generally entered into the atmosphere by the processes like comminution, heating, landfilling, combustion, improper recycling and recovery methods, including grinding and melting. In particularly, the hazardous ingredients of E-waste enter into the atmosphere by the abovementioned activities through dust, fumes and smoke (Zheng et al., 2016; Dai et al., 2017; Zhang et al., 2017). The detailed outline about the probable sources of E-waste hazardous ingredients is presented in the Figure 5.1B. The extent of hazardous property of E-waste depends on chemical composition of e-wastes, more specifically the age and type of discarded electronic gadgets. The different E-waste associated air contaminants are tabulated in Table 5.2. Among them, heavy metals are utilized during the electronic gadgets production, while others, like different organic pollutants namely polycyclic aromatic hydrocarbons (PAHs), dioxin related substances, etc., are generated either by E-waste improper recycling or recovery methods. For example, E-wastes burning emits toxic gases and fumes. Open burning of circuit boards for recovery of metals through mercury amalgamation and/or acid bath emits lead, copper, etc., into the air.

TABLE 5.2
Hazardous Air Pollutants Originated from E-waste

Hazardous Pollutants	Source of E-waste
PCB (polychlorinated biphenyls)	Condensers, Transformers
PVC (polyvinyl chloride)	Cable insulation and computer housing
CFC (Chlorofluorocarbon)	Cooling unit, insulation foam
Cadmium (Cd)	Chips resistors and semiconductors, rechargeable Ni-Cd-batteries, fluorescent layer (CRT screen), printer inks and toners.
Chromium VI	Data tapes, floppy-disks, Corrosion protection of untreated and galvanized steel plates
Lead (PB)	Cathode ray tube (CRT) screens, batteries, solder in printed wiring/ circuit boards, glass panels and gaskets in computer monitors.
Mercury (Hg)	Relays and switches, printed circuit boards, alkaline batteries
Arsenic (As)	Small quantities in the form of gallium arsenide within light emitting diodes.
Nickel (Ni)	Rechargeable Ni-Cd-batteries or NiMH batteries, electron gun in CRT
Lithium (Li)	Li-batteries

5.4 AIR POLLUTION BY E-WASTE-DERIVED CONTAMINANTS

Clean air is very vital for human sustenance. But the air is severely contaminated by diverse range of pollutants. Contamination of air by E-waste associated pollutants has recently gained utmost concern among the scientific community in the globe. E-waste pollutants not only contaminate air but also aggravate the climate change issues. The occurrences of air pollution by E-waste pollutants are described in below:

A variety of contaminants (ranging from hazardous to non-hazardous contaminants) are released into the air due to thermal processing, unregulated burning and dismantling processes of electronic trash, creating smoke. For instance, whenever it burned, polyvinyl chloride (PVC) releases a significant amount of hydrogen chloride (HCl) gas. When HCl and water combine, they produce hydrochloric acid, which is harmful. Furthermore, PVC burning emits furans, dioxin-related compounds (DRCs) and dioxins, including polychlorinated dibenzofurans (PCDFs) and dibenzo-p-dioxins (PCDDs), polybrominated dibenzo-p-dioxins/dibenzofurans (PBDD/Fs) and DL-PCBs as well as their mixed chlorinated/brominated homologues (PXDD/Fs) as they cover plastic computer casings and copper wires (Grause et al. 2011; Duan et al. 2011). Combustion of E-waste has caused an enhanced air level of total PBDEs in Guiyu up to 16,575 pg/m^3 is 300 times more than neighboring Hong Kong (Deng et al. 2007). Guiyu's air contained over 11, 000 pg/m^3 PBDEs during day and it is around 5,000 pg/m^3 at night-time (Chen et al. 2009). An analysis of the airborne levels of PBDEs (BDE-17, 28, 47, 49, 66, 85, 99, 100, 153, and 154) from E-waste collection facilities in China and Thailand revealed that PBDEs amount in out-and indoor air of storage center in Thailand has been noted to be 8–150 and 46–350 pg/m^3, respectively, but were lower than the China outdoor and indoor air PBDEs level (Muenhor et al. 2010). In contrast to the reference site, an investigation from a storage center in Canada revealed that ambient environment of work stations was adulterated with Tris (2-chloroethyl) phosphate (59 ng/m^3) and BDE-209 (156 ng/m^3), whereas the dust has been polluted with Triphenyl phosphate (47,00 ng/g) and BDE-209 (96,300 ng/g) (Nguyen et al. 2019). High quantities of total suspended particles (TSP) in surrounding atmosphere, particularly for tiny pieces of 2.5 mm diameter ($PM_{2.5}$), have indeed been connected to higher risks of mortality (Deng et al. 2006; Awasthi et al. 2016; Park et al. 2002). Particle pollution is another name for particulate matter (PM). It serves as the primary ambient air quality indicator. The E-waste recycling today plays a sizable part in the worsening of ambient air status (Awasthi et al. 2016). Monitoring of ambient PM levels is a crucial component of global air monitoring, establishing a causal association between PM exposures and adverse health effects (Al Katheeri et al. 2012). However, a lot of study looked into the ambient PM_{10} levels. Because there is substantial associations between the exposures to these are PM fractions and harmful health impacts. scientists have only recently begun to concentrate on tiny particles like PM_{10} and $PM_{2.5}$. Owing to the existence of hazardous compounds and trace metals in ambient air and their detrimental environmental effects, PM pollution is a severe environmental problem (Suvarapu and Baek, 2017). Atmospheric heavy metal pollution around recycling facilities of E-waste and subsequent surrounding areas have become a hot topic now-a-day. The transport and survivability of E-waste dust contaminants emitted during recycle processing are mediated by the air, among the most significant sources. According to studies by Eguchi et al. (2012), Ejiogu (2013), and Bi et al. (2010), excessive heavy metals values in ambient air harm the environment. As per findings of Deng et al. (2006), enriched metals Zn (1.038 $\mu g/m^3$) and Cr (1.161 $\mu g/m^3$) had greater amounts of TSP than Mn (0.0606 $\mu g/m^3$), Pb (0.444 $\mu g/m^3$) and Cu (0.483 $\mu g/m^3$) in the air pollution surrounding e-waste recycling centers. Nevertheless, lead, mercury and cadmium can be released into the air when electronic garbage is burned (ESDO 2014). A 14-inch monitor tube containing between 2.5 and 4 kg of lead. If the tube is smashed and burns, it releases poisonous vapors into the atmosphere, causing air pollution (Ramachandra and Saira, 2004). In Guiyu, the detected HMs levels, such as Cu, Cr and Zn at 483, 1161 and 1038 ng/m^3, respectively, were around 4 to 33 times greater than other Asian countries. PBDD/Fs level of 8.12–61 pg/m^3 and PAHs at 40–347 and 22.7–263 ng/m^3 in TSP and $PM_{2.5}$, respectively, have also been found at similar location *i.e.*, Guiyu (Deng et al. 2006). At 64.9–2365 pg/m^3, Guiyu declared highest environmental air PCDD/Fs level in the world. Around e-waste recycling facilities, ambient phthalates esters (PAE) levels range from 200 to 1200 ng/m3, that is 1–2 times greater than vicinity urban areas levels (Li et al. 2007; Gu et al. 2010; Huang et al. 2013). According to reports, the atmosphere surrounding recycling facility in Bangalore, India, included substantial proportions of Ti, Cu, Mn, Cr, In, Bi, Co, Sn, and Sb and Pb than the reference location (Sepulveda et al. 2010). During inappropriate e-waste dismantling and recycling processes, including grinding, incineration, landfilling and melting of e-waste like discarded refrigerators and air-conditioners emits chlorofluorocarbons (CFCs), which is another prime air pollutant and finally enter into the atmosphere and caused air pollution. The incidence of CFCs emission reflected as climate change in the tune of ozone layer depletion and global warming.

5.5 LEGISLATIVE FRAMEWORK FOR ENVIRONMENTAL PROTECTION

The Basel Convention is the first international agreement to reduce the cross-border movement of electronic trash. However, the European Union is a leader in implementing e-waste directives to control and handle electronic waste. Thus far, two directives have been issued to safeguard the environment, specifically the air: "Directive 2002/95/EC (RoHS-Restriction of Hazardous Substances)" and "directive 2002/96/EC on waste electrical and electronic equipment." Switzerland was the first country to establish a formal e-waste management program, the 'Ordinance on the Return, the Taking Back and the Disposal of Electrical and Electronic Appliances.' The law mandated two operational e-waste recycling schemes. The "brown" electronic devices, including televisions, computers, and radios, are overseen by the Swiss Association for Information, Communication, and Organizational Technology.

On the other hand, "white" electronic devices, such as refrigerators, washing machines, and ovens, are regulated by the Stiftung Entsorgung Schweiz System (Gaidajis et al., 2010). The EU has recently enacted E-waste management rules/laws (WEEE Directive 2012/19/EU) by utilizing its WEEE directives in 2012. These laws have been active since February 14, 2014. The comprehensive regulatory measures for preventing E-waste contamination are outlined in Table 5.3.

TABLE 5.3
Regulatory Provisions for E-waste Management

Sl. No.	Regulations
1	The Basel Convention 1998; Stockholm Convention on Persistent Organic Pollutants
2	The Bamako Convention
3	The Environmental Protection Agency Act, 1994 (Act 490).
4	The E-waste (Management and Handling) Rules 2011 by MoEF (India)
5	Regional Platform on Electronic Waste in Latin America and Caribbean (RELAC); Agreement on Environmental Cooperation of North America (ACAAN) (Mexico)
6	General Law for the Prevention and Management of Solid Waste; Regulation of the General Law for the Prevention and Management of Solid Waste.
7	National Environmental (Electrical/Electronic Sector) Regulations 2011 (Nigeria)
8	The Mercosur Policy Agreement from 2006 (Brazil).
9	National Solid Waste Policy of 2010 (Brazil)

5.6 FUTURE CONCERNS AND RECOMMENDATIONS

Our civilization is filled with electronic devices and associated garbage. They stand out for having a complicated chemical make-up and being challenging to quantify both locally and globally. E-waste has sparked the production and emission of a variety of dangerous compounds, particularly POPs and heavy metals due to improper recycling and recovery practices. These hazardous pollutants significantly harmed different environmental segments by their erratic management. In conclusion, it is crucial to separate e-waste from other types of solid garbage and recycle it in order to recover valuable raw ingredients and vital metals. Management program must be logically planned so that the necessary resources utilization for system operations do not offset the environmental advantages, including financial benefits from recuperation.

REFERENCES

Al Katheeri, E., Al Jallad, F., & Al Omar, M. (2012). Assessment of gaseous and particulate pollutants in the ambient air in Al Mirfa City, United Arab Emirates. *Journal of Environmental Protection*, *3*(7), 640–647. https://doi.org/10.4236/jep.2012.37077

Alabi, O. A., Adeoluwa, Y. M., Huo, X., Xu, X., & Bakare, A. A. (2021). Environmental contamination and public health effects of electronic waste: An overview. *Journal of Environmental Health Science and Engineering*, *19*, 1209–1227. https://doi.org/10.1007/s40201-021-00654-5

Andeobu, L., Wibowo, S., & Grandhi, S. (2021). A systematic review of e-waste generation and environmental management of Asia Pacific countries. *International journal of environmental research and public health*, *18*(17), 9051. https://doi.org/10.3390/ijerph18179051

Ansari, M., Ehrampoush, M. H., Farzadkia, M., & Ahmadi, E. (2019). Dynamic assessment of economic and environmental performance index and generation, composition, environmental and human health risks of hospital solid waste in developing countries; A state of the art of review. *Environment international*, *132*, 105073. https://doi.org/10.1016/j.envint.2019.105073

Awasthi, A. K., Zeng, X., & Li, J. (2016). Environmental pollution of electronic waste recycling in India: A critical review. *Environmental pollution*, *211*, 259–270. https://doi.org/10.1016/j.envpol.2015.11.027

Bi, X., Simoneit, B. R., Wang, Z., Wang, X., Sheng, G., &Fu, J. (2010). The major components of particles emitted during recycling of waste printed circuit boards in a typical e-waste workshop of South China. *Atmospheric Environment*, *44*(35), 4440–4445. https://doi.org/10.1016/j.atmosenv.2010.07.040

Bridgen K, Labunska I, Santillo D, Allsopp M (2005). Recycling of electronic waste in China & India: Workplace & Environmental Contamination. *Greenpeace international, Amsterdam.* http://www.greenpeaceto/publications.htm

Chen, D., Bi, X., Zhao, J., Chen, L., Tan, J., Mai, B., … & Wong, M. (2009). Pollution characterization and diurnal variation of PBDEs in the atmosphere of an E-waste dismantling region. *Environmental Pollution, 157*(3), 1051–1057. https://doi.org/10.1016/j.envpol.2008.06.005

Dai, Y., Huo, X., Zhang, Y., Yang, T., Li, M., & Xu, X. (2017). Elevated lead levels and changes in blood morphology and erythrocyte CR1 in preschool children from an e waste area. *Science of the Total Environment, 592*, 51–59. https://doi.org/10.1016/j.scitotenv.2017.03.080

Deng, W. J., Louie, P. K. K., Liu, W. K., Bi, X. H., Fu, J. M., & Wong, M. H. (2006). Atmospheric levels and cytotoxicity of PAHs and heavy metals in TSP and PM2. 5 at an electronic waste recycling site in southeastChina.*AtmosphericEnvironment*,*40*(36),6945–6955.https://doi.org/10.1016/j.atmosenv.2006.06.032

Deng, W. J., Zheng, J. S., Bi, X. H., Fu, J. M., Wong, M. H., 2007. Distribution of PBDEs in air particles from an electronic waste recycling site compared with Guangzhou and Hong Kong, South China. *Environment International, 33* (8), 1063–1069. https://doi.org/10.1016/j.envint.2007.06.007.

Duan, H., Li, J., Liu, Y., Yamazaki, N., & Jiang, W. (2011). Characterization and inventory of PCDD/Fs and PBDD/Fs emissions from the incineration of waste printed circuit board. *Environmental science & technology, 45*(15), 6322–6328.https://doi.org/10.1021/es2007403

Eguchi, A., Nomiyama, K., Devanathan, G., Subramanian, A., Bulbule, K. A., Parthasarathy, P., …&Tanabe, S. (2012). Different profiles of anthropogenic and naturally produced organohalogen compounds in serum from residents living near a coastal area and e-waste recycling workers in India. *Environment international, 47*, 8–16. https://doi.org/10.1016/j.envint.2012.05.003

Ejiogu, A. R. (2013). E-waste economics: a Nigerian perspective. *Management of Environmental Quality: An International Journal, 24*(2), 199213. https://doi.org/10.1108/14777831311303083

ESDO.(2014). Magnitude of the Flow of E-waste in Bangladesh. 1–40.

Forti, V., Balde, C. P., Kuehr, R., & Bel, G. (2020). The Global E-waste Monitor 2020: Quantities, flows and the circular economy potential.

Gaidajis, G., Angelakoglou, K., & Aktsoglou, D. (2010). E-waste: environmental problems and current management. *Journal of Engineering Science and Technology Review, 3*(1), 193–199.

Grause, G., Karakita, D., Ishibashi, J., Kameda, T., Bhaskar, T., & Yoshioka, T. (2011). TG–MS investigation of brominated products from the degradation of brominated flame retardants in high-impact polystyrene. *Chemosphere, 85*(3), 368–373. https://doi.org/10.1016/j.chemosphere.2011.06.104

Gu, Z., Feng, J., Han, W., Wu, M., Fu, J., & Sheng, G. (2010). Characteristics of organic matter in PM2. 5 from an e-waste dismantling area in Taizhou, China. *Chemosphere, 80*(7), 800–806. https://doi.org/10.1016/j.chemosphere.2010.04.078

Huang, D. Y., Zhou, S. G., Hong, W., Feng, W. F., & Tao, L. (2013). Pollution characteristics of volatile organic compounds, polycyclic aromatic hydrocarbons and phthalate esters emitted from plastic wastes recycling granulation plants in Xingtan Town, South China. *Atmospheric Environment, 71*, 327–334. https://doi.org/10.1016/j.atmosenv.2013.02.011

Kaya, M., (2016). Recovery of metals from electronic waste by physical and chemical recycling processes. *International Journal of Chemical, Molecular, Nuclear, Materials and Metallurgical Engineering*10 (2), 232–243.

Li, H., Yu, L., Sheng, G., Fu, J., & Peng, P. A. (2007). Severe PCDD/F and PBDD/F pollution in air around an electronic waste dismantling area in China. *Environmental Science & Technology, 41*(16), 5641–5646. https://doi.org/10.1021/es0702925

Muenhor, D., Harrad, S., Ali, N., & Covaci, A. (2010). Brominated flame retardants (BFRs) in air and dust from electronic waste storage facilities in Thailand. *Environment international, 36*(7), 690–698. https://doi.org/10.1016/j.envint.2010.05.002

Nguyen, L. V., Diamond, M. L., Venier, M., Stubbings, W. A., Romanak, K., Bajard, L., …&Arrandale, V. H. (2019). Exposure of Canadian electronic waste dismantlers to flame retardants. *Environment international, 129*, 95–104. https://doi.org/10.1016/j.envint.2019.04.056

Park, J. S., Wade, T. L., & Sweet, S. T. (2002). Atmospheric deposition of PAHs, PCBs, and organochlorine pesticides to Corpus Christi Bay, Texas. *Atmospheric Environment, 36*(10), 1707–1720. https://doi.org/10.1016/S1352-2310(01)00586-6

Ramachandra, T. V., & Saira, V. K. (2004). Environmentally sound options for e-wastes management. *Envis Journal of Human Settlements, 5*.

Sepúlveda, A., Schluep, M., Renaud, F. G., Streicher, M., Kuehr, R., Hagelüken, C., & Gerecke, A. C. (2010). A review of the environmental fate and effects of hazardous substances released from electrical and electronic equipments during recycling: Examples from China and India. *Environmental impact assessment review, 30*(1), 28–41. https://doi.org/10.1016/j.eiar.2009.04.001

Suvarapu, L. N., & Baek, S. O. (2017). Determination of heavy metals in the ambient atmosphere: A review. *Toxicology and industrial health*, *33*(1), 79–96. https://doi.org/10.1177/0748233716654827

Zhang, B., Huo, X., Xu, L., Cheng, Z., Cong, X., Lu, X., & Xu, X. (2017). Elevated lead levels from e-waste exposure are linked to decreased olfactory memory in children. *Environmental Pollution*, *231*, 1112–1121. https://doi.org/10.1016/j.envpol.2017.07.015

Zheng, X., Xu, X., Yekeen, T. A., Zhang, Y., Chen, A., Kim, S. S., …& Huo, X. (2016). Ambient air heavy metals in PM2. 5 and potential human health risk assessment in an informal electronic-waste recycling site of China. *Aerosol and Air Quality Research*, *16*(2), 388–397. https://doi.org/10.4209/aaqr.2014.11.0292

6 Electronic Wastes and Contamination of Water

Laxmi Kant Bhardwaj
Amity Institute of Environmental Toxicology, Safety and Management, Amity University, Noida, India

Prangya Rath
Amity Institute of Environmental Sciences, Amity University, Noida, India

Moharana Choudhury
Environmental Research and Management Division, Voice of Environment (VoE), Guwahati, India

America Metzdorff
Facultad de Ciencias, Instituto de Ciencias Químicas, Universidad Austral de Chile, Valdivia, Chile

Pankaj Kanti Jodder
Department of Geography and Earth Science, University of North Carolina at Charlotte, Charlotte, North Carolina, USA

6.1 IMPORTANCE OF WATER FOR SUSTAINING LIFE

Water is a precious gift from nature for all flora and fauna (Bhardwaj and Sharma, 2021a; Bhardwaj, 2022a). Water is the medium through which, nature, weaves the connection between biotic and abiotic components. Every living organism on this planet depends on water for survival. The ability to act as a universal solvent, temperature buffer, and medium for cellular metabolism, makes, water, essential for sustaining life. The fundamental role of water is to facilitate the growth of living organisms, serve as a habitat or reproductive environment for various microorganisms, and offer the necessary nutrients and minerals for sustaining life. Water serves as a medium for the transportation of oxygen, minerals, nutrients, and waste products to and from the cells (Bhardwaj et al., 2025). Water provides lubrication to the mucous membranes of the respiratory and gastrointestinal tracts, which are essential for the correct functioning of the digestive system. Water is the medium through which the body generates energy through metabolic processes and other chemical reactions, and also acts as a shock absorber or padding material between organs. Water covers three-fourths of the surface of the Earth, of which, approximately 97 % is salt water and the remaining percentage is freshwater, which is frozen in ice caps and glaciers. Living organisms can only use less than 1 % of the water on Earth to meet daily needs. Human beings rely heavily on groundwater for drinking purposes. Groundwater is an important reserve of high-quality water and contains 90 % of the freshwater resources. Unfortunately, urbanization has led to significant increases in the destruction of water resources. Contamination of water bodies through e-wastes, Micro-Nano Plastics along with other hazardous chemicals has become a global concern. Several techniques have emerged for the

DOI: 10.1201/9781003582311-6

treatment of water, but nanotechnology is a cheaper and more efficient technique as compared to other techniques (Bhardwaj et al., 2022). In the present chapter, the author focuses on the ways by which e-wastes contaminate water, its impact on aquatic ecosystems, specifically phytoplankton and aquatic microorganisms, and the ways by which plant and animal species and human beings are getting affected. This chapter also provides a concise description of ways by which pollutants from e-wastes can be identified and quantified from water bodies.

6.2 ELECTRONIC WASTES AND CONTAMINATION OF WATER

When electronic devices such as computers, printers, scanners, fax machines, calculators, compact discs (CDs), televisions, medical equipment, transceivers, VCRs, battery cells, monitors, iPods, refrigerators, smart watches, tablets, video games, laptops, camera, xerox machines, cellular phones, air conditioners, and washing machines (to name a few) are no longer fit for human use or have passed their expiration date, they become e-waste (Table 6.1 and Figure 6.1A). The primary factor behind the rapid increase in e-waste production is the swift replacement of old and existing electronic equipment with newer and advanced ones. Due to this tendency of people to switch to new models, the lifespan of equipment has decreased significantly. A number of industries, including information technology (IT), pharma, food, chemical, electrical, telecom, business process outsourcing (BPO), and the environment, have seen an increase in the use of electronic equipment (Iyer, 2018). The generation of e-wastes is directly proportional to technological advancement and the exponential rise in the need for system upgradation (Kiddee et al., 2013) and management of E-waste is important (Agarwal et al., 2023). The hazardous chemical compounds generated from e-wastes include organic compounds such as Brominated Flame Retardants (BFRs), Chlorinated Benzenes (CBs), Polycyclic Aromatic Hydrocarbons (PAHs), Polybrominated Diphenyl Ethers (PBDEs), Polychlorinated Biphenyls (PCBs), Polybrominated Biphenyls (PBBs), Hexabromocyclododecane (HBCDs), Polychlorinated Dibenzo-p-dioxins (PCDDs), Polybrominated Dibenzo-p-dioxins (PBDDs), Polychlorinated Dibenzofurans (PCDFs), and Polybrominated Dibenzofurans (PBDFs) and inorganic compounds such as Lead (Pb), Cadmium (Cd), Arsenic (As), Mercury (Hg), Nickel (Ni), Selenium (Se), Zinc (Zn), Cobalt (Co), Chromium (Cr), Copper (Cu), and Tin (Sn) to name a few, contaminates the air, water, and soil (Panwar and Ahmed, 2018). The e-wastes (having the potential to cause a public health crisis), present in the environment, enter the food chain through bio-accumulation (Luo et al., 2011).

TABLE 6.1
Different Types of Wastes Produced by Different Electrical Devices

S. No.	Electronic Devices	Waste Produced from Electrical Devices
1	LCD screens	Mercury (Hg), Lead (Pb)
2	Computer	Plastic housing, screws, small plastic components, small metal parts (ferrous-containing and non-ferrous containing groups), circuit board (metals, precious metals, and fiberglass), clips, Lead, Cadmium, hexavalent Chromium, monitors, Beryllium, Brominated Flame Retardants (BFRs), Mercury (Hg)
3	Mobile phones	Arsenic (As), Cadmium (Cd), Palladium (Pd), Lead (Pb)
4	Optical fibers	Lead (Pb), Copper (Cu), Zirconium (Zr), Yttrium (Y), Fluorine (F)
5	Television	Mercury (Hg), Lead (Pb), plastic, glass, circuit board, small metal parts
6	Rechargeable batteries	Cadmium (Cd), Lithium (Li), Nickel (Ni)
7	Laptop	Small metal parts, Lead (Pb), Nickel (Ni), plastics, Lithium (Li), Cadmium (Cd), circuit board, Mercury (Hg)
8	Computer mouse	Small metal parts, plastic, circuit board
9	Electric motor	Copper coil, metal parts, Lead (Pb), Mercury (Hg), plastic, different wires
10	Electric bulb	Tungsten filament, wire, small metal part, glass, low-pressure inert gas
11	Keyboard	Small metal parts, circuit board, plastic, wire

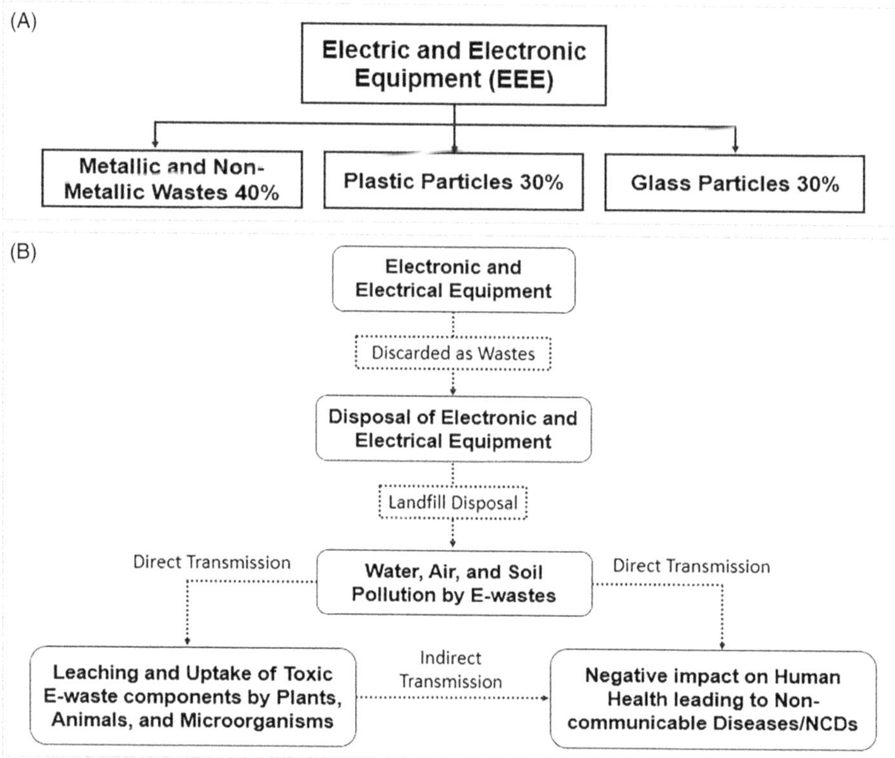

FIGURE 6.1 (A) Composition percentage of electronic wastes. (B) Fate of E-wastes from generation to abiotic sources to biotic sources.

Heavy metals pose a major threat because of their mobility, toxicity, and inability to break down in the environment (Alam et al., 2023; Bhardwaj et al., 2023). These metals can be found in dissolved, particulate, and colloidal states (Adepoju-Bello et al., 2009). As a result of leaching, these metals are more concentrated in the water and soil of areas with e-waste recycling facilities (Marcovecchio et al., 2007). Researchers have reported a high concentration of Pb, Cu, and Cd in paddy fields, which were far from e-waste recycling facilities and were primarily transported by irrigation water (Wu et al., 2015). The small fragments of plastics can enter the environments of freshwater from wastewater effluent and degraded plastic wastes and act as carrier of heavy metals (Bhardwaj and Sharma, 2021b; Bhardwaj, 2022b; Bhardwaj, 2023; Bhardwaj et al., 2024).

6.3 EFFECTS OF ELECTRONIC WASTES ON WATER QUALITY

Frequent burning, dumping, and recycling of e-waste impart a number of negative effects are on the environment. Due to the illegal importation of e-waste from other nations, developing nations are confronted with significant environmental challenges. Over the course of decades, the water that was used to wash the e-waste was dumped into nearby water bodies, resulting in heavy metal contamination. Likewise, unsalvageable waste produced subsequent to washing was placed in the dirt, accordingly, the unsafe substances can drain into the groundwater by precipitation. The use of such stream water for irrigation may result in heavy metal contamination of the vegetation and paddy fields. Researchers have reported that the recycling of electronic waste had no effect on shallow groundwater and Nigerian well water (Wu et al., 2014). Studies have shown that well water, an important source of potable water in Nigeria is less affected by the recycling activities of e-wastes (Alabi and Bakare, 2015). The lesser concentration of heavy metals was found in the ponds and was due to transfer from the surface soil through rainfall, prominently in wet seasons (Rahman et al., 2012). Cd, Cr, Mn, Ni, Pb, and Cu were reported in riverine water in Guiya (China) which was situated near the e-waste recycling site (Wong et al., 2007). Higher concentrations of Pb, Cd, Cr, Zn, PAHs, PCBs, and PBDEs were reported in the edible plants which were growing near e-waste markets in Nigeria (Alabi et al., 2021). The presence of these pollutants in plants is because of open burning and indiscriminate disposal of e-waste. It has also been demonstrated that the e-waste dumpsite leachates produced during rainy seasons contain high levels of heavy metals, PBDEs, PCBs, and PAHs. Underground water is the primary source of potable water for most people in most places. Approximately 70% of surface water in India is polluted (Rao and Mamatha, 2004). There are two ways that e-waste can contaminate water bodies: First, Dumping e-wastes in landfills can contaminate the groundwater and surface water because the hazardous substances can leach into water supplies. And second, Improper recycling generates toxic by-products which can contaminate the residential water supply after that, these by-products may lead to further contamination by entering the surface water namely ponds and rivers.

6.4 EFFECTS OF ELECTRONIC WASTES ON PHYTOPLANKTON AND AQUATIC MICROORGANISMS

The toxic substances emitted from electronic wastes can cause harmful impacts on the physiology, reproduction, development, metabolism, growth, and behavior of several aquatic creatures. These persistent toxins do not degrade over time and are transferred from one creature to another. Individuals with a significant reliance on fish in their diet are more susceptible to the adverse effects of hazardous chemicals. However, data are needed to determine the possible detrimental impacts of electronic waste on aquatic creatures. However, numerous researchers have documented various forms of unsafe chemicals derived from diverse creatures. Heavy metals (e.g., Pb, Cd, Cu, and Zn) and organic pollutants (e.g., PBDEs and PCDD/Fs) were reported in the sediments of resident water bodies of Ghana (Huang et al., 2014) and these contaminants were hazardous to aquatic species. High levels of PCBs, PCDDs, and PCDFs were reported in eight aquatic biota species in China (Zhu et al., 2015). Bisphenol A-bis diphenyl phosphate (BPA-BDPP) (< 0.002–0.21 ng/l in surface water; 0.005–89 ng/g in sediment; 0.012–0.52 ng/g in biota species) and 1,3–phenylene bis diphenyl phosphate (PBDPP) (< 0.004–1.2 ng/g in sediment; < 0.002–0.22 ng/g in aquatic organisms) were reported in Taihu Lake, China (Zhao et al., 2019). Cu (1657.14 ± 259.3 μg/g) and PBDE (7831.32 ± 1273.86 ng/g) were reported in the leachate and they might be playing a partial role in producing deleterious effects on Daphnia Magna (Bao et al., 2020).

6.5 CONTAMINATION OF WATER AND ITS EFFECT ON PLANTS AND ANIMALS

Electronic waste pollutes water, which subsequently infiltrates the food chain. Plant roots and other portions of the plant absorb heavy metals like Pb, Ni, Fe, Zn, Mn, and As (Sharma and Agrawal, 2005). Both animals and humans ingest these plants. Upon entering the body, these metals gradually accumulate and have the potential to harm vital organs such as the liver and kidneys (Okereafor et al., 2020). E-waste pollution in water bodies leads to the modification of nutrient composition in aquatic ecosystems, hence promoting the growth of algal blooms. Excessive proliferation of algae results in eutrophication and reduced amounts of oxygen in water (Newtech Recycling, 2020). Metals' toxicity hampers the growth of roots and shoots, disrupts the process of photosynthesis, and induces oxidative stress in plants (Rizwan et al., 2019). The toxicity of Cu and Pb in plants results in delayed germination, reduced seedling length, and changes in the quantity of lateral roots (Zulfiqar et al., 2019). The ingestion of plants growing in close proximity to e-waste landfills can lead to various health ailments (Vongdala et al., 2019). Research has demonstrated that contaminants, such as PBDE and heavy metals, have the potential to impact soil fertility negatively (Vu et al., 2019). Toxic substances, such as Barium (Ba) and Lithium (Li), those are produced from electronic trash (e-waste) have been discovered to have detrimental effects on several parts of the human body, including the nervous system, heart, liver, kidneys, and reproductive system. The bacterial activity in e-waste produces toxic methylated mercury, which then enters the human food chain and spreads through the processes of bioaccumulation and biomagnification (Luhar and Luhar, 2019). The noxious compounds included in electronic wastes have been identified as the causative agents causing gill hypertrophy, gill necrosis, intestinal mucosal necrosis, and necrotizing hepatitis in aquatic organisms (Ibrahim et al., 2020). India showed a significantly higher quantity of Pb (17–247 times higher) in wastewater, which was attributed to e-waste activities (Alabi et al., 2021). Groundwater near the e-waste market was found to have a high content of Cr, As, Pb, and Cd, exceeding the permissible limits for drinking water by a wide margin (Alabi et al., 2021).

6.6 CONTAMINATION OF WATER AND ITS EFFECT ON HUMAN HEALTH

Electronic trash presents a significant hazard to humans, animals, and the ecosystem. The majority of electronic waste is commonly disposed of worldwide through methods such as open burning or acid baths. Substances such as dioxins, furans, and heavy metals are discharged into the environment, which leads to their toxicity and negatively impacts the health of workers and individuals residing in close proximity to landfills. Individuals of all genders and age groups, including men, women, and children, are engaged in basic and environmentally harmful recycling methods to recover metals, inks, and plastics from electronic garbage such as computers. Developing nations employ more detrimental and costly techniques, such as incinerating electrical devices to liquefy polymers and retrieve metals. As a result, carcinogens and neurotoxins are emitted into the atmosphere, contributing to the formation of smog. For instance, the combustion of fibers can release brominated and chlorinated dioxins, which have the potential to cause cancer in people. E-waste is a growing concern in India, posing risks to human health and the environment. This issue is mostly attributed to a lack of awareness and poor levels of education (Kwatra et al., 2014). Workers are unknowingly involved in the recycling, processing, and disposal of electronic garbage. Workers failed to adhere to the correct usage of personal protective equipment (PPE), perhaps leading to additional hazardous exposure from e-waste. According to Monika (2010), some 25,000 workers in Delhi are employed in raw e-waste treatment facilities, where they manually handle 10,000–20,000 tonnes of e-waste without any protective equipment. Workers engage in the illicit disposal of electronic trash outside of established procedures, which can result in significant health risks (Park and Soni, 2019). Exposure to toxic substances can happen either through direct contact with the skin, breathing in tiny particles, or consuming dust that is contaminated. Exposure to electronic trash (e-waste) can result in various health issues, such as alterations in thyroid function, adverse birth outcomes, miscarriage, stillbirth, early birth, and stunted child growth (Grant et al., 2013).

Various adverse effects of exposure were also reported, such as behavioral changes, decreased lung function, and DNA damage. Pregnant women and toddlers are susceptible to health hazards posed by electronic waste. Recycling electronic debris has resulted in elevated levels of lead in youngsters' bloodstreams. Exposure to e-waste has been associated with abortion, premature birth, and a decrease in birth weight. Tap water from residential areas near landfills contains heavy metals, including Pb, Cr, Ni, Zn, and Cd (Olafisoye et al., 2013). The concentrations of heavy metals were higher during the summer compared to the wet season. Female animals subjected to Pb toxicity (Bhardwaj et al., 2023) have shown an increased incidence of miscarriages and deaths. Exposure to excessive levels of lead, which is a form of metal poisoning, can cause severe symptoms like abdominal discomfort, irritability, exhaustion, and lethargy. Pregnant women and children are susceptible to health hazards associated with electronic waste. Recycling electronic debris has led to higher levels of lead in the blood of youngsters. Exposure to e-waste has been associated with abortion, premature birth, and a decrease in birth weight. Tap water from residential areas near landfills contains heavy metals, including Pb, Cr, Ni, Zn, and Cd (Olafisoye et al., 2013). During the summer, the concentrations of heavy metals were higher compared to the rainy season. Female animals subjected to Pb toxicity have shown an increased incidence of miscarriages and deaths, as reported by Bhardwaj et al. in 2023. Exposure to excessive levels of lead, which is a form of metal poisoning, can result in significant symptoms like abdominal discomfort, irritability, weariness, lethargy, and anemia (Cecil et al., 2008). Pb toxicity has also been reported as a possible human carcinogen. Ni has been observed to seep from the soil into groundwater. It causes damage to the skin epidermis and produces asthma symptoms for approx. 10 to 20 % of the population is directly affected by Ni toxicity (Lenntech, 2009). Workers exposed to the most Ni dust suffered from bronchitis and lung damage. Nickel hydroxide has been shown to act as a carcinogen. The incineration of e-waste in landfills is one of the causes of nitrogen exposure to the atmosphere (Lenntech, 2009). Zinc is essential for the physiology of plants and animals, but high levels of Zn in water have been reported to cause problems such as bitter taste, acrid taste, and milky color (Lenntech, 2009). The presence of used ZnS in the glass of the display is one of the causes of the waste of Zn electrons. Zn exposure can also result from burnt and improperly disassembled computer system monitors. Direct exposure or ingestion of Zn is very harmful and leads to skin and lung problems (Adaramodu et al., 2012). Cd is mainly found in infrared detectors, semiconductors, and chip resistors. Previous screen samples contained approx. 5–10 g Cd. Ni-Cds are used in the manufacture of several types of batteries. Cds enter the environment through many electronic elements. The presence of lower amounts of Cd in the waters can be attributed to the reuse of infrared detectors, semiconductors, and chip resistors. Exposure to Cd occurs through breathing and ingesting contaminated food and water. Cr (VI) is a toxic form of Cr metal. It is used to provide a layer of protection in electronic items. Prolonged exposure to Cr causes high blood pressure, kidney damage, asthma and bronchitis, and damages DNA. Landfill leachate or fly ash are the main causes of burning Cr (VI). The various hazardous substances and their sources and associated health effects are presented in Table 6.2. The direct and indirect effects of e-waste are shown in Figure 6.1B.

TABLE 6.2
Chemical Compounds, with Their Sources and Health Effects

S. No.	Name of the Compounds	Sources	Health Effects
1	Cadmium	PCBs, rechargeable batteries, switches, CRTs, light-sensitive resistors, and UV stabilizers in older PVC cables	Pulmonary, carcinogen, bone structure, deficits in learning, and kidney damage.
2	Polyvinyl chloride (PVC)	Insulation of electrical cables	After burning dioxin is produced which is an endocrine disrupter
3	Lead	Glass panels, CRT monitor glass, lead-acid batteries, PCBs, and gaskets in computer monitors	Damage to the C.N.S (central nervous system) and kidney, miscarriage, lower IQ in children
4	Antimony	Trace metals	Carcinogen to humans
5	Barium	CRTs	Muscle weakness, damage to the heart, liver, and spleen
6	Phthalates	Plastics especially PVC	Carcinogenic and toxic to reproduction
7	Beryllium	Circuit boards	Lung cancer
8	Mercury	Switches, fluorescent tubes, lead acid batteries	Impairment of sensory organs, dermatitis, loss of memory, weakness in muscles, failure of the nervous system particularly in children
9	Chloro-benzenes	Formed during the combustion of the chlorinated plastics PVC	Hazardous impacts on the thyroid, liver, and C.N.S (central nervous system)
10	Hexavalent chromium	Metal coatings to guard against corrosion	Carcinogenic, inhibit cell proliferation, and cause DNA single-strand breaks
11	Sulfur	Lead-acid batteries	Damage to the liver, kidneys, heart, and irritation in the eyes and throat
12	Brominated flame retardants (BFRs)	Most of the electronic items	Thyroid and liver problems

6.7 DETERMINATION OF ELECTRONIC WASTES IN WATER BODIES

E-waste contamination is a major cause of groundwater pollution, which degrades the quality of water for drinking and other purposes (Gupta and Nath, 2020). High concentrations of heavy metals such as Cd (0.002–0.005 mg/l), Cr (0.001–0.008 mg/l), Ni (0.001–0.011 mg/l), Pb (0.007–0.020 mg/l), and Zn (0.003–0.007 mg/l) were reported in well water samples (Otache et al., 2014). Very high concentrations of dioxin and Pb have been observed in local water pools situated near e-waste dumping sites (Arya and Kumar, 2020). The presence of heavy metals in groundwater is responsible to play role in the electrical conductivity (EC) of ions and salts present in groundwater. The hardness and TDS of groundwater increased due to the presence of bicarbonates, chloride, and sulphates of calcium and magnesium (Khan et al., 2021). The water bodies of Delhi have been contaminated with the leaching of heavy metals present in e-wastes (Khan et al., 2021). CPCB and the state government together signed a memorandum to determine the e-waste in water bodies in Bhopal in 2019 and set the country's first e-waste clinic in Bhopal (Bhopal News-Times of India, 2019). Traceable amounts of PBDE and >16 times higher heavy metals such as Cd, Cr, Pb, etc. were reported in the samples of water bodies near e-waste dumping sites in Delhi (Khan et al., 2021). Furans present in e-wastes get accumulated in water bodies and form complexes. These complexes can be detected by FTIR techniques (Lockwood et al., 2018). Similarly, researchers reported Zn and Cu in water bodies and stated that they were >29 times higher than water standards and >8 times higher than referenced water levels (Khan et al., 2021). The different parameters such as pH, temperature, turbidity, TDS, BOD, COD, conductivity, total alkalinity, sulphates, nitrates, Zn, Cu, Fe, Ni, Hg, Pb, etc. were observed to be more in the water samples obtained from contaminated areas (Ma et al., 2020). A novel chromogenic extractant method was used for the determination of the level of Cu (II) in water samples (Kharade et al., 2022).

6.8 FUTURE CONCERNS AND RECOMMENDATIONS

Approximately two-thirds (66%) of the world's population are subject to e-waste regulations, although only one-fifth (20%) of global e-waste is recycled yearly. Most electronic wastes are disposed of at landfills located near industrial zones. Consequently, the harmful compounds emitted from these electronic wastes have the potential to contaminate water bodies through leaching. This water becomes unsuitable for consumption and other uses. Research has shown that the level of pollutants is higher near garbage sites and decreases as the distance between the waste sites and water bodies increases. The governments of both wealthy and developing countries face a significant forthcoming challenge in managing electronic trash (e-waste). The exponential growth of e-waste poses a substantial threat to public health. The escalating issues linked to e-waste necessitate a heightened focus on endeavors about e-waste recycling and advancements in e-waste management. Keeping all the above things in mind, we are giving some recommendations which are as follows:

- Physical and chemical techniques have to be used to prevent future pollutant dissemination.
- Insights into solutions containing plants and microbes' combinations for remediating highly contaminated sites are required.
- It is advised that the government encourages new entrepreneurs to develop solutions for removing e-waste by providing them with necessary financial support and technical guidance.
- To minimize illegal activities related to e-waste recycling procedures, the government needs to ensure initiatives to build more legal e-waste disposal centers.
- For the management of e-wastes, the government could declare incentives that help in ensuring compliance across electronics industries. It could be either in the form of concessions in tax or provide rebates.

- The associated competent bodies in developing as well as transition countries are required to frame mechanisms that could handle and treat e-waste in a safer and more sustainable procedure.
- Activities like increasing awareness campaigns, capacity building, etc. are critically required for promoting e-waste management programmes that are also environment friendly.
- It is strongly recommended that future research should focus more on examining a broader scale and long-term approaches for reducing the levels of contamination in the biota in the nearby water bodies.

REFERENCES

Adaramodu, A. A., Osuntogun, A. O., & Ehi-Eromosele, C. O. (2012). Heavy metal concentration of surface dust present in e-waste components: the Westminister Electronic Market, Lagos case study. *Resour Environ*, *2*(2), 9–13.

Adepoju-Bello, A. A., Ojomolade, O. O., Ayoola, G. A., & Coker, H. A. B. (2009). Quantitative analysis of some toxic metals in domestic water obtained from Lagos metropolis. *The Nigeria Journal of Pharmacy*, *42*(1): 57–60.

Agarwal, S., Darbar, S., Saha, S., Choudhury, M., & Singh, R. P. (2023). E-waste management using different cost-effective, eco-friendly biological techniques: an overview. *Waste Management and Resource Recycling in the Developing World*, 205–235, https://doi.org/10.1016/B978-0-323-90463-6.00026-9

Alabi, O. A., Adeoluwa, Y. M., Huo, X., Xu, X., & Bakare, A. A. (2021). Environmental contamination and public health effects of electronic waste: An overview. *Journal of Environmental Health Science and Engineering*, *19*(1), 1209–1227.

Alabi, O. A., & Bakare, A. A. (2015). Perceived public health effects of occupational and residential exposure to electronic wastes in Lagos, Nigeria. *Zoologist*, *13*, 62–71.

Alam, S., Bhardwaj, L.K., Mallick, R., & Rai, S. (2023). Estimation of Heavy Metals and Fluoride Ion in Vegetables Grown Nearby the Stretch of River Yamuna, Delhi (NCR), India. *Indian Journal of Environmental Protection*, *43* (1), 64–73.

Arya, S., & Kumar, S. (2020). E-waste in India at a glance: Current trends, regulations, challenges and management strategies. *Journal of Cleaner Production*, *271*, 122707.

Bao, S., Pan, B., Wang, L., Cheng, Z., Liu, X., Zhou, Z., & Nie, X. (2020). Adverse effects in Daphnia magna exposed to e-waste leachate: Assessment based on life trait changes and responses of detoxification-related genes. *Environmental Research*, *188*, 109821.

Bhardwaj, L. K. (2022a). Evaluation of Bis (2-ethylhexyl) Phthalate (DEHP) in the PET Bottled Mineral Water of Different Brands and Impact of Heat by GC–MS/MS. *Chemistry Africa*, *5*, 929–942.

Bhardwaj, L. K., Rath, P., & Choudhury, M. (2022). A Comprehensive Review on the Classification, Uses, Sources of Nanoparticles (NPs) and Their Toxicity on Health. *Aerosol Science and Engineering*, *7*, 69–86.

Bhardwaj, L. K., & Sharma, A. (2021a). Estimation of physico-chemical, trace metals, microbiological and phthalate in PET bottled water. *Chemistry Africa*, *4*(4), 981–991.

Bhardwaj, L. K. (2022b). Microplastics (MPs) in drinking water: uses, sources and transport. *Futuristic Trends in Agriculture Engineering & Food Sciences, IIP Proceedings*, volume 2, book 9, part 1, chapter 28. ISBN: 978-93-95632-65-2.

Bhardwaj, L. K., & Sharma, A. (2021b). Microplastics (MPs) in Drinking Water: Uses, Sources & Transport. *Preprints* 2021, 2021040498. https://doi.org/10.20944/preprints202104.0498.v1)

Bhardwaj, L. K., Sharma, S., & Jindal, T. (2023). Estimation of Physico-Chemical and Heavy Metals in the Lakes of Grovnes & Broknes Peninsula, Larsemann Hill, East Antarctica. *Chemistry Africa*, *6*, 2677–2694

Bhardwaj, L. K., Rath, P., Rehman, F., Yadav, P., Massahi, T., Mittal, R., & Jain, H. (2025). Status of Micro/Nano-Plastic (MNP) Pollution in the Aquatic System. In *Societal and Environmental Ramifications of Plastic Pollution* (pp. 43–62). IGI Global.

Bhardwaj, L. K. (2023). Occurrence of microplastics (MPs) in Antarctica and its impact on the health of organisms. *Maritime Technology and Research*, 6(2), 265418. https://doi.org/10.33175/mtr.2024.265418

Bhopal News-Times of India (2019). India's first e-waste clinic in Bhopal soon. *The Times of India*. Retrieved from https://timesofindia.indiatimes.com/city/Bhopal/countrys-first-e-waste-clinic-in-bhopal-soon/articleshow/71342753.cms

Bhardwaj, L. K., Rath, P., Yadav, P., & Gupta, U. (2024). Microplastic contamination, an emerging threat to the freshwater environment: a systematic review. *Environmental Systems Research*, *13*(1), 8. https://doi.org/10.1186/s40068-024-00338-7

Cecil, K. M., Brubaker, C. J., Adler, C. M., Dietrich, K. N., Altaye, M., Egelhoff, J. C., ... & Lanphear, B. P. (2008). Decreased brain volume in adults with childhood lead exposure. *PLoS Medicine*, *5*(5), e112.

Grant, K., Goldizen, F. C., Sly, P. D., Brune, M. N., Neira, M., van den Berg, M., & Norman, R. E. (2013). Health consequences of exposure to e-waste: a systematic review. *The Lancet Global Health*, *1*(6), e350–e361.

Gupta, N., & Nath, M. (2020). Groundwater contamination By E-waste and its remedial measure-a literature review. In *Journal of Physics: Conference Series* (Vol. *1531*, No. 1, p. 012023). IOP Publishing.

Huang, J., Nkrumah, P. N., Anim, D. O., & Mensah, E. (2014). E-waste disposal effects on the aquatic environment: Accra, Ghana. *Reviews of environmental contamination and toxicology*, *229*, 19–34. https://doi.org/10.1007/978-3-319-03777-6_2

Ibrahim, A., Narayanan, M., Salman, K., & Mohamed, S. (2020). The effect of freshwater acidification in the tissues of Cyprinus carpio. *Global Journal of Zoology*, *5*(1), 009–015.

Iyer, L. S. (2018). Knowledge, Attitude and Behaviour (KAB) of Student Community Towards Electronic Waste-A Case Study. *Indian Journal of Science and Technology*, *11*(10), 1–10.

Khan, J., Kumar, A., Giri, A., Pal, D. B., Tripathi, A., & Giri, D. D. (2021). Impact of electronic waste pollutants on underground water. *Groundwater Geochemistry: Pollution and Remediation Methods*, 265–281, https://doi.org/10.1002/9781119709732.ch13.

Kharade, S. B., Chougale, R. K., Barache, U. B., Sanadi, K. R., Rathod, K. C., Gaikwad, S. H., ... & Kamble, G. S. (2022). Design and optimization of sensitive analytical spectrophotometric method for micro determination of copper (II) from e-waste by using of novel chromogenic extractant. *Spectrochimica Acta Part A: Molecular and Biomolecular Spectroscopy*, *267*, 120502.

Kiddee, P., Naidu, R., & Wong, M. H. (2013). Electronic waste management approaches: An overview. *Waste management*, *33*(5), 1237–1250.

Kwatra, S., Pandey, S., & Sharma, S. (2014). Understanding public knowledge and awareness on e-waste in an urban setting in India: A case study for Delhi. *Management of Environmental Quality: An International Journal*, *25*(6), 752–765, https://doi.org/10.1108/MEQ-12-2013-0139.

Lenntech, W. T. (2009). Chemical properties, health and environmental effects of copper. *Lenntech Water Treatment and Purification Holding BV* www.lenntech.com/periodic/elements/cu.htm

Lockwood, S. P., Fuller, T. G., & Newby, J. J. (2018). Structure and spectroscopy of Furan: H_2O complexes. *The Journal of Physical Chemistry A*, *122*(36), 7160–7170.

Luhar, S., & Luhar, I. (2019). Potential application of E-wastes in construction industry: A review. *Construction and Building Materials*, *203*, 222–240.

Luo, C., Liu, C., Wang, Y., Liu, X., Li, F., Zhang, G., & Li, X. (2011). Heavy metal contamination in soils and vegetables near an e-waste processing site, south China. *Journal of hazardous materials*, *186*(1), 481–490.

Ma, J., Wu, S., Shekhar, N. V., Biswas, S., & Sahu, A. K. (2020). Determination of physicochemical parameters and levels of heavy metals in food waste water wit environmental effects. *Bioinorganic Chemistry and Applications*, https://doi.org/10.1155/2020/8886093.

Marcovecchio, J. E., & Botté, S. E. (2007). Heavy metals, major metals, trace elements. In *Handbook of water analysis*, 289–326.

Monika, J. K. (2010). Best Fiber Laser Cutting Machine| Best Fiber Laser Cutting Machine Factory-MORN LASER. *Indian Journal of Community Medicine: Official Publication of Indian Association of Preventive & Social Medicine*, *35*(3), 382.

Newtech Recycling (2020). The Effects of E-Waste on Water. Retrieved from https://newtechrecycling.com/the-effects-of-e-waste-on-water/

Okereafor, U., Makhatha, M., Mekuto, L., Uche-Okereafor, N., Sebola, T., & Mavumengwana, V. (2020). Toxic metal implications on agricultural soils, plants, animals, aquatic life and human health. *International journal of environmental research and public health*, *17*(7), 2204.

Olafisoye, O. B., Adefioye, T., & Osibote, O. A. (2013). Heavy Metals Contamination of Water, Soil, and Plants around an Electronic Waste Dumpsite. *Polish journal of environmental studies*, *22*(5), 1431–1439.

Otache, M. Y., Musa, J. J., Animashaun, I. M., & Oji, D. M. (2014). Evaluation of the Effects of Electronic Wasteon Topsoil and Groundwater, *International Journal of Science, Engineering and Technology Research (IJSETR)*, *3*(12).

Panwar, R. M., & Ahmed, S. (2018). Assessment of contamination of soil and groundwater due to e-waste handling. *Current Science*, *114*(1), 166–173.

Park, M., & Soni, L. (2019). India's two-million-tonne e-waste problem has deadly consequences. *Quartz India*, https://qz.com/india/1553483/indias-two-million-tonne-e-waste-problem-has-deadly-consequences

Rahman, S. H., Khanam, D., Adyel, T. M., Islam, M. S., Ahsan, M. A., & Akbor, M. A. (2012). Assessment of heavy metal contamination of agricultural soil around Dhaka Export Processing Zone (DEPZ), Bangladesh: implication of seasonal variation and indices. *Applied sciences*, *2*(3), 584–601.

Rao, S. M., & Mamatha, P. (2004). Water quality in sustainable water management. *Current science*, *87*(7), 942–947.

Rizwan, M., Ali, S., & Maqbool, A. (2019). A critical review on the effects of zinc at toxic levels of cadmium in plants. *Environmental Science and Pollution Research*, *26*(7), 6279–6289.

Sharma, R. K., & Agrawal, M. (2005). Biological effects of heavy metals: an overview. *Journal of environmental Biology*, *26*(2), 301–313.

Vongdala, N., Tran, H. D., Xuan, T. D., Teschke, R., & Khanh, T. D. (2019). Heavy metal accumulation in water, soil, and plants of municipal solid waste landfill in Vientiane, Laos. *International journal of environmental research and public health*, *16*(1), 22.

Vu, H. L., Ng, K. T. W., & Bolingbroke, D. (2019). Time-lagged effects of weekly climatic and socio-economic factors on ANN municipal yard waste prediction models. *Waste Management*, *84*, 129–140.

Wong, C. S., Duzgoren-Aydin, N. S., Aydin, A., & Wong, M. H. (2007). Evidence of excessive releases of metals from primitive e-waste processing in Guiyu, China. *Environmental Pollution*, *148*(1), 62–72.

Wu, C., Luo, Y., Deng, S., Teng, Y., & Song, J. (2014). Spatial characteristics of cadmium in topsoils in a typical e-waste recycling area in southeast China and its potential threat to shallow groundwater. *Science of the Total Environment*, *472*, 556–561.

Wu, Q., Leung, J. Y., Geng, X., Chen, S., Huang, X., Li, H., … & Lu, Y. (2015). Heavy metal contamination of soil and water in the vicinity of an abandoned e-waste recycling site: implications for dissemination of heavy metals. *Science of the Total Environment*, *506*, 217–225.

Zhao, H., Liu, L., Li, Y., Zhao, F., Zhang, S., Mu, D., … & Hu, J. (2019). Occurrence, bioaccumulation, and trophic transfer of oligomeric organophosphorus flame retardants in an aquatic environment. *Environmental Science & Technology Letters*, *6*(6), 323–328.

Zhu, C., Wang, P., Li, Y., Chen, Z., Li, W., Ssebugere, P., … & Jiang, G. (2015). Bioconcentration and trophic transfer of polychlorinated biphenyls and polychlorinated dibenzo-p-dioxins and dibenzofurans in aquatic animals from an e-waste dismantling area in East China. *Environmental Science: Processes & Impacts*, *17*(3), 693–699.

Zulfiqar, U., Farooq, M., Hussain, S., Maqsood, M., Hussain, M., Ishfaq, M., … & Anjum, M. Z. (2019). Lead toxicity in plants: Impacts and remediation. *Journal of environmental management*, *250*, 109557.

7 Electronic Wastes and Contamination of Soil

Jyoti Sarwan, Lakshika, Richa Chaurasia, and Jagadeesh Chandra Bose K
University Institute of Biotechnology, Chandigarh University, Chandigarh, India

7.1 IMPORTANCE OF SOIL FOR SUSTAINING LIFE

Any electrical or electronic equipment that has been disposed of by the user with no intention of using it again is referred to as E-waste. With increased development and industrialization have come increased variety and a number of electronic gadgets. Electronic and electrical equipment for sure have made life more convenient to live but increase in the number of gadgets has also risen the E-waste amount generated to a high level and it has become a cause of worry. All sorts of equipment which come under electrical and electronic equipment, including mobiles, scanners, earphones, printers, computers, etc. are considered as E-waste. Toxic substances like cadmium, brominated flame retardants, chromium, (PCBs), and lead are included in E-waste components (WHO, 2019). These toxic substances can take a heavy toll on the health of human beings as well as on environment. The health effects include skin disorders, respiratory problems, asthma, bronchitis, reduced lung development, coughing, and high blood pressure. If we try to estimate, one person contributes to 6 kgs of hazardous E-waste and in total 44.7 m tons as in 2016 (WHO, 2019; Baldé et al., 2017). In countries like India, Nigeria, and Ghana which are still in developing stage and even in developed countries like China rudimentary processing of E-waste is done which generates a lot of toxic residues as well as emissions to soil, water, and air (Ogunseitan 2013). Figure 7.1 shows ways of dismantling of E-waste in developed and developing countries. E-waste has become a major concern source as it has a growth rate of around 3–5% per annum which is actually thrice the rate of growth in municipal solid waste (Ogungbuyi et al. 2012). A big chunk of electronic waste is generated in developed countries like Europe, United States, and China. In countries like Nigeria, due to the production of more attractive models to keep up with the trends, the lifespan of electronics has substantially shortened which is a bigger concern as managing normal wastes there already is a pretty difficult task due to the lack of proper stands of waste disposal (Kiddee, 2013). Most locations in Nigeria are used as Ewaste-processing sites without considering the effects it is going to have on the environment which contributes as a huge health and environmental hazard to the residents exposed to it (Shahabuddin et al., 2023).

Soil is considered to be the top-most layer of lithosphere and responsible for providing nutrition to plants for growth and is a home to various animals and microorganisms which are necessary for proper functioning of the biosphere. Electronic waste is composed of metals including copper, aluminum, nickel, lead, tin, manganese and iron, which are protected by an outer covering of glass, fiber, plastic, ceramics, and metal oxides (Ilankoon et al., 2018). Virtually almost every E-waste consists of some valuable metals out of which copper is mostly observed. An old cathode ray tube computer contains 17.3% electronic parts, 43.7% metals, 15% glass and 23.3% plastic (Berkhout and Hertin, 2004). A study confirms the presence of tenfold higher concentrations of metals that hold value in a waste printed circuit board as to that of metals that are mined commercially (Ilankoon

 DOI: 10.1201/9781003582311-7

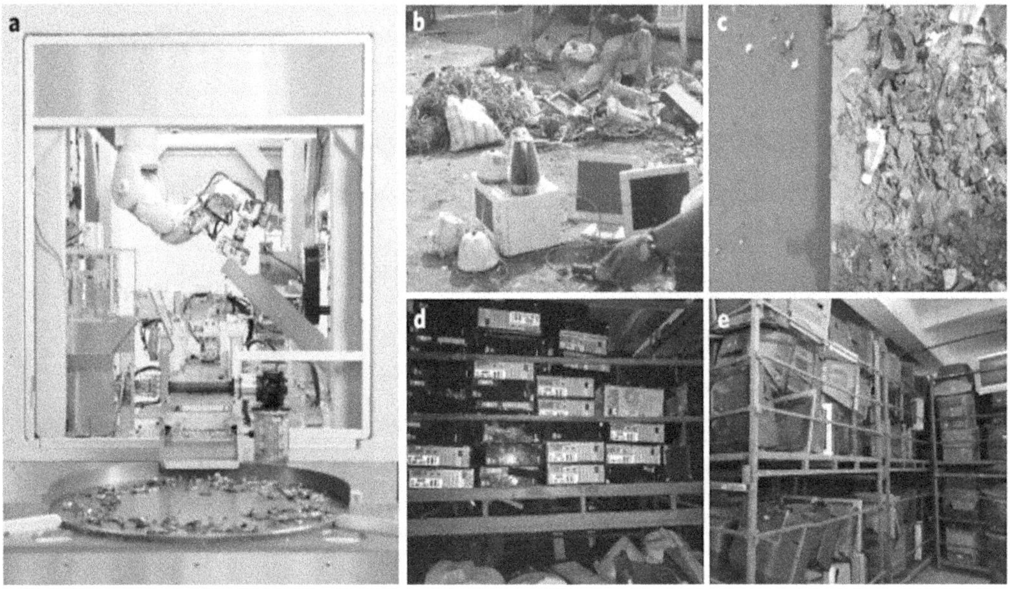

FIGURE 7.1 (a) Formal dismantling of E-waste by daisy robot which belongs to Apple; (b) Manual dismantling in Agbogbloshie market in Accra and similar places; (c) Environmental pollution caused by informal E-waste recovery; (d and e) Piles of E-waste in government facilities in China and Beijing.

et al., 2018). An estimate of 24 kg gold, 9 kg palladium, 9000 kg copper, and 250 kg silver can be generated from one million mobile phones (Ilankoon et al., 2018). Lithium, Beryllium, Cerium, Gallium, and several other metals that form complex mixture of hazardous pollutants can be found in batteries, LCD monitors, and conductive materials (Ladou and Lovegrove, 2008).

7.2 SCENARIO OF SOIL CONTAMINATION BY ELECTRONIC WASTES

On the basis of research conducted previously on E-waste from 2015 to 2022, some countries (Indonesia, India, Australia, China, and Malaysia) of Asia Pacific are the largest manufacturer of electronic waste (Forti, V et al., 2020; Schumacher K.A et al., 2021). As per the UNEP report titled "Recycling from Electronic Waste to Resources," in the forthcoming decade the E-waste amount generated from mobile phones and computers could rise by 500% in countries like India. The statistical analysis shows that the United States assembles 3-million-ton E-waste each year (world leader in E-waste production) and China assembles approximately 2.3-million-ton E-waste. India has become the fifth massive electronic waste creator in the world. Eighty thousand people in India are employed in reprocess province out of E-waste, for example, town Seelampur where enormous bulk of E-waste gets recycled. The approximate E-waste abandoned by India in the year 2016 was 1.85 milliontons. The city which produced utmost amount of E-waste was Mumbai (24%) going along with New Delhi (21.2%), Bangalore (10.1%), Chennai (9.1%), and New Delhi (21.2%). About 10 divergent categories of E-waste acknowledged globally. E-waste can be classified on the basis of numerous factors such as protectorate of occupation and households on EEE (electrical & electronic equipment), socio-economic conditions, and consumer behaviors (Balde et al., 2017). Household component has greater consequences (42.1%) followed by telecommunication and IT (33.9%), consumer devices (13.7%), and small household implements (4.7%) (European 2012). The main dominating consistency in developing nations comprises TVs, mobile phones, and computers. Figure 7.2(A) shows distribution of E-waste in different countries and Figure 7.2(B) shows distribution according to the zones in India. E-waste accumulates in the environment, can be toxic and

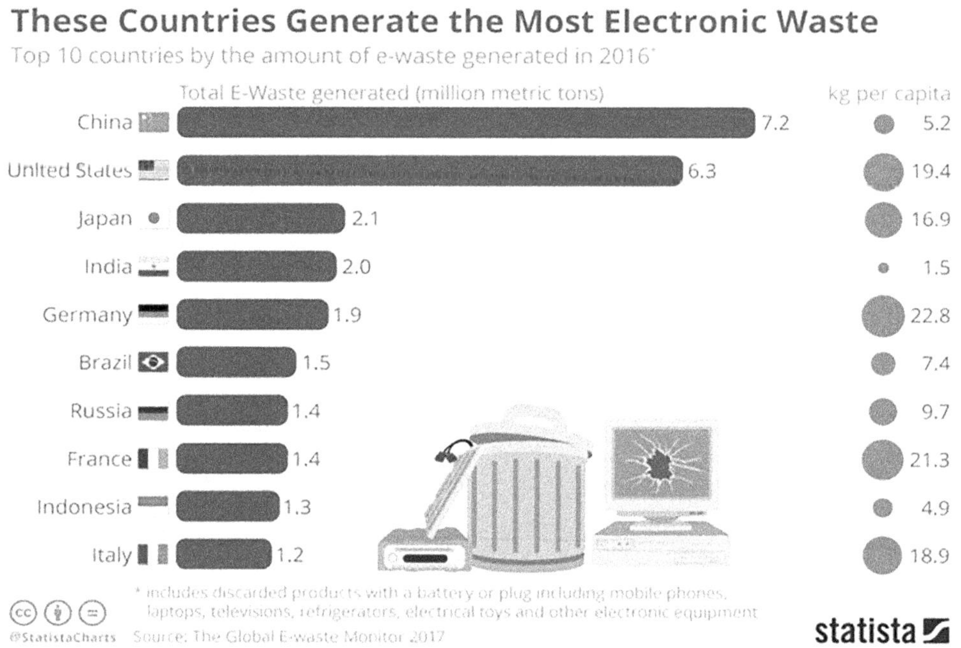

(A)

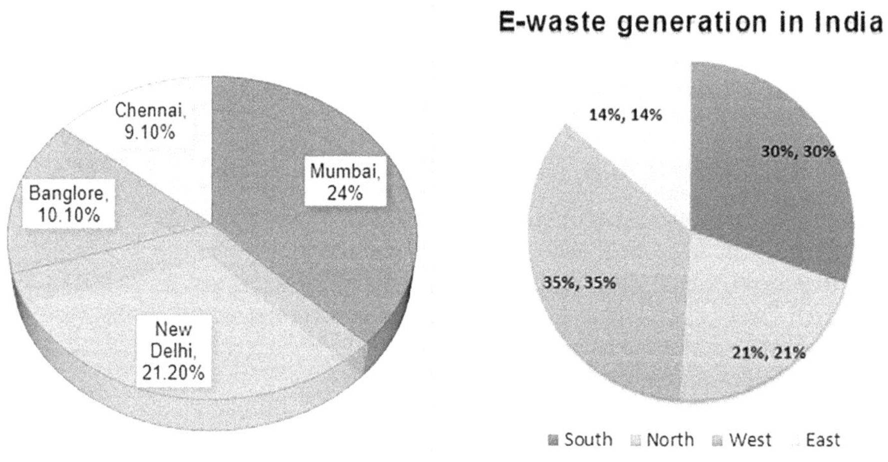

(B)

FIGURE 7.2 (A) Electronic waste distribution in India; (B) Distribution according to the zones. This chart ranks the top 10 countries generating the highest amount of electronic waste in 2016.

non-biodegradable in the soil, air, water, and living things. Improper destruction of E-waste in landfills or other non-disposing areas can cause harm worldwide to the environment, polluting ecosystems for generations to come. Toxic chemicals let out due to improper landfills affect the earth's soil, water, air, and most importantly human health. Several alternative practices are under investigation for printed motherboard recycling. Proper dumping or reusing of electronic gadgets reduces health complications and greenhouse-gas emissions. The extreme part of E-waste gets transported to

E-Waste Component	Process Used	Potential Environmental Hazard
Cathode ray tubes (used in TVs, computer monitors, ATM, video cameras, and more)	Breaking and removal of yoke, then dumping	Lead, barium and other heavy metals leaching into the ground water and release of toxic phosphor
Printed circuit board (image behind table – a thin plate on which chips and other electronic components are placed)	De-soldering and removal of computer chips; open burning and acid baths to remove metals after chips are removed.	Air emissions and discharge into rivers of glass dust, tin, lead, brominated dioxin, beryllium cadmium, and mercury
Chips and other gold-plated components	Chemical stripping using nitric and hydrochloric acid and burning of chips	PAHs, heavy metals, brominated flame retardants discharged directly into rivers acidifying fish and flora. Tin and lead contamination of surface and groundwater. Air emissions of brominated dioxins, heavy metals, and PAHs
Plastics from printers, keyboards, monitors, etc.	Shredding and low temp melting to be reused	Emissions of brominated dioxins, heavy metals and hydrocarbons
Computer wires	Open burning and stripping to remove copper	PAHs released into air, water and soil.

Prem Baboo,Sr. Manager(Prod), National Fertilizers Ltd. India

FIGURE 7.3 Sources of dumping electronic wastes and hazards relating to it. The table shows the comparison of various E-waste regulations around the world.

developing countries (Brazil, India, Nigeria, etc.), only 35% of which would be around 3.3 million tons of 9.5 million tons of used E-waste treated within EU. Figure 7.3 shows sources of dumping E-waste and has a relation to it.

7.3 EFFECTS OF ELECTRONIC WASTES ON SOIL QUALITY

Soil is referred to as a mixture of organic matter, minerals, liquids, and gases which is also a home to several microorganisms and animals. The quality of soil not only determines the survival of these microbes and animals but also of the flora that is native to that particular area (Gogoi et al., 2024). The amount of minerals present in soil whether toxic or useful are transferred directly or indirectly to humans via a food chain which begins with the vegetation growing there. The dumping of E-waste in soil has led to not only a decrease in the microorganisms and animals living in the soil but also has a severe impact on the growth of vegetation. Dumping of the electronic waste has also led to an increase in amount of Pb, Cd, Mn, and compounds like PCBs which has resulted in increase in acidity of the soil making it unfit for the plants to grow there. Wastes s from computers and electronic gadgets are dumped in the form of landfills which produce leachates and ultimately result in the contamination of ground water. Also, the computer chips that are being dumped generate sludge and acids which further acidify the soil. Due to contamination of ground water, there have been a shortage of water in places like Hong Kong's Guiyu which is a result of unlawful electronic waste recycling. Figure 7.4 shows cycle of electronic waste in soiland how it is a global ecological threat. The sale of Ni-Cd batteries has been banned by the European Union which contain at least 6–8% cadmium which can seep deep into the soil and not only cause damage to microorganisms, but if left incorrectly recycled can also disrupt the ecological setup of the soil (Sankhla et al.,2016). Toxins like organic pollutants are non-biodegradable which if accumulated can lead to long-term health risks. One of the informal techniques to recycle the E-waste is open air burning which is used for component separation for solder and copper recovery from electric cables. It leads to release of

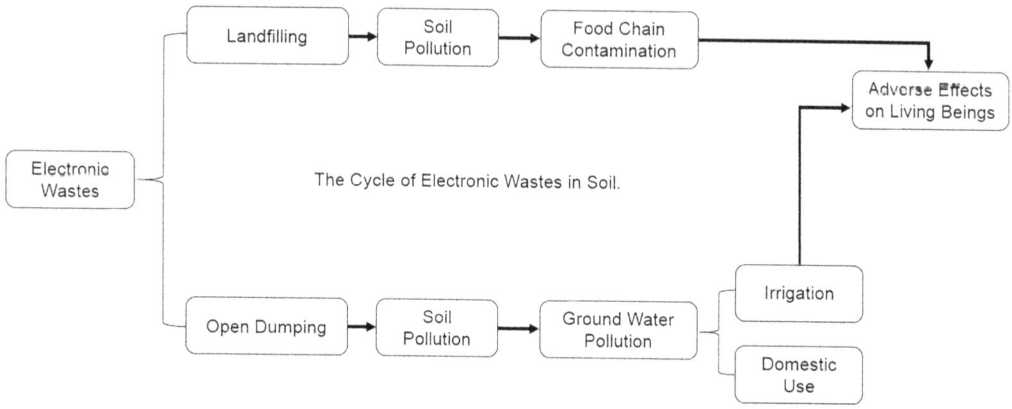

FIGURE 7.4 The cycle of electronic wastes in soil. E-waste and their leachates impact on human health and environment (Global ecological threat and management).

several harmful substances in air and accumulation of others in soil and water resources. Research was carried out on surface soils and river sediments in areas around Vietnam and in WEEE places high levels of toxins of dioxin-like compounds were found as a result of open burning (Kumari et al., 2023).

7.4 LEACHATES PRODUCED FROM E-WASTES

Leachate is a by-product generated from E-waste processing plants, landfills, and incineration plants which are highly toxic in nature. Leachates are composed of lethal organic pollutants, aromatic compounds, and ammonia nitrogen compounds. The leachate composition depends upon the site of dumping, landfill time, type of waste, retention time in incineration plants, and source that is contributing to the waste whereas the quantity is based mostly upon the location, timespan, moisture and the process of treatment. The leachates produced from landfills are mainly a result of precipitation, ground water infiltration, free water from the waste, and produced from that during decomposition of organic matter (Hussein and Ibrahim, 2023) (Datta,2023).

7.5 SOIL CONTAMINATION AND IMPACT ON HUMAN HEALTH

E-waste generated effects the quality of soil being used in various activities such as for production of food which impacts human health. It causes cancerous disease, lungs damage, kidney failure, etc. Leachate is a by-product generated from E-waste-processing plants, landfills, and incineration plants which are highly toxic in nature. Leachates are composed of lethal organic pollutants, aromatic compounds. and ammonia nitrogen compounds. The composition of leachate depends on site of dumping, landfill time, properties of the waste, retention time of holding cell in incineration plants and source of waste whereas the quantity is based mostly upon the location, timespan, moisture, and the process of treatment (Ghulam and Abushammala, 2023). The leachates produced from landfills are mainly a result of precipitation, ground water infiltration, free water from the waste, and from that produced during decomposition of organic matter.

7.6 TOXIC EFFECTS OF E-WASTES ON SOIL AND MICROBIAL COMMUNITY

Presence of a healthy soil is necessary for functioning of both natural and managed ecosystems. Microorganisms act as an important part of soil and help in maintenance of the soil quality and also act as the indicators of the number of pollutants as well as contaminants that could be present inside the soil. Microorganisms have a vital role in decomposition of organic matter which is essential for maintaining the soil quality and nutrient cycling. E-waste is composed of harmful toxic compounds that affect the soil microbial populations and proper management of E-waste (Agarwal et al., 2023). It has been proved that HMs reduce the microbial populations in soil. It has been shown by the studies that metal exposure on a regular basis can cause the microbial populations to undergo metal tolerance. Isolation of many species like Acinetobacter sp, Pseudomonas sp, and Sphingomonas sp shows tolerance and resistance to HMs and zinc, copper, cadmium metals etc (Yan et al., 2023; Li et al., 2023).

7.7 FUTURE CONCERNS AND RECOMMENDATIONS

Soil is an essential part of our environment. To help prevent more E-waste contamination in soil and further soil degradation, better disposal and recycling of E-waste needs to be done. Adding chemicals that may help in degrading the metals already present in soil may be helpful, and a significant reduction in E-waste creation would benefit this situation. E-waste is considered to be the most hazardous waste of all. There are numerous negative impacts on air, soil, human health, and all living organisms. Somehow, E-waste is highly responsible for disturbing many activities in the ecosystem and that particular environment. Not only animals and humans are in consideration for facing the impacts of E-waste, but even the microbial communities have to suffer the toxic effects of E-waste. Although various strategies have already been introduced in the recycling process, more is needed to reduce the poisonous impacts on humans, microbes, plants, and the environment. Hence, each country must introduce strategies to sustain a healthy environment.

REFERENCES

Agarwal, S., Darbar, S., Saha, S., Choudhury, M., & Singh, R. P. (2023). E-waste management using different cost-effective, eco-friendly biological techniques: an overview. *Waste Management and Resource Recycling in the Developing World*, 205–235.

Baldé, C. P., Forti, V., Gray, V., Kuehr, R., & Stegmann, P. (2017). *The global e-waste monitor 2017: Quantities, flows and resources*. United Nations University, International Telecommunication Union, and International Solid Waste Association.

Berkhout, F., & Hertin, J. (2004). De-materialising and re-materialising: digital technologies and the environment. *Futures*, *36*(8), 903–920.

Forti, V., Balde, C. P., Kuehr, R., & Bel, G. (2020). The Global E-waste Monitor 2020: Quantities, flows and the circular economy potential.

Datta, M. G. (2023). Electronic Waste and Their Management Strategies. In *Microbial Technology for Sustainable E-waste Management* (pp. 45–61). Cham: Springer International Publishing.

Ghulam, S. T., & Abushammala, H. (2023). Challenges and Opportunities in the Management of Electronic Waste and Its Impact on Human Health and Environment. *Sustainability*, *15*(3), 1837.

Gogoi, N., Sarma, A., Choudhury, M., Samanta, P., Faizan, M., & Sadhak, S. (2024). Soil heavy metal pollution and ecological risk assessment in disturbed and undisturbed soil of Morigaon, Assam. *Discover Environment*, 2(1), 13. https://doi.org/10.1007/s44274-024-00039-z

Hussein, O. A., & Ibrahim, J. A. A. (2023). Leachates Recirculation Impact on the Stabilization of the Solid Wastes–A Review. *Journal of Ecological Engineering*, *24*(4), 172–183.

Ilankoon, I. M. S. K., Ghorbani, Y., Chong, M. N., Herath, G., Moyo, T., & Petersen, J. (2018). E-waste in the international context–A review of trade flows, regulations, hazards, waste management strategies and technologies for value recovery. *Waste Management*, *82*, 258–275.

Kiddee, P., Naidu, R., & Wong, M. H. (2013). Electronic waste management approaches: An overview. *Waste management*, *33*(5), 1237–1250.

Kumari, B., Singh, G., Gautam, A., Sinam, G., Pal, S., Mishra, K., & Mallick, S. (2023). Enhancement in Ni–Cd Phytoremediation Efficiency of Helianthus annuus L. from Battery Waste Contaminated Soil by Bacterial Augmentation, Isolated from E-Waste Contaminated Sites. *International Journal of Environmental Research*, *17*(1), 18.

Ladou, J., & Lovegrove, S. (2008). Export of electronics equipment waste. *International journal of occupational and environmental health*, *14*(1), 1–10.

Li, J., Ma, D., Tian, J., Sun, T., Meng, Q., Li, J., & Shan, A. (2023). The responses of organic acid production and microbial community to different carbon source additions during the anaerobic fermentation of Chinese cabbage waste. *Bioresource Technology*, *371*(1), 128624.

Ogungbuyi O., Nnorom I. C., Osibanjo O., Schluep M. (2012) May. e-Waste Country Assessment Nigeria, e-Waste Africa project of the Secretariat of Basel Convention.

Ogunseitan, O. A. (2013). The Basel Convention and e-waste: translation of scientific uncertainty to protective policy. *The lancet global health*, *1*(6), e313–e314.

Sankhla, M. S., Kumari, M., Nandan, M., Kumar, R., & Agrawal, P. (2016). Heavy metals contamination in water and their hazardous effect on human health-a review. *Int. J. Curr. Microbiol. App. Sci*, *5*(10), 759–766.

Schumacher, K. A., & Agbemabiese, L. (2021). E-waste legislation in the US: An analysis of the disparate design and resulting influence on collection rates across States. *Journal of Environmental Planning and Management*, *64*(6), 1067–1088.

Shahabuddin, M., Uddin, M. N., Chowdhury, J. I., Ahmed, S. F., Uddin, M. N., Mofijur, M., & Uddin, M. A. (2023). A review of the recent development, challenges, and opportunities of electronic waste (e-waste). *International Journal of Environmental Science and Technology*, *20*(4), 4513–4520.

Yan, K., You, Q., Wang, S., Zou, Y., Chen, J., Xu, J., & Wang, H. (2023). Depth-dependent patterns of soil microbial community in the E-waste dismantling area. *Journal of Hazardous Materials*, *444*, 130379.

8 Electronic Wastes, Its Impact on Wildlife and Biodiversity

Pallabi Roy Chowdhury
Baranagar Baghajatin Social Welfare Organization, Baranagar, India

Srijan Goswami
Department of Biotechnology, University of North Bengal, Darjeeling, India

Rakesh Choudhary
Department of Civil Engineering, National Institute of Technology Delhi, Delhi, India

Moharana Choudhury
Environmental Research and Management Division, Voice of Environment (VoE), Guwahati, India

8.1 IMPORTANCE OF BIODIVERSITY AND WILDLIFE IN ECOLOGICAL BALANCE

Biodiversity, or biological diversity, encompasses the full range of life on Earth, encompassing species diversity, genetic variation within species, and the ecosystems they create. Biodiversity refers to the collective presence of all genes (genetic diversity), species (species diversity), and ecosystems (ecosystem diversity) within a specific geographical region. Ecological balance, sometimes called environmental equilibrium or ecological stability, is the condition of steady balance within an ecosystem. In an ecologically balanced system, the interactions between living organisms (biotic factors) and their surroundings (abiotic factors) are in a state of relative harmony, leading to a stable and sustainable ecosystem. Biodiversity and wildlife are integral aspects of ecology. Additional elements encompass biotic interdependencies, nutrient circulation, and energy transfer. Biodiversity and animals are essential for preserving the well-being and equilibrium of ecosystems. The ecosystem provides a range of functions, such as agricultural pollination, water purification, climate regulation, and disease prevention. Moreover, biodiversity and wildlife have inherent worth, enhancing the cultural, artistic, and recreational dimensions of human existence (Bratman, 2019) (Isbell et al., 2021) (IPBES, 2019) (Milner-Gulland et al., 2020) (Maffi et al., 2020).

Biodiversity and ecological balance are closely interlinked, and their relationship is fundamental to the health and sustainability of ecosystems. Biodiversity contributes to the stability and resilience of ecosystems. Diverse ecosystems are better able to withstand and recover from environmental changes, disturbances, and stressors. Different species play unique roles, and the interactions between them create a web of dependencies that helps maintain balance. Biodiversity provides functional diversity within ecosystems. Each species has a specific role or function, such as nutrient

DOI: 10.1201/9781003582311-8

cycling, pollination, pest control, or decomposition. The variety of functions performed by different species contributes to the overall balance and efficiency of the ecosystem. Biodiversity is essential for the provision of ecosystem services that directly or indirectly benefit humans. These services include air and water purification, climate regulation, soil fertility, and the production of food and medicine. A diverse range of species is necessary for the proper functioning of these services, contributing to ecological balance. Within biodiversity, genetic diversity at the species level is crucial for adaptation and resilience. A diverse gene pool enables species to evolve and adapt to changing environmental conditions, ensuring the long-term survival of populations. The various components of biodiversity (genetic, species, and ecosystem diversity) are interconnected and influence each other. Changes in one aspect of biodiversity can have cascading effects on others, leading to either a positive or negative feedback loop. This interconnectedness helps maintain a dynamic equilibrium within ecosystems. Biodiversity is directly linked to human well-being. Many resources are essential for human survival and development, such as food, medicine, and clean water, are derived from diverse ecosystems. The loss of biodiversity can compromise these resources, affecting the balance between human societies and their environment (Bratman, 2019) (Isbell et al., 2021) (IPBES, 2019) (Milner-Gulland et al., 2020) (Maffi et al., 2020).

Biodiversity loss can have profound and far-reaching consequences on ecological balance, disrupting the intricate relationships and functions within ecosystems. The consequences are numerous and can affect various aspects of ecosystem dynamics. Biodiversity loss can compromise the ability of ecosystems to provide essential services such as pollination, water purification, pest control, and climate regulation. This can lead to a decline in agricultural productivity, increased vulnerability to natural disasters, and a reduction in the overall well-being of human populations. Biodiverse ecosystems tend to be more resilient to environmental changes and disturbances. As species are lost, ecosystems become less adaptable to new conditions, making them more susceptible to disruptions such as extreme weather events, diseases, or invasive species. Biodiversity loss can disrupt nutrient cycling within ecosystems. Different species play key roles in nutrient uptake, decomposition, and recycling. The loss of specific species can lead to imbalances in nutrient availability, affecting the health and productivity of plants and other organisms in the ecosystem. Biodiversity loss can result in the proliferation of certain species, including pests and disease vectors. In diverse ecosystems, natural predators and competitors help regulate the populations of potential pests. As biodiversity declines, these regulatory mechanisms weaken, leading to increased risks of pest outbreaks and the spread of diseases. Biodiversity loss also involves the loss of genetic diversity within populations. Reduced genetic diversity can make species more vulnerable to diseases, environmental changes, and other stressors. It limits their ability to adapt and evolve, which can ultimately lead to population decline or even extinction. The removal of specific species can lead to shifts in the structure of ecosystems. For example, the loss of top predators can result in an increase in herbivore populations, leading to overgrazing and changes in vegetation composition. These shifts can have cascading effects on other species and disrupt the balance of the entire ecosystem. Biodiversity loss can destabilize food webs by affecting the availability and quality of resources for different trophic levels. The disappearance of key species in a food web can lead to population declines in species at other trophic levels, ultimately affecting the overall structure and functioning of the ecosystem.: Biodiverse ecosystems often have stronger defenses against invasive species. However, as biodiversity declines, ecosystems become more susceptible to the establishment and spread of invasive species, further disrupting native communities and ecological balance. Biodiversity loss can result in the disappearance of unique and culturally significant species, impacting the aesthetic and cultural value of ecosystems. Many societies rely on diverse ecosystems for traditional practices, spiritual beliefs, and cultural heritage (Bratman, 2019) (Isbell et al., 2021) (IPBES, 2019) (Milner-Gulland et al., 2020) (Maffi et al., 2020).

Addressing biodiversity loss requires conservation measures, habitat restoration, sustainable resource management, and global efforts to mitigate the drivers of biodiversity decline. Recognizing the consequences of biodiversity loss on ecological balance is crucial for informed decision-making and effective conservation strategies.

8.2 E-WASTES AND DESTRUCTION OF BIODIVERSITY AND WILDLIFE

Electronic garbage, often known as e-waste, encompasses discarded electronic devices such as computers, smartphones, televisions, and other electronic gadgets. Improper management and disposal of electronic trash can cause significant harm to biodiversity. Electronic equipment frequently includes toxic substances such as lead, mercury, cadmium, and brominated flame retardants. Improper disposal or incineration of e-waste can result in the release of harmful compounds into the soil and water, leading to the contamination of ecosystems. These contaminants can cause detrimental impacts on the health of plants, animals, and microbes, resulting in a reduction in biodiversity. Electronic debris that is disposed of in landfills or through informal recycling methods has the potential to pollute the nearby soil (Paul et al., 2019). Electronic gadgets emit hazardous compounds that can build up in the soil, leading to negative impacts on plant growth and soil-dwelling species. This disturbance can have a domino effect on the entire ecosystem, affecting the variety of plant and animal species. Improper e-waste disposal, such as dumping or insufficient recycling, can result in the emission of toxic compounds into water bodies. The presence of pollutants in the water can have detrimental effects on aquatic ecosystems, causing harm to fish, amphibians, and other creatures that inhabit water environments. Alterations in water quality can disturb the equilibrium of marine ecosystems and lead to the depletion of biodiversity. Incinerating electronic garbage, a prevalent practice in certain areas emits hazardous vapors into the atmosphere. Airborne contaminants, such as heavy metals and organic compounds, can accumulate on plants and in the soil. Aerial pollution can have detrimental effects on the well-being of plants and cause disruptions in the food chain, hence impacting multiple species in terrestrial ecosystems. Habitat destruction can result from the mining and exploitation of raw minerals used in electronic products. Mining operations can lead to the removal of forests, marshes, and other natural habitats, with a direct negative influence on the biodiversity of these ecosystems. E-waste pollution and disruption can result in population decreases of diverse plant and animal species. Species that exhibit higher susceptibility to environmental changes may encounter heightened risks, while those with specific habitats or dietary requirements may experience difficulties in adapting to modified environments. Certain hazardous compounds present in electronic waste can gradually build up in the tissues of living creatures. Bioaccumulation is the process by which contaminants accumulate in animals at higher trophic levels, resulting in increased concentrations. Biomagnification is the process by which larger concentrations of substances are transferred to predators, which can potentially have an impact on top predators in ecosystems (Wu, 2021) (Jayaraj et al., 2020) (Baldé, 2020) (Lebreton, 2021).

E-waste contamination can affect soil microbial communities. Microorganisms play crucial roles in nutrient cycling and soil health. Disruption of these microbial communities can have cascading effects on plant health and the overall functioning of terrestrial ecosystems. To address the negative impact of e-waste on biodiversity, it is crucial to promote responsible e-waste management practices, including proper recycling, safe disposal, and the development of more sustainable electronic product design. Global initiatives and regulations that encourage the recycling and safe handling of e-waste are essential to mitigate its impact on ecosystems and biodiversity (Wu, 2021) (Jayaraj et al., 2020) (Baldé, 2020) (Lebreton, 2021).

8.3 E-WASTES AND TERRESTRIAL ECOSYSTEM

Constant exposure to electronic waste (e-waste) can have significant negative impacts on terrestrial ecosystems and wildlife. Improper disposal of e-waste, such as dumping in landfills or open burning, can lead to soil contamination. Toxic substances like heavy metals (lead, mercury, cadmium) and hazardous chemicals from e-waste can leach into the soil, affecting soil quality. This contamination can disrupt nutrient cycling, soil structure, and the health of plants, ultimately impacting the entire terrestrial ecosystem. E-waste components can leach into the soil and, eventually, reach nearby water bodies. Rainwater can wash contaminants into rivers and streams, affecting aquatic ecosystems (Mondal et al., 2023). This runoff may introduce toxic substances to freshwater habitats, potentially harming aquatic plants, invertebrates, and fish. The interconnectedness of terrestrial and aquatic ecosystems means that pollution originating from e-waste can have cascading effects. Toxic substances from e-waste can enter the food chain. In the soil, plants may absorb these contaminants, leading to bioaccumulation. As herbivores consume contaminated plants, the concentration of pollutants can increase in their tissues. This process continues up the food chain, with predators at higher trophic levels accumulating higher levels of contaminants through biomagnification. This can lead to health issues and population declines in terrestrial wildlife. E-waste pollutants can negatively affect plant growth and development. The presence of heavy metals and toxic chemicals in the soil can lead to reduced plant productivity, altered nutrient uptake, and changes in plant community composition. This can have cascading effects on herbivores and other organisms dependent on plant resources. The mining and extraction of raw materials for electronic devices contribute to habitat destruction. This activity often involves clearing forests, disrupting ecosystems, and displacing wildlife. The loss of natural habitats can lead to declines in biodiversity, with certain species losing their homes and facing increased threats. Wildlife may suffer direct health effects from exposure to e-waste pollutants. For example, animals that ingest contaminated soil or water may experience physiological and reproductive issues. The presence of toxic substances can compromise the immune systems of wildlife, making them more susceptible to diseases. Wildlife may mistakenly ingest e-waste components, mistaking them for food. This can occur when animals scavenge for food in landfills or other areas where e-waste is improperly disposed of. Ingestion of plastic components, heavy metals, or other harmful materials can lead to internal injuries, poisoning, or death. The presence of e-waste in natural habitats can disrupt the behavior of wildlife. Animals may encounter unfamiliar and potentially hazardous materials, altering their feeding, mating, and nesting behaviors. This disruption can have long-term consequences on population dynamics and ecosystem functioning (Akinyemi, 2021) (Das et al., 2020) (He et al., 2021) (Hu et al., 2021).

To mitigate these impacts, it is essential to promote responsible e-waste management practices, including proper recycling, safe disposal, and the reduction of hazardous materials in electronic devices. Additionally, awareness and education about the environmental consequences of e-waste can contribute to more sustainable consumer choices and industrial practices.

8.4 E-WASTES AND MARINE ECOSYSTEM

Constant exposure to electronic waste (e-waste) can have detrimental effects on marine ecosystems and organisms. Improper disposal of e-waste can lead to the release of toxic substances into water bodies, either directly or indirectly. Rainwater can carry pollutants from e-waste dumping sites to rivers and oceans, leading to contamination. This contamination can include heavy metals, persistent organic pollutants (POPs), and other hazardous chemicals, affecting the water quality in marine ecosystems. Contaminants from e-waste can enter marine food chains. Phytoplankton, marine plants, and small organisms absorb these pollutants, leading to bioaccumulation in their tissues. As larger organisms are consume smaller ones, the concentration of contaminants increases through biomagnification. Top predators in marine ecosystems, such as large fish and marine mammals, may accumulate high levels of toxic substances, posing health risks to these species. Marine organisms, including fish, shellfish, and marine invertebrates, can be directly affected by the toxic substances found in e-waste. These contaminants can cause developmental abnormalities, reproductive issues, impaired immune systems, and increased susceptibility to diseases. Some pollutants may also interfere with the endocrine systems of marine organisms, disrupting hormonal functions. E-waste pollution in coastal areas and marine environments can lead to habitat degradation. Dumping sites and the leaching of toxic substances into the water can negatively impact the health of coral reefs, seagrass beds, and other marine habitats. This degradation can result in the loss of critical habitats for various marine species. E-waste often contains plastic components, and improper disposal can contribute to plastic pollution in the oceans. Plastics can break down into smaller particles known as microplastics, which are ingested by marine organisms. The ingestion of microplastics can lead to physical harm, digestive issues, and potential transfer of contaminants up the food chain. Marine species already facing conservation challenges, such as sea turtles, marine mammals, and certain fish species, may be further threatened by exposure to e-waste. These species may encounter and ingest hazardous materials, leading to additional stress on their populations. E-waste pollution can disrupt the balance and functioning of marine ecosystems. Changes in the abundance and distribution of marine species, as well as alterations in community structures, can have cascading effects on ecosystem dynamics. This disruption can impact the overall health and resilience of marine ecosystems. Due to ocean currents and water circulation patterns, contaminants from e-waste can be transported over long distances. This global distribution means that even regions far from e-waste disposal sites can be affected, highlighting the interconnected nature of marine ecosystems. To address these issues, it is crucial to promote responsible e-waste management, including proper recycling, waste reduction, and the safe disposal of electronic devices. International cooperation and awareness are essential to address the global nature of e-waste pollution and its impact on marine ecosystems and organisms (Li et al., 2021) (Woodring et al., 2020) (Cao et al., 2021) (Tang et al., 2021).

8.5 E-WASTES AND CONTAMINATION OF FOOD CHAIN

Electronic waste (e-waste) can disrupt food chains in various ways, affecting both terrestrial and aquatic ecosystems. E-waste often contains hazardous materials such as heavy metals (e.g., lead, mercury, and cadmium) and persistent organic pollutants (POPs). When these substances enter the environment through improper disposal or recycling processes, they can accumulate in the tissues of organisms. As organisms at lower trophic levels (such as plants or small invertebrates) absorb these contaminants, the concentrations can increase through a process known as bioaccumulation. Higher trophic level organisms, such as predators, feed on organisms at lower trophic levels, incorporating the accumulated toxins in their bodies. The process of biomagnification occurs as the concentrations of these contaminants increase up the food chain. Top predators in the food chain, such as large fish, marine mammals, or birds of prey, may accumulate particularly high levels of toxic substances, leading to potential health risks for these species. E-waste pollutants, when leached into the soil or water, can negatively affect primary producers such as plants, algae, or phytoplankton. These organisms form the base of the food chain, and any disruption in their health and productivity can affect the entire trophic structure. Contaminated primary producers may transfer toxins to herbivores, starting the process of bioaccumulation. The toxic substances from e-waste can affect the health and survival of certain species within the food chain. Some species may be more sensitive to contaminants, leading to declines in their populations. This can result in imbalances and shifts in the composition of species within the food chain. E-waste pollution can contribute to the decline of certain species, leading to a reduction in biodiversity within ecosystems. Biodiversity loss can disrupt the intricate interactions between species and compromise the stability of food webs (Shi et al., 2021) (Chen et al., 2020) (Zhang et al., 2021) (Luo et al., 2021).

Herbivores, which feed on plants and algae, may suffer from the direct effects of e-waste contaminants present in their food sources. Additionally, omnivores that consume both plants and animals may face exposure to toxins through multiple pathways, further contributing to the disruption of their feeding habits and health. E-waste pollutants can impact the reproductive success of species within the food chain. Reproductive processes such as egg development, hatching success, and the survival of offspring may be compromised due to the presence of contaminants, affecting the long-term sustainability of populations. Apex predators, at the top of the food chain, are particularly vulnerable to the effects of e-waste pollution. These species may experience both direct exposure to contaminants and indirect effects through the consumption of contaminated prey. Negative impacts on apex predators can have cascading effects on the entire ecosystem. To address the disruption of food chains by e-waste, it is essential to implement proper e-waste management practices, including responsible disposal and recycling. Additionally, efforts to reduce the use of hazardous materials in electronic devices and promote sustainable manufacturing practices can help mitigate the environmental impact on food chains. Public awareness and policy measures are crucial components in addressing this environmental challenge (Shi et al., 2021) (Chen et al., 2020) (Zhang et al., 2021) (Luo et al., 2021).

8.6 CONCLUSION

In conclusion, the pervasive impact of electronic waste (e-waste) on biodiversity and wildlife underscores the urgent need for comprehensive action to mitigate its detrimental effects. E-waste, laden with hazardous substances such as heavy metals, flame retardants, and persistent organic pollutants, infiltrates ecosystems through improper disposal and recycling practices. This pollution threatens the stability and resilience of ecosystems, leading to habitat degradation, loss of biodiversity, and adverse health effects on wildlife populations.

From soil and water contamination to bioaccumulation in organisms across the food chain, the repercussions of e-waste pollution reverberate through terrestrial, freshwater, and marine environments. Wildlife, from terrestrial mammals and birds to aquatic species, faces heightened risks of reproductive abnormalities, immune suppression, and neurological disorders due to exposure to e-waste-derived pollutants. Moreover, habitat destruction and fragmentation exacerbate biodiversity loss and disrupt critical ecosystem functions, further imperiling wildlife populations.

Addressing the impacts of e-waste on biodiversity and wildlife requires concerted efforts at multiple levels, including policy interventions, technological innovations, and public awareness campaigns. Strengthening regulatory frameworks for e-waste management, promoting sustainable recycling practices, and fostering international cooperation are essential steps in mitigating e-waste pollution. Additionally, investing in research on the long-term effects of e-waste on ecosystems and wildlife health is imperative for informed decision-making and effective conservation strategies.

Ultimately, safeguarding biodiversity and wildlife from the impacts of e-waste pollution necessitates a holistic approach that prioritizes ecosystem health, sustainability, and the well-being of present and future generations. By recognizing the interconnectedness of human activities and natural systems, we can strive toward a harmonious coexistence where biodiversity thrives and wildlife flourishes in balanced ecosystems, free from the pervasive threat of electronic waste contamination.

REFERENCES

Akinyemi, F. O., Aworunse, O. S., & Owojori, O. J. (2021). Assessment of the concentration of heavy metals in soils around electronic waste disposal sites in Lagos, Nigeria. *Environmental Monitoring and Assessment*, 193(2), 66.

Baldé, C. P., Wang, F., Kuehr, R., & Huisman, J. (2020). *The Global E-waste Monitor 2020: Quantities, Flows and the Circular Economy Potential*. United Nations University (UNU), International Telecommunication Union (ITU), and International Solid Waste Association (ISWA).

Bratman, G. N., Anderson, C. B., Berman, M. G., Cochran, B., de Vries, S., Flanders, J., & Kahn Jr, P. H. (2019). Nature and mental health: An ecosystem service perspective. *Science Advances*, 5(7), eaax0903.

Cao, S., He, P. J., & Xu, X. R. (2021). Potential toxicological effects of e-waste-derived pollutants on marine organisms: A critical review. *Journal of Hazardous Materials*, 407, 124417.

Chen, H., Zhou, Y., Wang, X., Wu, X., Zhang, Y., Gu, Q., & Li, X. (2020). The bioaccumulation of heavy metals in soil–plant system: A review. *Ecotoxicology and Environmental Safety*, 201, 110812.

Das, S., Sarkar, B., & Mukherjee, S. (2020). Impact of electronic waste on plant growth, physiology and metabolic profiling. *Environmental Pollution*, 266(Pt 1), 115122.

He, Z., Ye, X., Zhang, C., Wang, X., Shen, J., & Kong, X. (2021). Environmental exposure of wildlife to emerging e-waste pollution in East China: A systematic review and meta-analysis. *Environment International*, 146, 106199.

Hu, X., Zhang, C., Zhang, Y., Li, M., & Liu, Y. (2021). Phytoremediation potential and mechanisms of heavy metal pollution in e-waste-contaminated soil: A review. *Environmental Pollution*, 276, 116601.

IPBES (Intergovernmental Science-Policy Platform on Biodiversity and Ecosystem Services). (2019). Global Assessment Report on Biodiversity and Ecosystem Services. IPBES Secretariat, Bonn, Germany.

Isbell, F., Cowles, J., Dee, L. E., & Loreau, M. (2021). The biodiversity–stability relationship in ecological communities: a plethora of patterns and mechanisms. *Ecology Letters*, 24(1), 220–238.

Jayaraj, R. L., Megha, P., & Sreedev, P. (2020). Review on the toxicological effects of electronic waste on the environment and human health. *Journal of Applied Toxicology*, 40(1), 8–19.

Lebreton, L., Fischer, E., Jambeck, J., & Rocca, J. (2021). E-waste governance in the United States and China: the need for a critical raw material security approach. *Journal of Environmental Management*, 278(Pt 1), 111424.

Li, J., Li, X., Zhang, J., Zhang, L., Li, X., & Yang, L. (2021). E-waste pollution in the coastal zone: A review of sources, distribution, and potential effects. *Environmental Pollution*, 284, 117195.

Luo, X., Zhang, W., Liu, R., & Wu, F. (2021). Bioaccumulation of polycyclic aromatic hydrocarbons (PAHs) in the food chain and human health risk assessment from the areas of E-waste recycling site in East China. *Chemosphere*, 263, 127939.

Maffi, L. (2020). Biocultural diversity and sustainability: Perspectives, insights, and real-world solutions. *Proceedings of the National Academy of Sciences*, 117(3), 1452–1460.

Milner-Gulland, E. J., Mcgregor, J. A., Agarwala, M., Atkinson, G., Bevan, P., Clements, T., & Williams, D. R. (2020). Accounting for the impact of conservation on human well-being. *Conservation Biology*, 34(3), 626–636.

Mondal, T., Choudhury, M., Kundu, D., Dutta, D., & Samanta, P. (2023). Landfill: An eclectic review on structure, reactions and remediation approach. *Waste Management*, 164, 127–142. https://doi.org/10.1016/j.wasman.2023.03.034

Paul, S., Choudhury, M., Deb, U., Pegu, R., Das, S., & Bhattacharya, S. S. (2019). Assessing the ecological impacts of ageing on hazard potential of solid waste landfills: A green approach through vermitechnology. *Journal of cleaner production*, 236, 117643. https://doi.org/10.1016/j.jclepro.2019.117643

Shi, G., Chen, Z., Bi, X., Li, Y., Teng, J., Wang, L., … & Xu, S. (2021). Spatial distribution and health risk assessment of heavy metals and persistent organic pollutants in agricultural soils around e-waste recycling areas in China. *Environmental Pollution*, 278, 116839.

Tang, Y., Li, Z., Lin, X., & Li, W. (2021). E-waste management strategies and technology for environmental protection and sustainable development: A critical review. *Journal of Cleaner Production*, 289, 125768.

Woodring, J., Tarrant, A. M., & Webster, N. S. (2020). Ingestion of microplastics and natural fibres in the benthic coral reef sponge Carteriospongia foliascens. *Journal of Hazardous Materials*, 384, 121396.

Wu, J., Li, M., Bi, X., Yu, L., Liu, X., Zhu, F., … & Fu, J. (2021). Heavy metal contamination in soils and plants around a large-scale e-waste recycling site: implications for potential health risks. *Environmental Geochemistry and Health*, 43(2), 713–728.

Zhang, Y., Wang, X., Yan, Z., Gao, Y., & Zeng, X. (2021). Risk assessment of heavy metal and polycyclic aromatic hydrocarbon pollution in poultry products around e-waste recycling sites in China. *Environmental Pollution*, 278, 116864.

9 Electronic Wastes, Its Impact on Insects

Kamidi Rahul
Central Sericultural Research & Training Institute, Berhampore, India
Industrial Insect and Sericulture Division, National Institute of Agricultural Sciences, RDA, Wanju-gun, Republic of Korea

Pooja Makwana
Central Sericultural Research & Training Institute, Berhampore, India

Katsuhiko Ito
Graduate School of Agriculture, Tokyo University of Agriculture and Technology, Tokyo, Japan

Bindu Subhadra
Department of Biological and Earth Sciences, McEver Hall, Russellville, Arkansas, United States

Dhiraj Kumar
Department of Zoology, Hansraj College, University of Delhi, Delhi, India

Anil Pappachan
P2 Basic Seed Farm, National Silkworm Seed Organization (NSSO), Yelagiri Hills, India

9.1 UNDERSTANDING THE INSECT AND E-WASTE CONNECTION

Insects are without a doubt the most prevalent and fascinating creatures on earth, with more than one million species having been documented so far (Stork, 2018). They are unparalleled to any other multicellular life forms on this planet in terms of abundance, diversity, and ecological significance. Insects play vital roles in providing an array of indispensable services, many of which are essential for human sustenance and well-being (Miličić et al., 2021). Insects are involved in pollination of a variety of crops with commercial importance as well as numerous wild plants, dispersal of seeds, breakdown of organic matter, recycling of nutrients, fostering fertility of soil and controlling pests (Gogoi et al., 2024). They also serve as sources of sustenance to some species of birds, fish and amphibians, indicators of environmental pollution, climate change and model organisms in scientific experiments to study various biological processes. Often, it is also assumed that insects cause crop destruction, infrastructure damage, serve as carriers of infectious diseases, and sting or bite humans and other animals. However, the species that fulfill the aforesaid description only make up to a very small proportion of all known insect species (Sollai and Solari, 2022). Rise in population, rapid urbanization and industrialization are contributing to the deliberate or accidental discharge of diverse and noxious pollutants into the environment. The last decade also witnessed a tremendous

 DOI: 10.1201/9781003582311-9

surge in disposal of a specific group of pollutants, known as the e-waste pollutants (Murthy and Ramakrishna, 2022). These pollutants when enter the terrestrial or aquatic realms, acts directly on the life forms inhabiting them or indirectly by altering the physical environment leading to loss in biodiversity and subsequently, the functioning of an ecosystem. Insects being ubiquitous are vital in the sustainability and functioning of any ecosystem in which they are distributed. Owing to their immense diversity, insects serve as a key link in a food web connecting plants and other animals. The issue at hand is about the role of insects in aquatic and terrestrial habitats contaminated with e-waste and this chapter summarizes about the same.

9.2 BIOACCUMULATION AND BIOMAGNIFICATION OF E-WASTE POLLUTANTS – ROLE OF INSECTS

The onsite/field level evaluation studies related to bioaccumulation and/or biomagnification of e-waste pollutants involving insects were mostly performed by researchers in the vicinity of e-waste recycling facilities in China. The published research supports the fact that insects bioaccumulate e-waste constituents and also act as vectors for transmission of the same to the upper-order members of the food chain in terrestrial as well as aquatic habitats. Before continuing, let's acquaint ourselves with the phrases "bioaccumulation" and "biomagnification", which are two distinct processes that are assessed to depict the degree of pollutant transmission within food webs. The term bioaccumulation refers to the gradual increase in concentration of a pollutant over time in a specified organism upon uptake from the surrounding environment. Biomagnification is a phenomenon in which pollutants tend to concentrate as they pass from one trophic level to another. In other words, the concentration of a pollutant increases as it progress through the food chain (Miller, Hamann and Kroon, 2020; Daley, Corkum and Drouillard, 2011). Another important term "biomagnification factor (BMF)" is frequently used to assess whether or not pollutants are being biomagnified in predators from their prey, with a BMF more than 1 is an indication that biomagnification actually occurred. In broad terms, by definition BMF is the ratio of the concentration of a specific pollutant in a predator (consumer) to that in its prey (dietary source). BMFs are considered more appropriate for assessment in terrestrial ecosystems and it is important to keep in mind that concentration of the pollutant in predator may not result exclusively from its diet but also from its consumption from external contaminated milieu (Chen et al., 2022; Franklin, 2016).

9.3 DECHLORANE PLUS (DP)

Both the syn- and anti- forms of the stereoisomers of DP are widely used in a diverse range of electrical appliances as a flame retardant. Insects (prey) belonging to different taxa as well as frogs (predators) were sampled from a same habitat (agricultural field) near to an e-waste recycling facility in South China and assessed for the presence of *anti*-DP, *syn*-DP, and *anti*-Cl_{11}-DP (DP dechlorinated compound). The ΣDP in insects was found to be 87.1 ng/g lipid weight (lw). All the aforementioned three compounds were found to be biomagnified in frogs from insects (BMF > 1). It is also stated that female frogs have passed these chemical pollutants to their eggs through maternal transmission (Wu et al., 2018).

9.4 HEXABROMOCYCLODODECANE (HBCDD)

A study was undertaken from a pond which was contaminated by e-waste and its vicinity located at Longtang, South China to determine the presence of an additive flame retardant HBCDD used in electronics (Liu et al., 2020a). The samples included beetles, butterflies, crickets, dragonflies, grasshoppers, mantises, mole-crickets, and water stinkbugs as well as other biotic and abiotic samples. Aquatic insects such as water stinkbugs (1.1 ng g^{-1} lw) and dragonfly larvae (6 ng g^{-1} lw) showed higher concentrations of HBCDD compared to terrestrial insects (0.31-0.69 ng g^{-1} lw).

The researchers attributed the same to the pond ecosystem that was greatly contaminated with e-waste pollutants in comparison to terrestrial habitat. It has been also reported that BMF of α-HBCDD in between grasshopper (predator) and grass (prey) was 1.52; fish and aquatic insects (6.25); waterbird eggs and aquatic insects (26.9); toads and terrestrial insects (0.29), frogs and terrestrial insects (0.85); lizards and terrestrial insects (0.63) and terrestrial birds and terrestrial insects (2.03) clearly indicating the role of insects in transfer of pollutants from one trophic level to other in aquatic as well as terrestrial food webs (Liu et al., 2020a).

9.5 POLYCHLORINATED BIPHENYLS (PCBs) AND POLYBROMINATED DIPHENYL ETHERS (PBDEs)

PBDEs, another flame retardant in electronics were detected in insect as well as frog samples collected from a polluted site at Qingyuan, South China, which is world's second largest e-waste recycling territory. Among the eighteen PBDEs tested in the samples, thirteen of them (BDEs 85, 99, 100, 153, 154, 183, 196, 197, 203, 205, 207, 208, and 209) have biomagnified in frogs from insects (BMF > 1). The concentration of different BDEs varied in between 0.01-6.32 ng/g wet weight (ww) in insect samples. Transfer of the contaminants from mother frogs to eggs was also reported (Wu et al., 2009). Nie et al. (2015) sampled various terrestrial species which included insects from localities around e-waste burning site in Longtang, South China and determined for the congener profiles, concentrations, and magnitude of biomagnification of PBDEs. The mean concentrations of Σ_{21}PBDEs in grasshopper, ant, dragonfly and butterfly ranged in between 474-1096 ng g^{-1} lw. BMFs of Σ21PBDEs in between grasshopper-turtledove, ant-turtledove, dragonfly-turtledove, and butterfly-turtledove were 1.7, 3.6, 2.2 and 1.6, respectively. It is very interesting to note that BMFs of individual congeners among the aforesaid trophic relations varied.

The presence of PBDEs and PCBs in different parts of rice plants, soil and grasshopper (*Oxya chinensis*) collected from an agricultural field in the vicinity of an e-waste processing locality at Niutang, China were assessed by Han et al. (2017). The concentration of above contaminants was higher in soils nearer to e-waste processing sites when compared to housing neighborhoods. It is also reported that the pollutants were bioaccumulated in rice plants from the contaminated soils and biomagnified from rice plants to the grasshoppers (Han et al., 2017). At the Xilin Gol grassland, Inner Mongolia, Bute et al. (2020) investigated the distribution of PBDEs in a food chain of grassland ecosystem. The researchers did mention that lizards, snakes, and toads (heterothermic organisms) had augmented levels of Σ_{10} PBDEs than insects. The authors indicated that insects serve as major carriers of pollutants to life forms at upper trophic levels. It is also observed that the BMF initially increased as the trophic level ascended before remaining constant in the grass-insect-lizard-snake food chain whereas BMF continued to enhance as the trophic level rose in the grass-insect-mouse-weasel food chain. The presence and concentrations of persistent organic pollutants (POPs) in insects including chafer, cricket, locust, moth, mantis, stinkbug (terrestrial) as well as dragonfly, water scavenger beetle (aquatic) collected across one square kilometer area near a deserted e-waste recycling site in South China were assessed. The biomagnification of the target contaminants in songbirds from terrestrial and aquatic invertebrates was also investigated. The highest concentrations of PDBEs and PCBs were detected in dragonfly when compared to the other insect species tested. It is also reported that the majority of POPs detected in birds originated from aquatic insects suggesting that the pollutants are bio transferred from aquatic to riparian habitat (Wu et al., 2022). Amphibians and insect species belonging to six and eight taxa respectively were sampled from a deserted e-waste recycling unit, South China and BMFs of POPs from insects to amphibians was assessed. The median concentrations of PBDEs, PCBs, 1,2-bis(2,4,6-tribromophenoxy)ethane (BTBPE), decabromodiphenyl ethane (DBDPE) in insects ranged in between not detected (nd)-66.5, 43.4-526, nd-108 and nd-81.9 ng/g lw, respectively. The authors evaluated the dietary compositions of amphibians using quantitative fatty acid signature analysis. The most significant contribution to the diet of amphibians was made by dragon flies and BMFs resultant from quantitative amphibian/

insect relationships exhibited more reliable results when compared to BMFs based on amphibian/each insect or amphibian/combined prey interactions (Ding et al., 2022). Luo et al. (2022) evaluated *Asota caricae* and leaves of its host plant *Ficus hispida* as well as *Aloa lactinea* collected from woods surrounding e-waste processing locality at Longtang, South China for the presence of PCBs, polycyclic aromatic hydrocarbons (PAHs), PBDEs. All the assessed contaminants in varying amounts were detected in the insects as well as in *F. hispida* samples. The authors mentioned that a negative correlation was observed between BMFs measured from *A. caricae* larvae to *F. hispida* and log K_{OW} for all compounds which is in stark contrast to results from earlier studies. The trend reversed upon metamorphosis, implying that pollutants have undergone biotransformation during metamorphosis (Luo et al., 2022). Butterflies, dragonflies, grasshoppers, litchi stinkbugs and moths collected from in and around 50 m radius of a pond heavily contaminated with e-waste in Longtang, South China were assessed for the presence of persistent halogenated organic pollutants (HOPs). ΣHOPs [ΣPCBs, ΣPBDEs, ΣDBDPE, ΣDP, Σbrominated flame retardants (BFRs), Σdichlorodiphenyltrichloroethanes (DDTs)] levels in ng/g lw were found to be highest in dragonfly followed by butterfly and moth, grasshopper and litchi stinkbug. The concentrations of individual HOPs excepting DDT also followed a similar pattern as mentioned above. Significant positive correlations were observed between bioaccumulation factors for dragonfly larvae; BMF for moth, grasshopper larvae and octanol-water partition co-efficient of the chemicals. The authors also indicated that shedding of exuviae during metamorphosis as well as excretory processes may be involved in reduction of the contaminant burden by the insects (Liu et al., 2018). Liu et al. (2020b) reported that HOPs accumulated more readily in homeotherms than in poikilotherms, as evidenced by the fact that the BMFs of HOPs (excluding DBDPE) in insect-bird food chains were considerably greater in comparison to insect-toad and insect-lizard food chains.

9.6 OTHER BROMINATED FLAME RETARDANTS

Alternative brominated flame retardants (ABFRs) were also detected in insects as well as in other biota sampled from an e-waste recycling unit at Qingyuan, China. ABFRs were found to be highest in dragonfly (*Aeshnidae rambur*) larvae when compared to Chinese bitterling, northern snakehead, mud carp, crucian carp, shrimp, tilapia, and water beetle. Fatty acid signature analysis was used by the authors to interpret the interactions between predators and their prey. It is reported that pentabromotoluene (PBT) and pentabromoethylbenzene (PBEB) were biomagnified in northern snakehead (Tao et al., 2019). Chen et al. (2022) performed studies related to biomagnification of the various novel brominated flame retardants (NBFRs) in Xilingol grassland food chain, Inner Mongolia, China. The researchers depicted that BMF of pentabromotoluene (PBT), pentabromobenzene (PBBz), hexabromobenzene (HBB), pentabromoethylbenzene (PBEB), and tetrabromo-*p*-xylene (pTBX) in between insect and grass food chain is less than 1 signifying that there was no biomagnification. All the aforementioned NBFRs were found to be biomagnified in bird (Cuckoo) from insects whereas only pTBX, PBBz, PBEB biomagnified in another bird (Swallow). Biomagnification of pTBX (toad-insect) as well as pTBX and PBBz (lizard-insect) was also reported (Chen et al., 2022).

9.7 SHORT-CHAIN CHLORINATED PARAFFINS (SCCPs) AND MEDIUM-CHAIN CHLORINATED PARAFFINS (MCCPs)

Insects including mantis, dragonfly (*Aeshnidae rambur, Libellulidae rambur*), grasshopper, cricket, mole cricket, and beetle collected from locations surrounding 500 m radius of an e-waste dump site in Longtang, China were assessed for the presence of SCCPs and MCCPs. The levels of the aforementioned target contaminants were highest in dragonflies, mantises (carnivorous); followed by crickets, grasshoppers, mole crickets (burrowing); and beetles (phytophagous). The evaluated contaminants ranged in between 2200-12000 ng/g lw (SCCPs) and 990-19000 ng/g lw (MCCPs) in the tested insect species. Furthermore, biomagnification of both the contaminants (trophic magnification factors of ΣSCCPs: 2.08; ΣMCCPs: 2.45) in the experimented terrestrial food web was reported (Liu et al., 2020c). In another study, SCCPs in the ranges of 28-170 ng/g ww were detected in native as well as migratory birds (long-tailed shrike, white wagtail, grey-backed thrush, great tit, and red-flanked bluetail) that are insectivorous around a facility which processes e-waste at Longtang, South China (Luo et al., 2015).

9.8 ORGANOPHOSPHORUS FLAME RETARDANTS (PFRs) AND PLASTICIZERS

Wild insects belonging to different orders Coleoptera, Hemiptera, Lepidoptera, Odonata and Orthoptera were collected from localities extensively contaminated with e-waste constituents in Longtang, China and analyzed for the presence of twelve PFRs and plasticizers. The most prevalent compounds detected in the order were tris(2-ethylhexyl) phosphate > triphenyl phosphate >tris(1-chloro-2-propyl) phosphate>tris(2-chloroethyl) phosphate> 2-ethylhexyldiphenyl phosphate > tri-cresyl phosphate. The average concentrations of the above mentioned abundant contaminants in insects ranged in between 0.1 and 5.8 ng/g ww. Fifty percent of the total PFRs pollutants were distributed in insects of order Odonata, 18% in Lepidoptera, 16% in Orthoptera, 13% in Hemiptera and 3% in Coleoptera. The level of contamination was found to be in the given order dragonfly nymphs > moth adults > terrestrial stinkbug adults. The remaining insect species (dragon fly adults, mole-cricket adults, cricket adults, grasshopper nymphs, aquatic beetle adults, aquatic stinkbug adults, grasshopper adults, terrestrial beetle adults, moth larvae) accounted for less than five percent of the total PFR contamination. The differences in accumulation of pollutants may be attributed to variations in feeding habits of insects as well as habitats (Poma et al., 2019). Insects of different orders (butterfly, dragonfly, grasshopper, litchi stinkbug, moth, shining leaf chafer and water scavenger beetle) were gathered from an e-waste disposal region in Guangdong, South China and traces of PFRs (2.3 to 91 ng/g ww) and plasticizers (420-15600 ng/g ww) were detected. In view of differing habitats as well as dietary behaviors, the levels as well as compositions of the aforementioned e-waste pollutants varied greatly in the seven insect taxa. Grasshopper and litchi stinkbug larvae had considerably greater PFR levels than their adults and the same trend was otherwise in case of butterflies, dragonflies and moths. In addition, the authors also signified that ecdysis is a crucial mechanism through which PFRs and plasticizers were disposed of by dragonflies. Average BMF of PFRs between grasshopper-grass and moth larvae-guava was 10 and 0.67 respectively whereas the same in terms of plasticizers were 1.3 and 1 (Liu et al., 2021). More related studies authored by Prince, Taylor and Angelini, 2020; Zhang et al., 2011; Daley, Corkum and Drouillard, 2011; Mo et al., 2019 can also be referred for additional information.

9.9 CONCLUSION

POPs are most detected in biotic (including insects) and abiotic samples from e-waste-contaminated aquatic as well as terrestrial habitats. Insects belonging to diverse taxa responded differently when exposed to various e-waste pollutants. Most of the studies indicated that dragonflies are most sensitive to e-waste contaminants. Furthermore, insects play a critical role in accumulating and transferring e-waste pollutants to insectivores in diverse habitats.

REFERENCES

Bute, Chen, W., Li, T., Liu, Y., Wang, H., Zhao, P., Hu, J., & Jin, J. (2020). Distributions and biomagnification of polybrominated diphenyl ethers in a grassland ecosystem food chain. *Science of the Total Environment, 747*, 141141.

Chen, W., Bu, T., Li, T., Bao, J., Wang, Y., Hu, J., & Jin, J. (2022). Concentration, distribution and biomagnification of novel brominated flame retardant in grassland food chain and sheep from Inner Mongolia, China. *International Journal of Environmental Research and Public Health, 19*, 12785.

Daley, J. M., Corkum. L. D., & Drouillard, K. G. (2011). Aquatic to terrestrial transfer of sediment associated persistent organic pollutants is enhanced by bioamplification processes. *Environmental Toxicology and Chemistry*, *30*(9), 2167–2174.

Ding, Y., Zheng, X., Yu, L., Lu, R., Wu, X., Luo, X., & Mai, B. (2022). Occurrence and distribution of persistent organic pollutants (POPs) in amphibian species: Implications from biomagnification factors based on quantitative fatty acid signature analysis. *Environmental Science & Technology*, *56*(5), 3117–3126.

Franklin, J. (2016). How reliable are field-derived biomagnification factors and trophic magnification factors as indicators of bioaccumulation potential? Conclusions from a case study on per- and polyfluoroalkyl substances. *Integrated Environmental Assessment and Management*, *12*(1), 6–20.

Gogoi, N., Sarma, A., Choudhury, M., Samanta, P., Faizan, M., & Sadhak, S. (2024). Soil heavy metal pollution and ecological risk assessment in disturbed and undisturbed soil of Morigaon, Assam. *Discover Environment*, 2(1), 13. https://doi.org/10.1007/s44274-024-00039-z

Han, Z. X., Wang, N., Zhang, H. L., & Zhao, Y. X. (2017). Bioaccumulation of PBDEs and PCBs in a small food chain at electronic waste recycling sites. *Environmental Forensics*, *18*(1), 44–49.

Liu, Y., Luo, X. J., Huang, L. Q., Yu, L. H., & Mai, B. X. (2018). Bioaccumulation of persistent halogenated organic pollutants in insects: Common alterations to the pollutant pattern for different insects during metamorphosis. *Environmental Science & Technology*, *52*(9), 5145–5153.

Liu, Y. E., Luo, X. J., Liu, Y., Zeng, Y. H., & Mai, B. X. (2021). Bioaccumulation of legacy and emerging organophosphorus flame retardants and plasticizers in insects during metamorphosis. *Journal of Hazardous Materials*, *406*, 124688.

Liu, Y., Luo, X., Zeng, Y., Deng, M., Tu, W., Wu, Y., & Mai, B. (2020a). Bioaccumulation and biomagnification of hexabromocyclododecane (HBCDD) in insect-dominated food webs from a former e-waste recycling site in South China. *Chemosphere*, *240*, 124813.

Liu, Y., Luo, X., Zeng, Y., Tu, W., Deng, M., Wu, Y., & Mai, B. (2020b). Species-specific biomagnification and habitat-dependent trophic transfer of halogenated organic pollutants in insect-dominated food webs from an e-waste recycling site. *Environment International*, *138*, 105674.

Liu, Y., Luo, X., Zeng, Y., Wang, Q., Tu, W., Yang, C., & Mai, B. (2020c). Trophic magnification of short- and medium-chain chlorinated paraffins in terrestrial food webs and their bioamplification in insects and amphibians during metamorphosis. *Environmental Science & Technology*, *54*(18), 11282–11291.

Luo, X. J., Huang, L. Q., Guan, K. L., Qi, X. M., Zeng, Y. H., & Mai, B. X. (2022). Bioamplification and biomagnification of polycyclic aromatic hydrocarbons and halogenated organic pollutants in moths from an electronic waste recycling site. *Environmental Toxicology and Chemistry*, *41*(10), 2395–2403.

Luo, X. J., Sun, Y. X., Wu, J. P., Chen, S. J., & Mai, B. X. (2015). Short-chain chlorinated paraffins in terrestrial bird species inhabiting an e-waste recycling site in South China. *Environmental Pollution*, *198*, 41–46.

Miličić, M., Popov, S., Branco, V. V., & Cardoso, P. (2021). Insect threats and conservation through the lens of global experts. *Conservation Letters*, *14*, e12814.

Miller, M. E., Hamann, M., & Kroon, F. J. (2020). Bioaccumulation and biomagnification of microplastics in marine organisms: A review and meta-analysis of current data. *PLoS ONE*, *15*(10), e0240792.

Mo, L., Zheng, X., Zhu, C., Sun, Y., Yu, L., Luo, X., & Mai, B. (2019). Persistent organic pollutants (POPs) in oriental magpie-robins from e-waste, urban, and rural sites: Site-specific biomagnification of POPs. *Ecotoxicology and Environmental Safety*, *186*, 109758.

Murthy, V., & Ramakrishna, S. (2022). A review on global e-waste management: Urban mining towards a sustainable future and circular economy. *Sustainability*, *14*, 647.

Nie, Z., Tian, S., Tian, Y., Tang, Z., Tao, Y., Die, Q., Fang, Y., He, J., Wang, Q., & Huang, Q. (2015). The distribution and biomagnification of higher brominated BDEs in terrestrial organisms affected by a typical e-waste burning site in South China. *Chemosphere*, *118*, 301–308.

Prince, K. D., Taylor, S. D., & Angelini, C. (2020). A global, cross-system meta-analysis of polychlorinated biphenyl biomagnification. *Environmental Science & Technology*, *54*(18), 10989–11001.

Poma, G., Liu, Y., Cuykx, M., Tang, B., Luo, X. J., & Covaci, A. (2019). Occurrence of organophosphorus flame retardants and plasticizers in wild insects from a former e-waste recycling site in the Guangdong province, South China. *Science of the Total Environment*, *650*, 709–712.

Sollai, G., & Solari, P. (2022). An overview of "Insect Biodiversity". *Diversity*, *14*, 134.

Stork, N. E. (2018). How many species of insects and other terrestrial arthropods are there on earth? *Annual Review of Entomology*, *63*, 31–45.

Tao, L., Zhang, Y., Wu, J. P., Wu, S. K., Liu, Y., Zeng, Y. H., Luo, X. J., & Mai, B. X. (2019). Biomagnification of PBDEs and alternative brominated flame retardants in a predatory fish: Using fatty acid signature as a primer. *Environment International*, *127*, 226–232.

Wu, J. P., Chen, X. Y., Si-Kang, W., Sun, Y., Feng, W. L., Tao, L., Luo, X. J., & Mai, B. X. (2018). Dechlorane Plus flame retardant in a contaminated frog species: Biomagnification and isomer-specific transfer from females to their eggs. *Chemosphere*, *211*, 218–225.

Wu, J. P., Luo, X. J., Zhang, Y., Chen, S. J., Mai, B. X., Guan, Y. T., & Yang, Z. Y. (2009). Residues of polybrominated diphenyl ethers in frogs (*Rana limnocharis*) from a contaminated site, South China: tissue distribution, biomagnification, and maternal transfer. *Environmental Science & Technology*, *43*(14), 5212–5217.

Wu, X., Zheng, X., Yu, L., Lu, R., Zhang, Q., Luo, X. J., & Mai, B. X. (2022). Biomagnification of persistent organic pollutants from terrestrial and aquatic invertebrates to songbirds: Associations with physiochemical and ecological indicators. *Environmental Science & Technology*, *56*(17), 12200–12209.

Zhang, X. L., Luo, X. J., Liu, H. Y., Yu, L. H., Chen, S. J., & Mai, B. X. (2011). Bioaccumulation of several brominated flame retardants and dechlorane plus in waterbirds from an e-waste recycling region in South China: associated with trophic level and diet sources. *Environmental Science & Technology*, *45*(2), 400–405.

10 Electronic Wastes and Its Impact on Microorganisms

Talluri Rameshwari K R and Rakshitha Rani N
JSS Academy of Higher Education and Research, Mysuru, India

Pratapa M G
ICAR-Central Institute of Fisheries Education, Mumbai, India

10.1 INTRODUCTION

Rapid population growth, industrialization, and urbanization have increased the need for electrical and technological gadgets. Because of the present quick advancements in technology and innovation, new products, superior and smarter electronic and electrical equipment (EEEs), are being manufactured by industries at a shocking rate. The rising production, acquisition, use, and disposal of electronic and electrical equipment (EEE) goods (WEEEs) result in the production of EEE garbage (Tansel et al., 2017). Electronic garbage is also known as "e-Trash" and encompasses little and large products such as smartphones, LED bulbs, freezers, printers, driers, temperature exchangers, and state-of-the-art computing gadgets (Bhattachraya et al., 2016). Due to the tremendous market expansion and quickening pace of technological advancement, most of these electric and electronic materials have comparatively short lifespans, which has resulted in the fastest buildup of e-waste (Sodha et al., 2019). Outdated electrical and electronic equipment with no intention of being reused is commonly referred to as "e-waste." The literature contains statistics indicating that the global production of electronic garbage in 2014 was close to 41.8 million tonnes. Based on an average growth rate of 4–5% per year, 49.8 million tonnes of electronic waste were produced in 2018 (Cui et al., 2016). The enormous volume of compound and varied e-waste produced have an adverse effect on people and the atmosphere when e-waste is not properly accomplished. Inappropriately, there isn't a reliable mechanism in place to store, transport, collect, handle, and dispose of e-waste. As a result, safeguarding the atmosphere from the negative consequences of inadequately treated e-waste has gained significant attention from businesses, governmental and non-governmental organizations, and even the general public (Kaya et al., 2016, Agarwal et al., 2023). The catastrophic ecological effects of e-waste, which are well known to most of us, have just recently come to light. People who are concerned have been forced to seek appropriate treatment as a result of these grave concerns. Base, precious, and rare earth metal contents in e-waste have been steadily increasing over time. In addition, these have several harmful organic pollutants in them (Shah et al., 2014). Printed circuit boards (PCBs) are the basic component of all electronic and many electrical equipment. They are necessary for the efficient, rapid, and functions of tiny to large electronic gadgets in a meaningful way. PCBs include numerous organic contaminants in addition to a wide range of metals. It cannot be burned or disposed of in a landfill as a result (Mondal et al., 2023, Paul et al., 2019). However, the most popular e-waste management techniques involve burning and land filling, which release toxic compounds and dangerous metals into the environment along with soil and groundwater (Choudhury et al.,2022). These pollutants will accumulate and spread throughout plant systems, eventually reaching animals and humans (Awasthi et al., 2016; Awasthi & Li, 2017). Because of increased awareness, e-waste management is now a major area of research on a global scale. The focus of this

DOI: 10.1201/9781003582311-10

chapter is on the effects of electronic e-waste on microorganisms, specifically on the application of microbial technology to extract metals from spent PCBs. The various types of microbes involved in the metal solubilization process are illustrated, as are the primary techniques for extracting metals (Dave et al., 2016, Selin et al., 2006).

10.2 PRINTED CIRCUIT BOARDS (PCBs) AND E-WASTE

China and India, where they belong to Asian countries, are the primary recipients of worldwide electronic trash for recycling since they have laxer regulations governing disposal (Dave et al., 2016). "Any unwelcome or damaged electrical and electronic products or equipment" is the basic definition of e-waste, while there are other meanings as well. As specified by the European Union (Selin et al., 2006), e-waste is the term used to describe any electric appliance that has reached the end of its useful life. Printed circuit boards, or PCBs, are an essential part that regulate the operation of most electric equipment and all electronic items. Precious metals are found to be present in significant amounts in printed circuit boards (Kiddee et al., 2013). Usually, 80–1500 grams of gold and 160–210 kilograms of copper are found in one metric tons of PCBs, it surpasses the amount of gold and copper found in its ores by 40–800 times. Twelve PCBs contain additional expensive base and other rare earth elements. PCBs are described as valuable secondary raw materials for these metals (Pradhan & Kumar 2012). They contain roughly 60 metals from the periodic table (Pradhan & Kumar 2012). Three weight percent of all e-waste is composed of PCBs. Because PCBs are such a rich and varied source of metals, they are seen as an enticing fraction for metal extraction, recycling, and recovery. As such, PCBs are seen as urban mines for recovering metals. PCBs are flat surfaces where electrical machineries viz., conductor chips, capacitors are mounted. Where, they connect the chips using copper sheets that are prevalent on PCB surfaces that usually covered in plastic and act as a structural element (Hadi et al., 2015, Erust et al., 2013). Different PCB kinds are produced using different FR-standards (flame retardants). Printed wiring boards are another name for PCBs (PWBs). PCBs have many different physical appearances in terms of size and shape. Less than 1–2 cm to a few cm is the size range of PCBs, depending on the apparatuses or gadgets that they utilize. PCBs from a phone, computer, television, bathtub light, and trash PCBs that were defective during manufacture. In addition, they are available in red, blue, green, and black color coatings. There is a wide range in the amount of metal included in PCBs, depending on the kind and the devices that use them. According to Dave et al. (2016), nickel, copper, and precious metals like silver and gold are demonstrated to be good suppliers in PCBs. However, because of the uneven mixing of heavy and dangerous metals, they present a challenging recovery problem. Therefore, despite PCBs' high metal content and the hazardous chemicals they contain, there are only a few commercially viable, environmentally benign separation procedures available.

10.3 ROLE OF MICROORGANISMS IN METAL EXTRACTION

Biohydrometallurgy is a process that involves the use of microorganisms to extract metals from solid wastes and concentrates. This method offers several advantages over traditional chemical extraction processes, such as decreased energy use and a smaller conservational effect, and the ability to treat low-grade ores that are not sparingly viable using conventional methods. Using microorganisms to extract metals from solid wastes and concentrates is known as biohydrometallurgy. In comparison to typical chemical extraction techniques, this technology has a number of benefits, including decreased energy consumption, less impact on the environment, and the capacity to handle low-grade ores that are not commercially feasible when using conventional methods. In the twenty-first century, solubilization of metals in the aqueous phase from ores, tailings, concentrates, and e-waste is receiving significant attention in both academic and practical research due to the abstraction of metals from solid wastelands and distillates using biohydrometallurgy. The metabolites of the bacteria are essential. Microbial technologies function in benign conditions at room temperature and pressure, making control and maintenance easier (Erust et al. 2013). The sulfide mines provided the first application of our understanding of the metal extraction mechanisms underpinning microbial activities, when the field-scale metal extraction method known as "bioleaching" or "biomining" from low-grade sulfide ores used microorganisms. Colmer and Hinkle (Finti et al., 2015) presented the first empirical proof of microbes' involvement in metal solubilization in 1947 and later isolated acidophilic autotrophic iron and sulfur-oxidizing bacteria from acid mine drainage (AMD). Colmer used microbes at Kennecott Copper Corporation in the early 1950s to recover copper from a mine dump. Using microbial technology, Zimmerley et al., 1958 recovered copper, gold, and cobalt on an industrial scale from their respective minerals. Excellent outcomes were obtained for the biomining method in the sulfides of the following metal groups, as well as for leaching copper and gold along with other metals such as, Ga, Ni, Mo, Cu, Zn, Pb and Platinum (Lee et al., 2012). The advancement of molecular biology paraphernalia and their custom in the biomining industry for the recognition and documentation of organisms have led to an increase in our understanding of metal microbial interaction and the function that microorganisms play in metal extraction and recovery. Microorganisms from all three domains of life—bacteria, eukaryotes, and archaea—have a potential role and application in biomining today because of advancements in methods for the identification, growth, and use of exacting microorganisms, as well as in the application of molecular biology techniques (Norris et al., 2007; Ehrlich et al., 2001, Vera et al., 2013). The primary microbes involved in the extraction of metals can be broadly classified as organic acid producers, iron and sulfur oxidizers, and cyanogenic organisms.

10.3.1 Iron and Sulfur Oxidizer

The extraction of metals from minerals and e-waste depends heavily on iron and sulfur oxidizers (Ehrlich et al., 2001, Vera et al., 2013). The most well known and well studied are *Leptospirillumferrooxidans*, *Acidithiobacillusferrooxidans*, and *Acidithiobacillusthiooxidans*. These are only a handful of the species that are known to be crucial in the solubilization of metals and to iron and sulfur dioxide. Many additional genera and species are believed to be actively involved in metal agitations at this time.

Most metal-leaching bacteria belong to four phylas:

1. Proteobacteria,
2. Nitrospirae,
3. Firmicutes, and
4. Actinobacteria

Leptospirillum belongs to the Nitrospirae family of bacteria, whereas Acidiphilum, Acidiferrobacter, Ferrovum, and Acidithiobacillus are Proteobacteria. Firmicutes include Alicyclobacillus and Sulfobacillus, and Actinobacteria include Ferrimicrobium, Acidimicrobium, and Ferrithrix. *Ferroplasmaacidiphilum* and *Ferroplasmaacidarmanus* are members of Euryarchaeota, whereas Sulfolobus, Acidianus, Metallosphaera, and Sulfurisphaera are members of the domain archaea and belong to Crenarchaeota (Vera et al., 2013). Based on their optimal development and activity temperature, these organisms are categorized as moderate thermophiles, mesophiles, or thermophiles. There aren't any clear temperature differences, though as species like *Acidimicrobium ferrooxidans* and *Acidithiobacilluscaldus* may thrive in the 25–55°C range without becoming mesophilic or moderately thermophilic. Based on the substrate preferences, they are divided into three groups: sulfur oxidizers, iron oxidizers, and a third group that oxidizes both sulfur and iron. *Acidithiobacillus ferrooxidans*, a bioleaching bacterium, is especially adaptive; it may live aerobically on iron or reduced inorganic sulfur compounds. When oxygen is not present, it keeps going and shows signs of anaerobic development by reducing ferric iron and oxidizing hydrogen and sulfur (Johnson et al., 2012). When ferric/ferrous ratios are elevated, the *Leptospirillum ferrooxidans* and *Leptospirillum ferriphillum* take the place of *Acidithiobacillus ferrooxidans* and play a dominant role in ferric iron generation to provide high ferric lixiviant, which is essential for the extraction of metals from waste PCBs. In contrast to iron and sulfuroxidizing organisms, PCBs do not use the pyrite or sulfur present in minerals as an energy source. There isn't a substrate that has the required iron or sulfur in it. There are two direct and indirect ways to explain *Acidithiobacillus ferrooxidans'* bioleaching of metal sulfides. Despite the general consensus that indirect mechanisms are the only way to extract metals from materials and that direct mechanisms do not exist, bioleaching is now better understood in terms of contact and noncontact processes—terms that are more generally used (Roowerder et al., 2007; Rawlings et al., 2002). Because extracellular polymeric substances (EPS) fill the void between the surface of the substrate and the organism's cell wall, their production is essential for contact processes. Specifically, PCB metals are discovered in e-waste as zero valent metals as Cu_0, Zn_0, Ni_0, etc. rather than as metal sulfides. Thus, inexplicable in Zn_0, Ni_0, and Cu_0 are converted to water-fathomable Cu_{2+}, Zn2+, and Ni_{2+} by biologically generated protons and/or ferric iron from reduced sulfur, ferrous, or sulfur-containing substances (Pradhan., 2012).

10.3.2 Cyanogenic Organisms

The cyanidation process has been used to remove silver along with gold and from minerals encompassing gold in alkaline environments since 1898 (Chauhan et al., 2018, Adrian et al., 1991). By combining with it to generate a water-soluble mixture, cyanide is used to extract gold. Using biogenic cyanide, alkaline bioleaching, often referred to as heterotrophic bioleaching, is an eco-friendly technique for extracting gold. Hydrocyanide is produced by a variety of heterotrophic organisms (HCN). The primary microorganisms known to synthesize HCN are *Chromobacterium violaceum*, *Pseudomonas fluorescens*, *Pseudomonas plecoglossicida*, and *Pseudomonas aeruginosa*. Additionally, it has been noted that fungi that make HCN include the polysphrous and clitocyde species of *Marsmiusoreades*. The recovery of gold involves the use of the enzyme HCN synthase. HCN is typically produced by Pseudomonas and *C. violaceum* at the beginning of the stationary phase, and by *C. violaceum* between the end of the exponential phase and the beginning of the stationary phase. During the late stationary phase, chromobacterium converts its HCN synthesis to cynoalanine, which detoxifies the system (Knowles et al., 1986) (Kumar et al., 2018).

10.4 A SINGLE-STEP BIOLEACHING METHOD

The inoculum from the exponential phase is transferred to the appropriate leaching medium along with the e-waste in a one-step bioleaching operation, and ferric iron or proton is produced by continuing the process, which gradually solubilizes the metals at the locations shown in equivalences. Hazardous chemical compounds and metals that are produced or present during the process impede the growth of organisms (Bosecker et al., 1997, Pradhan et al., 2012). The inoculum's growth and activities are thus inhibited or even terminated. These inhibitory effects limit the bioleaching process's ability to function to low pulp densities. Based on the statistics currently available in the literature, most processes use between 1 and 5% (w/v) or up to 10% (w/v) of e-waste (Brandl et al. 2008, Sabra et al. 2011, Liang et al. 2013). Low concentration of metal extraction medium is caused by low pulp density addition. The procedure's low metal extraction rates are also due to the difficulty of creating optimal conditions for both the organisms' development and metal extraction in a one-step process.

10.5 TWO-STEP BIOLEACHING PROCEDURE

The growth of the necessary organisms is the first stage of a two-step bioleaching process in which the appropriate medium under their ideal growth circumstances are noted. In the procedure's second phase, e-waste is introduced once the organism and lixiviant (ferric, proton, and cyanide) have grown to their maximum potential. The two-stage bioleaching procedure minimizes the toxicity of e-suppression wastes of the organisms and also offers the best conditions for development in the first phase and metal solubilization in the subsequent. The two-step bioleaching method is preferred because it yields more and faster metal extraction (Sand et al., 2001). In addition, the process can be conducted at a higher pulp density than a one-step bioleaching procedure. The second phase involves converting bio generated ferric iron to ferrous iron by using it in metal extractions. The created ferrous iron is moved to the initial stage of the process where it can be re-bio oxidized to produce ferric iron, which is used once more as a lixiviant for solubilizing metals. When e-waste is added to a two-step process, there is a considerable amount of ferric iron, proton, and cyanides already present. These substances react with the e-waste right away, solubilizing the metals at the quickest rate conceivable.

10.6 BIOLEACHING PROCESS

Bioleaching step-one, bioleaching step-two, and wasted medium method are used to eradicate metal from e-waste. The main distinction between e-wastes and sulfide minerals is this. In contrast to sulfide minerals, when metal sulfides, which are their form, are present coupled with pyrite or sulfur, e-waste contains metals in their zero valent state. For organisms that can oxidize sulfur and iron, pyrite, metal sulfides (or) sulfur serve as an vigor source (Sand et al., 2001). These are used as substrates during the step-one process by the leaching system's additional inoculum, which continuously produces proton or ferric as lixiviants. These lixiviants react with the sulfides metal, causing the metal to become soluble. No such substrates or energy sources are present in case of e-waste, so ferrous iron, pyrite, or sulfur must be added to the system of leaching, from which the organism produces the essential lixiviants (Shah et al., 2015, Ilyas et al., 2014).

10.7 FACTORS INFLUENCING E-WASTE BIOLEACHING

To recover metal from e-waste, a wide range of organisms are used. The synthesis of ferric iron, protonated organic acids, protons and HCN primarily controls the bioleaching process. The three main variables that affect metal ion extraction are pulp density, dissolved oxygen content, temperature, pH, iron/HCN absorption, e-Waste particle size, medium, type of culture, and inclusion of complexing agents and surfactants, irrespective of the lixiviant and the organisms employed in the bioleaching procedure (Shah et al., 2015). Since aerobic conditions are proven to be more favorable for all three categories of organisms, aeration or agitation is a key component of metal extraction. While it has been discovered that cyanogenic bacteria, which are acidic, pH is necessary for metal extractions mediated by fungi and iron and sulfur oxidizers, which are more active at alkaline pHs, ideally pH over 9.0. This is due to the produced HCN being converted to cyanide rather than being lost as vapor. However, it inhibits the organisms in some way, which limits the process above a particular pulp density. Increasing pulp density is sought for economically viable metal extractions. Less than 10% w/v pulp density was used in the majority of reported e-waste bioleaching methods. Shah et al. 2015 (Ilyas et al., 2014) evaluated the procedure at pulp densities of 15% w/v and higher. The pulp density limit may be partially removed by using spent medium bioleaching or two-step bioleaching. Along with highlighting the advantages of using ethylene diamine disuccinic acid (EDDS) over ethylene diamine tetra acetic acid, Lan et al., (Shah et al., 2015, Lan et al., 2009) have also reported on the effectiveness of individual chelating agents. They have also emphasized the advantages of including a complexing agent and a surfactant in e-waste metal extraction (EDTA). Depending on the target metal for solubilization, the influence of various additions differs. During the procedure, ferric iron acts as the lixiviant and the chelating agent acts as the ligand, leading to the ferric-chelate-metal complexation, which aids in maintaining more ferric iron in solution and improves extractions (Ylivainio et al., 2010). Smaller particles are said to allow for better metal extraction, but we've noticed that this isn't always the case with some types of e-waste, so choosing the right particle size is crucial. For faster and more effective metal extraction, these lixiviants, which aim for low absorption, should be preserved. Because the generation of lixiviants is what essentially controls e-waste bioleaching, concentrations of organic acids, ferric iron, HCN, and protons are important factors to consider (Pant et al., 2012). In order to better remove metals, bioleaching techniques from e-waste are now being investigated. With certain modifications, the one-step method is utilized; the total pulp density is added in fractions of two, three, or four lots rather than as one lot (fed batch process). In comparison to addition in a single lot, such fraction addition produces larger metal extractions (Shah et al., 2015, Segev et al., 2009). For high metal extraction, small-sized e-waste is typically favored; nevertheless, this needs grinding, which raises process costs, creates air pollution, makes it challenging to separate solids after leaching, and produces more sludge. The usage of PCBs without grinding increases the amount of metal extracted and lessens many of the issues that are mentioned as a result of using fine e-waste particles in the process. When using large PCBs without grinding, epoxy plates from the PCBs' glass can be used in a variety of applications after the leaching process (Abraham et al., 2002, An et al., 2011). Tenders of membrane technology may help to mitigate issues caused by the use of smaller e-waste particles. Cheaper membranes made of fabric that were tested by the author for capturing trash from the earth for bioleaching demonstrated improved metal extraction and made separating leached e-waste solids from the solution relatively simple (Perkins et al., 2014, Salam & Varma, 2019) (Su et al., 2013).

10.8 PRECIOUS METALS: MICROBIAL RECOVERY FROM E-WASTE POLLUTANTS

Compared to some ores, PCBs have higher amounts of some key elements. Thus, treating e-waste to prevent environmental contamination is equally important as microbiologically recovering valuable metals from it (Awashthi et al., 2000). The two steps in biometallurgy—the microbial recovery of precious metals—are bioleaching and biosorption. This approach is seen to be advantageous because of being less expensive, having minimal impact on the environment, and low energy consumption (Salam & Varma, 2019, Qiu et al., 2012). Cu may be successfully extracted, as can the precious metals Ag and Au. Microorganisms that can remove precious metals from used printed circuit boards (PCBs) could find profitable applications. For example, an efficient recovery of copper and gold from PCB was achieved using a two-stage bioleaching technique that combined *Pseudomonas putida* and *Pseudomonas fluorescens* with *Acidithiobacillus ferrivorans* and thiooxidans in the first and second steps, correspondingly (Abraham et al., 2002, Song et al., 2018). Moreover, reports exist of naturally occurring bacteria resistant to the toxicity of e-waste, such as *Pseudomonas balearica* SAE1, a bioleaching bacterium isolated from an e-waste recycling facility that has been the subject of numerous investigations (Kumar et al., 2018). It was able to extract silver and gold from PCB (An et al., 2011, Sobolev & Begonia, 2008). Researchers have investigated the potential use of Proteus hauseri ZMd44's unique copper oxidase in bioremediation and metal recovery. The copper oxidase gene was cloned and expressed in *E. coli*. The enzyme's remarkable capacity for Au adsorption was found, and it can be used as a biosensor or in the bio recovery of gold (Awashthi et al., 2016, Singh et al., 2017). In two hours, a set of bacteria utilized in effective bioleaching showed 100% copper recovery from 5 g/L PCBs. Using this method, the toxicity of PCBs to microorganisms has been inadequate. The microorganisms used were *Leptospirillum ferriphilum* and *Sulfobacillus thermosulfidooxidans*, which are capable of oxidizing iron and sulfur (Awashthi, 2000, Su et al., 2015, Suenaga et al., 2012).

10.9 METHODS FOR STUDYING SOIL MICROBIAL DIVERSITY

Researchers have been working on new techniques to explore the soil's microbial world and understand the roles that these organisms play in the ecosystem since the soil contains a wide range of non-cultivable bacteria and fungus. Between 2000-2004 and 8.3 million bacterial species are thought to exist per gramme of soil, and only 1% of this population can be grown in a lab setting using standard techniques (Awasthi et al., 2016, Bálint et al., 2016). According to this, it is estimated that there are roughly 1,500,000 species of fungi on the globe, yet the majority of them cannot be produced in a lab using conventional methods (Panwar et al., 2018, Perkins et al., 2014). Analysis of microbial diversity can be done using biochemical techniques such as plate counts, phospholipid fatty acid (PLFA), research on fatty acid methyl ester (FAME), and physiological profiling at the community level (CLPP), (Bhattacharya & Khare, 2016, Borja et al., 2005, Burd et al., (1998). These very specialized approaches produce very reproducible findings but are unable to identify nonculturable bacteria. Many molecular- techniques have been developed that provide important details regarding the makeup of a particular site's microbial community. A number of widely used molecular- techniques are as follows:

- Amplified ribosomal DNA restriction analysis (ARDRA),
- Denaturing and temperature gradient gel electrophoresis (DGGE and TGGE),
- Ribosomal intergenic spacer analysis (RISA)/automated RISA/RISA,
- Restriction fragment length polymorphism (RFLP), and
- Terminal restriction fragment length polymorphism (T-RFLP) (ARISA).

Nevertheless, despite the limitations of PCR biases, methods like DGGE/TGGE, T-RFLP, and ARISA offer a good picture of community profiles and can be used to compare variations in microbial communities (Salam & Varma, 2019, Needhidasan et al., 2014). One of the most effective methods for comprehending the functional variety of microbial communities is metagenomics, which entails directly removing DNA from environmental samples, cloning the DNA, and examining the genomic DNA retrieved from the complete community (Saritha et al., 2015a, 2015b; Segev et al., 2009). In light of this, several improved sequencing methods have been developed that provide read lengths sufficient for accurate assembly and investigation of the genetic diversity of microbes. With the help of enhanced techniques for metagenomics DNA analysis, such as the targeted slant of certain genes, the evolutionary and functional patterns of important genes from a variety of bacteria and habitats may be thoroughly comprehended (Cepa et al., 2007). Nowadays, high throughput sequencing (HTS) is frequently employed in biodiversity research. More focus goes to the microarray technology, where the dynamics and activities of hundreds to thousands of microbial populations in complex habitats may be simultaneously determined (Salam & Varma, 2019). Five categories of microarrays are used in microbial ecology research:

i. Using ribosomal rRNA genes as the basis for the probe, phylogenetic oligonucleotide arrays (POAs),
ii. Functional gene arrays (FGAs) are built using functional genes as their foundation,
iii. Community genome arrays (CGAs) based on the whole genome,
iv. Environmental DNA is used in metagenomic arrays (MGA), and
v. Open reading frames in whole genomes are the basis for whole genome open reading frame arrays (WGA) (Quideau et al., 2016, Rajapaksha et al., 2004, Roesch et al., 2007).

To be used as a standard instrument in microbial ecology research, microarray despite being a penetrating knowledge needs advancements in its technical aspects, such as in the collection of data analysis, pure samples, and automation. Numerous researches have made use of gene microarrays like Geo-Chip, which target particular genes involved in microbial activity. For instance, Geo-Chip (Awashthi et al., 2000) uses conventional gene families for a range of environmental stress responses, bacteriophages, and virulence mechanisms, as well as genes for organic cleanup, metal reduction, antibiotic resistance, and the carbon, nitrogen, sulfur, and phosphorous cycles (Cayumil et al., 2016, Lenart-Boroń et al., 2014, Liu et al., 2018).

Methods using gene microarrays may be constrained by the DNA fragments' non-specific binding to the probes (Singh et al., 2017, Sobolev et al., 2008, Taiwo et al., 2018, Segev et al., 2009).

The development of a method known as capture metagenomics enhanced methods for the investigation of microbial functional diversity and created probes to target many functional genes encoding for enzymes that break down organic compounds using carbohydrates. The web-based tools were used to create the probes, which successfully target a number of genes implicated in a certain pathway. For the examination of the functions of microbial communities in various contexts, numerous genes could be sequenced using this technique, which allowed it to detect non-specific binding and have higher resolution than the other approaches (Manoharan et al., 2015, Lal et al., 2006) The study of the organization of microbial communities has undergone a rapid development in recent years, and this has made it difficult to analyze the data. Balint and Boajă (2024), provide a thorough examination of the benefits and drawbacks of various statistical techniques and the integration of HTS data of marker genes for the analysis of microbial diversity and composition (2016). According to the current state of knowledge, a clear picture of the microbial community can be obtained by studying its structure and functions using both biochemical and molecular approaches (Tan et al., 2017, Tang et al., 2013).

10.10 FUTURE PROSPECTS AND CONCLUSION

Metal extraction from ore and concentrates using biomining is an ancient practice. In terms of the environment, it is a green technology. However, e-waste bioleaching research is less developed than other disciplines because handling a live system is more challenging than using chemical or mechanical methods (Guerin et al., 2002; Tchounwou et al., 2012). The current e-waste treatment techniques are hazardous and detrimental to humans and the biotic and abiotic environment (Salam & Varma, 2019; Needhidasan et al., 2014). Many technologies have been developed and are used to extract metal from e-waste. Although bio-hydro technology is one of these safest and acceptable techniques, there are still loopholes in the research that need to be filled. The two primary issues are that this technology can only handle tiny amounts of e-waste and that the toxicity of e-waste prevents the organisms utilized in the process from surviving for prolonged periods (Tu et al., 2014; Szanto et al., 2010). Therefore, it is vital to concentrate on creating novel strains, process optimization, and scaling up the current process for the management and treatment of e-waste. As a result, the area has to concentrate increasingly on various interdisciplinary topics such as microbiology, metallurgy, engineering, and biotechnology (Panwar et al., 2018; Perkins et al., 2014; Szanto et al., 2010). To reduce the environmental impact of the diverse harmful materials found in e-waste, it is necessary to rehabilitate the places where they were deposited effectively. The future of reducing the toxicity of e-waste dangers lies in microbiological hybrid techniques. Metals need to have any organic contaminants and colored epoxy coatings removed as a pre-treatment to make them vulnerable to microbial attack and enable faster and more efficient extraction. However, the effectiveness of biological treatments would be improved if they were employed in conjunction with various techniques, such as nanoparticles or certain non-polluting biodegraded substances. Furthering efforts on the "reduce, reuse, and recycle" method is also necessary to close the knowledge gap. The microbial population in the soil is impacted by heavy metals and hazardous organic pollutants, which also change the functional gene diversity of microbes. The development of microorganisms that can withstand the effects of toxic pollutants has improved methods for studying microbial diversity, mainly through a culture self-determining method, and current studies demonstrate that this has increased the diversity of the microbial functional community. Metagenomics and microarray technologies help examine the microbial community and functional diversity in contaminated environments. More research is necessary to fully understand how the microorganisms react to the increase in soil contamination caused by e-waste, the function of dominating microorganisms in ecosystems, and the primary channels via which organic contaminants degrade e-waste. To discover effective ways of remediating contaminated soils, rhizobial interactions with plants with phytoremediation capability must be investigated.

REFERENCES

Abraham, W. R., Nogales, B., Golyshin, P. N., Pieper, D. H., & Timmis, K. N. (2002). Polychlorinated biphenyl-degrading microbial communities in soils and sediments. *Current Opinion in Microbiology*, 5(3), 246–253.

Adrian G. Smith, & Mudder, T. (1991). *The chemistry and treatment of cyanidation wastes. Mining Journal Books Limited.*

Agarwal, S., Darbar, S., Saha, S., Choudhury, M., & Singh, R. P. (2023). E-waste management using different cost-effective, eco-friendly biological techniques: an overview. *Waste Management and Resource Recycling in the Developing World*, 205–235. https://doi.org/10.1016/B978-0-323-90463-6.00026-9

An, T., Zu, L., Li, G., Wan, S., Mai, B., & Wong, P. K. (2011). One-step process for debromination and aerobic mineralization of tetrabromobisphenol-A by a novel Ochrobactrum sp. T isolated from an e-waste recycling site. *Bioresource technology*, 102(19), 9148–9154.

Awashthi, S. K. (2000). *Central and state rules as amended for 1999: prevention of food adulteration act no 37 of 1954*. Ashoka Law House, New Delhi.

Awasthi, A. K., & Li, J. (2017). Management of electrical and electronic waste: A comparative evaluation of China and India. *Renewable and Sustainable Energy Reviews*, 76, 434–447.

Awasthi, A. K., Zeng, X., & Li, J. (2016). Environmental pollution of electronic waste recycling in India: A critical review. *Environmental Pollution*, 211, 259–270.

Bálint, M., Bahram, M., Eren, A. M., Faust, K., Fuhrman, J. A., Lindahl, B., .. & Tedersoo, L. (2016). Millions of reads, thousands of taxa: microbial community structure and associations analyzed via marker genes. *FEMS Microbiology Reviews*, 40(5), 686–700.

Balint, R. and Boajă, I.P. (2024), Assisted phytoextraction as a nature-based solution for the sustainable remediation of metal(loid)-contaminated soils. *Integrated Environmental Assessment and Management*, 20: 2003–2022. https://doi.org/10.1002/ieam.4907

Bhattacharya, A., & Khare, S. K. (2016). Sustainable options for mitigation of major toxicants originating from electronic waste. *Current Science*, 111(12), 1946–1954, https://doi.org/10.18520/cs/v111/i12/1946-1954.

Borja, J., Taleon, D. M., Auresenia, J., & Gallardo, S. (2005). Polychlorinated biphenyls and their biodegradation. *Process Biochemistry*, 40(6), 1999–2013.

Bosecker, K. (1997). Bioleaching: metal solubilization by microorganisms. *FEMS Microbiology Reviews*, 20(3–4), 591–604.

Brandl, H., Lehmann, S., Faramarzi, M. A., & Martinelli, D. (2008). Biomobilization of silver, gold, and platinum from solid waste materials by HCN-forming microorganisms. *Hydrometallurgy*, 94(1–4), 14–17.

Burd, G. I., Dixon, D. G., & Glick, B. R. (1998). A plant growth-promoting bacterium that decreases nickel toxicity in seedlings. *Applied and Environmental Microbiology*, 64(10), 3663–3668.

Cayumil, R., Khanna, R., Rajarao, R., Ikram-ul-Haq, M., Mukherjee, P. S., & Sahajwalla, V. (2016). Environmental impact of processing electronic waste–key issues and challenges. *E-waste in Transition—from Pollution to Resource. InTech*, 9–35, https://doi.org/10.5772/64139.

Cepa, C. E. P. A. (2007). *Canadian soil quality guidelines for the protection of environmental and human health.* Quebec, Canada: National Guidelines and Standards Office.

Chauhan, G., Jadhao, P. R., Pant, K. K., & Nigam, K. D. P. (2018). Novel technologies and conventional processes for recovery of metals from waste electrical and electronic equipment: challenges & opportunities–a review. *Journal of Environmental Chemical Engineering*, 6(1), 1288–1304.

Choudhury, M., Jyethi, D. S., Dutta, J., Purkayastha, S. P., Deb, D., Das, R., … & Bhattacharyya, K. G. (2022). Investigation of groundwater and soil quality near to a municipal waste disposal site in Silchar, Assam, India. *International Journal of Energy and Water Resources*, 6(1), 37–47. https://doi.org/10.1007/s42108-021-00117-5

Cui, H., & Anderson, C. G. (2016). Literature review of hydrometallurgical recycling of printed circuit boards (PCBs). *Journal of Advanced Chemical Engineering*, 6(1), 142–153.

Dave, S. R., Shah, M. B., & Tipre, D. R. (2016). E-Waste: Metal Pollution Threat or Metal Resource? *SOJ Biotech* 1(1), 14.

Ehrlich, H. L. (2001). Past, present and future of biohydrometallurgy. *Hydrometallurgy*, 59(2–3), 127–134.

Erüst, C., Akcil, A., Gahan, C. S., Tuncuk, A., & Deveci, H. (2013). Biohydrometallurgy of secondary metal resources: a potential alternative approach for metal recovery. *Journal of Chemical Technology & Biotechnology*, 88(12), 2115–2132.

Fonti, V., Amato, A., & Beolchini, F. (2015). Urban biomining: new challenges for a successful exploitation of WEEE by means of a biotechnological approach. *The Environment*, 329, https://doi.org/10.1201/b18124-14.

Guerin, T. F., Horner, S., McGovern, T., & Davey, B. (2002). An application of permeable reactive barrier technology to petroleum hydrocarbon contaminated groundwater. *Water Research*, 36(1), 15–24.

Hadi, P., Xu, M., Lin, C. S., Hui, C. W., & McKay, G. (2015). Waste printed circuit board recycling techniques and product utilization. *Journal of hazardous materials*, 283, 234–243.

Ilyas, S., Lee, J. C., & Kim, B. S. (2014). Bioremoval of heavy metals from recycling industry electronic waste by a consortium of moderate thermophiles: process development and optimization. *Journal of Cleaner Production*, 70, 194–202.

Johnson, D. B. (2012). Geomicrobiology of extremely acidic subsurface environments. *FEMS Microbiology Ecology*, 81(1), 2–12.

Kaya, M. (2016). Recovery of metals and nonmetals from electronic waste by physical and chemical recycling processes. *Waste Management*, 57, 64–90.

Kiddee, P., Naidu, R., & Wong, M. H. (2013). Electronic waste management approaches: An overview. *Waste Management*, 33(5), 1237–1250.

Knowles, C. J., & Bunch, A. W. (1986). Microbial cyanide metabolism. *Advances in Microbial Physiology*, 27, 73–111.

Kumar, A., Saini, H. S., & Kumar, S. (2018). Bioleaching of gold and silver from waste printed circuit boards by Pseudomonas balearica SAE1 isolated from an e-waste recycling facility. *Current Microbiology*, 75(2), 194–201.

Lal, R. (Ed.). (2006). *Encyclopedia of soil science (No. 11)*. CRC Press. https://books.google.co.in/books?hl=en&lr=&id=627Qopsj7bsC&oi=fnd&pg=PA931&dq=Lal,++R.+

Lan, Z., Hu, Y., & Qin, W. (2009). Effect of surfactant OPD on the bioleaching of marmatite. *Minerals Engineering*, 22(1), 10–13.

Lee, J. C., & Pandey, B. D. (2012). Bio-processing of solid wastes and secondary resources for metal extraction–a review. *Waste Management*, 32(1), 3–18.

Lenart-Boroń, A., & Boroń, P. (2014). The effect of industrial heavy metal pollution on microbial abundance and diversity in soils—a review. *IntechOpen*. 10.5772/57406

Liang, G., Tang, J., Liu, W., & Zhou, Q. (2013). Optimizing mixed culture of two acidophiles to improve copper recovery from printed circuit boards (PCBs). *Journal of Hazardous Materials*, 250, 238–245.

Liu, J., Chen, X., Shu, H. Y., Lin, X. R., Zhou, Q. X., Bramryd, T., .. & Huang, L. N. (2018). Microbial community structure and function in sediments from e-waste contaminated rivers at Guiyu area of China. *Environmental Pollution*, 235, 171–179.

Manoharan, L., Kushwaha, S. K., Hedlund, K., & Ahrén, D. (2015). Captured metagenomics: large-scale targeting of genes based on 'sequence capture' reveals functional diversity in soils. *DNA Research*, 22(6), 451–460.

Mondal, T., Choudhury, M., Kundu, D., Dutta, D., & Samanta, P. (2023). Landfill: An eclectic review on structure, reactions and remediation approach. *Waste Management*, 164, 127–142. https://doi.org/10.1016/j.wasman.2023.03.034

Needhidasan, S., Samuel, M., & Chidambaram, R. (2014). Electronic waste–an emerging threat to the environment of urban India. *Journal of Environmental Health Science and Engineering*, 12(1), 1–9.

Norris, P. R. (2007). Acidophile diversity in mineral sulfide oxidation. In *Biomining* (pp. 199–216). Springer, Berlin, Heidelberg.

Pant, D., Joshi, D., Upreti, M. K., & Kotnala, R. K. (2012). Chemical and biological extraction of metals present in E waste: a hybrid technology. *Waste Management*, 32(5), 979–990.

Panwar, R. M., & Ahmed, S. (2018). Assessment of contamination of soil and groundwater due to e-waste handling. *Current Science*, 114(1), 166–173.

Perkins, D. N., Drisse, M. N. B., Nxele, T., & Sly, P. D. (2014). E-waste: A global hazard. *Annals of global health*, 80(4), 286–295.

Pradhan, J. K., & Kumar, S. (2012). Metals bioleaching from electronic waste by Chromobacteriumviolaceum and Pseudomonads sp. *Waste Management & Research*, 30(11), 1151–1159.

Pradhan, S. C. (2012). Nonlocal finite element analysis and small scale effects of CNTs with Timoshenko beam theory. *Finite Elements in Analysis and Design*, 50, 8–20.

Paul, S., Choudhury, M., Deb, U., Pegu, R., Das, S., & Bhattacharya, S. S. (2019). Assessing the ecological impacts of ageing on hazard potential of solid waste landfills: A green approach through vermitechnology. *Journal of cleaner production*, 236, 117643. https://doi.org/10.1016/j.jclepro.2019.117643

Qiu, M., Chen, X., Deng, D., Guo, J., Sun, G., Mai, B., & Xu, M. (2012). Effects of electron donors on anaerobic microbial debromination of polybrominated diphenyl ethers (PBDEs). *Biodegradation*, 23(3), 351–361.

Quideau, S. A., McIntosh, A. C., Norris, C. E., Lloret, E., Swallow, M. J., & Hannam, K. (2016). Extraction and analysis of microbial phospholipid fatty acids in soils. *JoVE (Journal of Visualized Experiments)*, (114), e54360.

Rajapaksha, R. M. C. P., Tobor-Kapłon, M. A., & Baath, E. (2004). Metal toxicity affects fungal and bacterial activities in soil differently. *Applied and Environmental Microbiology*, 70(5), 2966–2973.

Rawlings, D. E. (2002). Heavy metal mining using microbes. *Annual Reviews in Microbiology*, 56(1), 65–91.

Roesch, L. F., Fulthorpe, R. R., Riva, A., Casella, G., Hadwin, A. K., Kent, A. D., … & Triplett, E. W. (2007). Pyrosequencing enumerates and contrasts soil microbial diversity. *The ISME Journal*, 1(4), 283–290.

Rohwerder, T., & Sand, W. (2007). Oxidation of inorganic sulfur compounds in acidophilic prokaryotes. *Engineering in Life Sciences*, 7(4), 301–309.

Sabra, N., Dubourguier, H. C., & Hamieh, T. (2011). Fungal leaching of heavy metals from sediments dredged from the Deûle Canal, France. 2(1), Article ID: 16702, 8 Pages, 10.4236/aces.2012.21001

Salam, M. D., & Varma, A. (2019). A review on impact of e-waste on soil microbial community and ecosystem function. *Pollution*, 5(4), 761–774.

Sand, W., Gehrke, T., Jozsa, P. G., & Schippers, A. (2001). (Bio) chemistry of bacterial leaching—direct vs. indirect bioleaching. *Hydrometallurgy*, 59(2–3), 159–175.

Saritha, V., Sunil Kumar, K. A., & Srikanth Vuppala, N. V. (2015a). Consumer attitudes and perceptions on electronic waste: An assessment. *Pollution*, 1(1), 31–43.

Saritha, K, Rajesh, A, Manjulatha, K, Setty, OH, Yenugu, S. (2015b). Mechanism of antibacterial action of the alcoholic extracts of Hemidesmus indicus (L.) R. Br. ex Schult, Leucas aspera (Wild.), Plumbago zeylanica L., and Tridax procumbens (L.) R. Br. ex Schult *Frontiers in microbiology*, 6, 577.

Segev, O., Kushmaro, A., & Brenner, A. (2009). Environmental impact of flame retardants (persistence and biodegradability). *International Journal of Environmental Research and Public Health*, 6(2), 478–491.

Selin, H., & VanDeveer, S. D. (2006). Raising global standards: hazardous substances and e-waste management in the European Union. *Environment: Science and Policy for Sustainable Development*, 48(10), 6–18.

Shah, M. B., Tipre, D. R., & Dave, S. R. (2014). Chemical and biological processes for multi-metal extraction from waste printed circuit boards of computers and mobile phones. *Waste Management & Research*, 32(11), 1134–1141.

Shah, M. B., Tipre, D. R., Purohit, M. S., & Dave, S. R. (2015). Development of two-step process for enhanced biorecovery of Cu–Zn–Ni from computer printed circuit boards. *Journal of Bioscience and Bioengineering*, 120(2), 167–173.

Singh, P., Borthakur, A., Singh, R., Awasthi, S., Pal, D. B., Srivastava, P., .. & Mishra, P. K. (2017). Utilization of temple floral waste for extraction of valuable products: A close loop approach towards environmental sustainability and waste management. *Pollution*, 3(1), 39–45.

Sobolev, D., & Begonia, M. F. (2008). Effects of heavy metal contamination upon soil microbes: lead-induced changes in general and denitrifying microbial communities as evidenced by molecular markers. *International Journal of Environmental Research and Public Health*, 5(5), 450–456.

Sodha, A. B., Qureshi, S. A., Khatri, B. R., Tipre, D. R., & Dave, S. R. (2019). Enhancement in iron oxidation and multi-metal extraction from waste television printed circuit boards by iron oxidizing Leptospirillumferiphillum isolated from coal sample. *Waste and Biomass Valorization*, 10(3), 671–680.

Song, M., Cheng, Z., Luo, C., Jiang, L., Zhang, D., Yin, H., & Zhang, G. (2018). Rhizospheric effects on the microbial community of e-waste-contaminated soils using phospholipid fatty acid and isoprenoid glycerol dialkyl glycerol tetraether analyses. *Environmental Science and Pollution Research*, 25(10), 9904–9914.

Su, X., Shen, H., Yao, X., Ding, L., Yu, C., & Shen, C. (2013). A novel approach to stimulate the biphenyl-degrading potential of bacterial community from PCBs-contaminated soil of e-waste recycling sites. *Bioresource Technology*, 146, 27–34.

Su, X.M., Liu, Y.D., Hashmi, M.Z., Ding, L.X. and Shen, C.F. (2015). Culture-dependent and culture-independent characterization of potentially functional biphenyl-degrading bacterial community in response to extracellular organic matter from Micrococcus luteus. *Microbial Biotechnology*, 8(3), 569–578.

Suenaga, H. (2012). Targeted metagenomics: a high-resolution metagenomics approach for specific gene clusters in complex microbial communities. *Environmental Microbiology*, 14(1), 13–22.

Szanto, A., Balint, B. L., Nagy, Z. S., Barta, E., Dezso, B., Pap, A., .. & Nagy, L. (2010). STAT6 transcription factor is a facilitator of the nuclear receptor PPARγ-regulated gene expression in macrophages and dendritic cells. *Immunity*, 33(5), 699–712.

Taiwo, M. O., Onatunde, O. O., Bamisile, O., Nwachukwu, B. C., & Sakariyau, A. O. (2018). Assessment of soil microorganisms, heavy metal levels and natural Radionuclei concentrations of three electronic waste dumpsites in Nigeria. *International Journal of Microbiological Research*, 9(3), 81–88.

Tan, S. I., Ng, I. S., & Yu, Y. J. (2017). Heterologous expression of an acidophilic multicopper oxidase in Escherichia coli and its applications in biorecovery of gold. *Bioresources and Bioprocessing*, 4(1), 1–10.

Tang, X., Qiao, J., Chen, C., Chen, L., Yu, C., Shen, C., & Chen, Y. (2013). Bacterial communities of polychlorinated biphenyls polluted soil around an e-waste recycling workshop. *Soil and Sediment Contamination: An International Journal*, 22(5), 562–573.

Tansel, B. (2017). From electronic consumer products to e-wastes: Global outlook, waste quantities, recycling challenges. *Environment International*, 98, 35–45.

Tchounwou, P. B., Yedjou, C. G., Patlolla, A. K., & Sutton, D. J. (2012). Heavy metals toxicity and the environment. EXS 101, 133Ã164.

Tu, Q., Yu, H., He, Z., Deng, Y. E., Wu, L., Van Nostrand, J. D., .. & Zhou, J. (2014). GeoChip 4: a functional gene-array-based high-throughput environmental technology for microbial community analysis. *Molecular Ecology Resources*, 14(5), 914–928.

Vera, M., Schippers, A., & Sand, W. (2013). Progress in bioleaching: fundamentals and mechanisms of bacterial metal sulfide oxidation—part A. *Applied Microbiology and Biotechnology*, 97(17), 7529–7541.

Ylivainio, K. (2010). Effects of iron (III) chelates on the solubility of heavy metals in calcareous soils. *Environmental Pollution*, 158(10), 3194–3200.

Zimmerley, S. R., Wilson, D. G., & Prater, J. D. (1958). U.S. Patent No. 2,829,964. Washington, DC: U.S. Patent and Trademark Office.

11 Electronic Wastes, Its Impact on Plants and Agricultural Crops

Rani Jha
Symbiosis Open Education Society, Pune, India

Avinash Sharma
Arunachal University of Studies, Namsai, India

Nguyen Tran Hai Bang
CUU Long Delta Rice Research Institute, Can Tho, Vietnam

Anwesha Gohain and Monjeet Sonowal
Arunachal University of Studies, Namsai, India

Zhang Shi
Shenyang Agricultural University, Shenyang, China

11.1 ROLE OF AGRICULTURE IN FOOD SAFETY

India is one of the world's largest agrarian societies, where over 70% of the population is dependent upon some form of agriculture and/or its allied sectors for livelihood. The scenario is more diverse in rural areas as compared to urban cities and contributes in share (%) of Gross Domestic Product (GDP). The growth in sustainable methods of agriculture has led to improved food security, employment generation, and agro-ecosystem technologies such as soil conservation, natural resource management, and agrobiodiversity conservation in a more holistic rural development (Karolina and Małgorzata, 2020). The agriculture-led allied sectors are transformed by the impact of the green, white, yellow, golden, red, brown, and blue revolutions (Dastagiri et al., 2014). The agricultural practices and agricultural industry are mostly dominated by rural villages due to the availability of land and manpower. Food is the most basic necessity for survival and is critical to the growth and development of a country and its people as a whole. Ensuring adequate nutrition for all is therefore considered a milestone under Goal 2 of the Sustainable Development Goals (SDG) by the United Nations (UN). Sustainable Development Goal 2 talks about providing food for all, especially the developing and disadvantaged in developing countries. Agriculture is the most important resource for maintaining food security. As an agriculture-dominated society, India can make a significant contribution to ending hunger (SDG 2). However, traditional agricultural practices of relocating cultivated land by the industrial sector have created significant competition and threatened the country's food security status quo (Singh et al., 2022). About 70% of India's population is engaged in agriculture. From an economic perspective, agriculture's contribution can be divided into increasing

DOI: 10.1201/9781003582311-11

gross national income, job creation, industry creation, agricultural inputs, and supply of raw materials (Gulati, Terway and Hussain, 2021). About 65% of the population is totally dependent on agriculture and related activities. In India, 41.49% of her workforce is employed in agro-related and agro-based industries (Verena and Navin, 2017). Most industries such as cotton fabric, jute, paper, and sugar rely on the extraction of raw materials to develop value-added products. Small-scale industries such as spinning, cotton ginning, presses, refineries, rice mills, fruit processing and agricultural activities, sericulture, beekeeping, beekeeping, agricultural machinery and tools, etc., are primarily involved in agricultural production and income (Bodie et al. 2019). Agriculture plays an important role in boosting state and national economies, primarily through direct and indirect taxes on agricultural inputs and outputs, such as the sale of farmland, taxes on agricultural products, and product processing fees where appropriate. About 50% of agricultural products such as tea, coffee, sugar, oilseeds, tobacco, spices, coconuts, jute, field crops, and horticultural crops are shipped worldwide (Anderson 2010). To overcome the hunger index and malnutrition, the accountability and availability of food provided to all groups of people is paramount. Agriculture provides grains and nutritious food and generates income around the world. It also maintains nutrient-rich crops, high yields, and better nutrition, alleviating poverty, hunger index, malnutrition, disease, illiteracy, environmental degradation, and discrimination against women (William et al. 2022). Agriculture improves food production and food diversity in the long term and repairs environmental degradation. Desertification, salinity, and climate change affect food availability and yields. Agricultural science includes intelligent plant breeding programs, such as in vitro techniques, molecular markers, genomics, and reverse genetics to develop high-yielding, high-value crops for domestic use and export markets, to increase income generation and contribute to socio-economic development (Gulati et al. 2021).

11.2 E-WASTES AND AGRO-ECOLOGICAL CONSEQUENCES

E-waste or electronic waste is scrap material from electronic devices. Classified electronic devices are used in society and institutions. The Environmental Protection Agency (EPA) has researched and noted categories of electronic materials such as common consumer electronics, electronic product types, etc., depending on the end use (Vats and Singh, 2014). The invention of new generation technology is forced to discard the old technology. In 2016, 44.7 million tons of waste electronic materials were generated worldwide. In 2018, e-waste was estimated at 50 million tons worldwide. In 2021, 57.4 million tons of e-waste were reported worldwide (Shashi and Sunil, 2020). Less than 20% of all e-waste collected is recycled by the industry. E-waste is harmful to ecosystems E-waste disposal pollutes the environment with soil, air, and water. Combustion of electronic materials releases vast amounts of hydrocarbons, causing harmful air pollution in open environments (Manisalidis 2020). Discarded electronic materials are decomposed underground. Waste electronic materials contain heavy metals (copper (Cu), lead (Pb), zinc (Zn), nickel (Ni), cadmium (Cd), chromium (Cr), iron (Fe), manganese (Mn), etc.), emits organic pollutants, i.e. hydrocarbons, aromatics, phenols, organochlorines, pesticides, and lithium filters pass through the soil and into groundwater and water bodies (Sathya 2022; Choudhury et al., 2022). The presence of these metals disrupts the habitat of zooplankton and phytoplankton communities in aquatic environments. The harmful effects of metals from e-waste remain in the ground for long periods of time and migrate into the facility over time. Morphological constituents such as roots, shoots, and leaves absorb organic pollutants such as the metals mentioned above, hydrocarbons, aromatics, phenols, organochlorines, and pesticides (Alengebawy 2021). The morphological parts of plants are divided into mechanical rhizome decomposition (metal degradation), plant filtration/plant extraction (metal assimilation), mobilization and immobilization (metal conversion), and plant stabilization (metal stabilization). It shows, removes, and stabilizes the aforementioned metallic and organic contaminants. Phytoremediation mechanisms alter plant anatomy and enrich metals and organic pollutants (Raklami 2022).

11.3 DISRUPTION OF SIGNALING AND BIOSYNTHETIC PATHWAYS

The toxicity of organic pollutants such as metals (Cu, Pb, Zn, Ni, Cd, Cr, Fe, Mn), hydrocarbons, aromatics, phenols, organochlorines, and pesticides causes physiological and anatomical abnormalities (Jaishankar 2014). The assimilation of organic pollutants such as the above metals, hydrocarbons, aromatics, phenols, organochlorines, and pesticides negatively affects the total chlorophyll content, chlorophyll a/b ratio, inhibits RuBisCO activity, and photosynthesis. Progressive reduction in the flow of electrons in the photochemical system is also evident (Haider 2021). The translocation of these metals declines abundance of proteins in Calvin cycle and large and small sub-units of RuBisCO (LSU and SSU) (Hossain and Komatsu, 2013). The assimilation of Cu, Pb, Zn, Ni, Cd, Cr, Fe and Mn, organic pollutants such as hydrocarbons, aromatic compounds, phenol, organic chlorinated, pesticides alters amount of substrate and enzymes of the TCA cycle, citric acid, and Krebs cycle (Singh 2017). The accumulation of heavy metals and organic molecules declines hydrolysis of sugars such as D-mannose, etc., mitigating water content and nutrient content in plants. The mitigation of sugars accumulation enhances water stress and inhibits photosynthetic and respiratory pathways (Meyer 2018). The assimilation of all the metals given above, organic pollutants such as hydrocarbons, aromatic compounds, phenols, organic chlorinated, and pesticides enhances biological nitrogen metabolism and are shown to inhibit protein synthesis and retards nitrate and nitrite reductase activity. It increases L-alanine, L-tryptophan, L-tyrosine and D-phenylalanine content and enhances cysteine content sulfur cycle (Roy 2021). The assimilation of these metals, organic pollutants such as hydrocarbons, aromatic compounds, phenol, organic chlorinated, and pesticides. has shown to increase the concentrations of phenylalanine, tyrosine, and tryptophan (Arora 2020). Tyrosine involved in the production of vitamin E and plastoquinone imposes stress resistance. Tryptophan produces secondary metabolites such as auxins, antitoxins, and glucosinolates (Parthasarathy 2018). Studies have also been done on translocation due to metals, organic pollutants such as hydrocarbons, aromatic compounds, phenols, organic chlorinated pesticides, mitigates metabolites such as flavonoids, anthocyanins, etc. (Pellaud and Mene-Saffrane, 2017) (Figure 11.1).

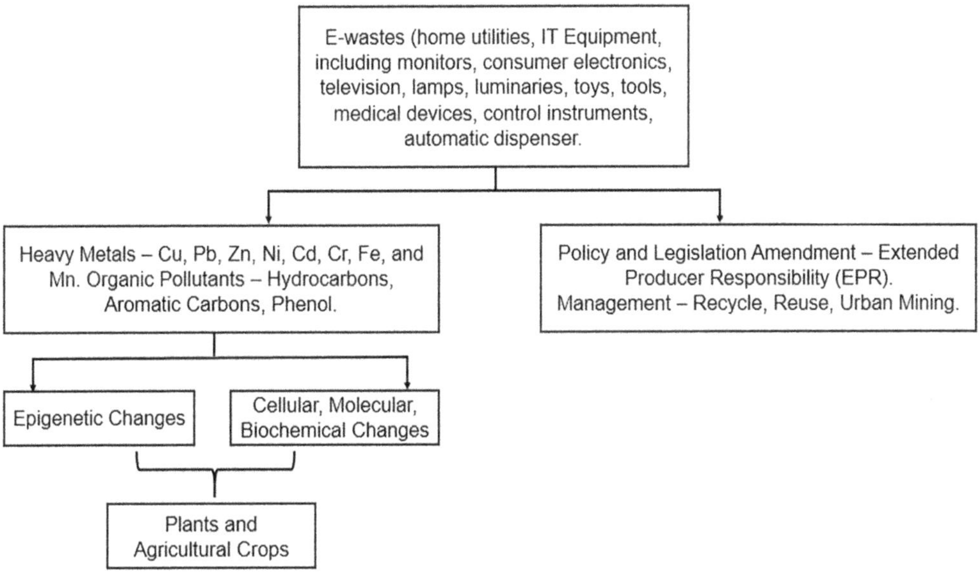

FIGURE 11.1 Epigenetic alteration and amelioration of electronic waste in plant.

11.4 OXIDATIVE DAMAGE TO PLANTS AND CROPS

The bioaccumulation of several of these metals mentioned before, organic pollutants such as hydrocarbons, aromatic compounds, and phenols, organic chlorinated, pesticides declines oxidative reactions and increase harmful oxidant molecule and ROS (Olaniran, Balgobind and Pillay, 2013). The metals induces two different oxidation mechanisms based on metal chemical properties a) Redox reaction: Fenton and Haber–Weiss reactions in presence of Cu^{+}/Cu^{2+} and Fe^{2+}/Fe^{3+} (formation of highly toxic hydroxyl radicals from hydrogen peroxide) ii) Non-redox reaction in presence of these metals cadmium, mercury and zinc (formation of reactive ROS). The formation of reactive oxygen species (ROS) mitigates the production of antioxidant molecule and enhances cell death in the plant (Pande 2022). The influx of metals mentioned before in this text, organic pollutants such as hydrocarbons, aromatic compounds, phenol, organic chlorinated, pesticides creates necrosis and lesion or Hypersensitive response in the plant (Briffa, Sinagra and Blundell, 2020). The production of ROS, superoxide, hydrogen peroxide, hydroperoxyl causes local cell death and restricts pest emergence. The ROS induces detrimental effect within morphological cells through lipid peroxidation, protein degradation, post protein degradation and DNA damage. The bioaccumulation of ROS in plant cells cleaves double bonds of DNA bases and forms di-, mono-, hydroxy-, and hydroxyl radicals, ring-saturated glycol, dehydrated, deaminated ring (Nita and Grzybowski, 2016). This glutathionylation initiates redox reaction in signal transduction and metabolic pathways. The conjugated mechanism of glutathionylation is transformed proteins structure and produces cysteinyl thiolate, sulfenic acid-protein alters signal and metabolic activity in the plant (Grek 2013). The post translational modifications such as carbonylation, nitrosylation produces ROS, superoxide, hydrogen peroxide, hydroperoxyl species induces reverse signal transduction in senescence, resistance and defense mechanisms, antioxidation and photorespiration and DNA binding-transcription factors (Xiong 2011).

11.5 GENETIC ANOMALIES-ABNORMAL REPRODUCTIVE RESPONSE OF PLANTS AND CROPS

The magnitude of copper, zinc, manganese etc. induces DNA damage and mutation in phanerogamic plant. The reactive species (ROS) damages all the molecular compositions of DNA such as cleavage of strand, transition of base, bond degradation and pyrimidine dimmers (Cadet and Wagner 2013). The heavy metal conjugates in the nucleus and DNA and exhibits bond and structural modification. The ROS and 8-oxoguanine are involved more in the DNA transformation through transversions mutation. The metals also alters genetic changes in chromatin structure, gene expression, DNA methylation, histone proteins and posttranslational modifications (Juan, 2021). The formation of lipid peroxidation, malondialdehyde and crotonaldehyde cleaves DNA sugars and double DNA strand restricting DNA replication. The bioaccumulation of metals, organic pollutants such as hydrocarbons, aromatic compounds, phenols, organic chlorinated, pesticides can directly impacts on transcription and translation alters with ROS-redox modifications and regulatory proteins causes cell death in the plant (Moretton and Loizou 2020). Metals such as cadmium and lead causes breakage of DNA strand. The chromium and arsenic induces retardation of DNA synthesis. The mitogen-activated protein-kinase (MAPK) enzyme produces regulatory protein induces certain genes translocate these metals in alfalfa, rice and *A. thaliana*. miRNAs provides signal of metal stress (Sinha et al. 2011). The micro RNAs were induced arsenic translocation in mustard (Ferragut, Udoh, and States, 2020). The high accumulation of metals causes epigenetic changes in the plants. The DNA damage of potato plant was imposed with the exposure of cadmium toxic caused yellowing of leaves, leaf distortion and growth inhibition in potato plants. Lead (Pb) causes gene mutation in *Brassica* and Barley seeds (Dutta 2018). Cadmium (Cd) causes DNA damage and mutations in barley and rice. Metals also induces DNA damage with production of ROS and oxidative stress.

The toxicity of several of these metals causes chromosomal changes and declines cell division (Bitto 2014). A lot of these metals causes nucleic acid damages in onion, broad bean, sunflower, potato and tobacco. For example copper (Cu) mitigates mitotic cell division and chromosome formation in root tip cells of *Allium cepa* var. *agrogarum* L. The seeds and seedlings of Litchi are damaged cell cycle and DNA with high doses of lead (Qin 2015).

11.6 DISEASED-CONTAMINATED PLANTS AND REDUCED CROP YIELD

Metals are also well known to affect the cellular and biochemical process leads to reduction of growth and yield. The bioaccumulation of metals, organic pollutants such as hydrocarbons, aromatic compounds, phenols, organic chlorinated, pesticides mitigates biomass, color pigments, yield, and organic and nutrients in the agricultural and horticultural crops (Okereafor 2020). The anthropogenic activities enhance the concentration of metals in the soil which impacts the crop productivity (Gogoi et al., 2024). The enormous accumulation of metals in the plant tissues transforms biochemical, physiological, and morphological functions in the crop productivity (Nguyen, Tran and Kumar, 2022). The high concentration of zinc and chromium adversely affected morphological, physiological and yield of wheat. The cadmium, chromium, lead, copper and zinc assimilation retards metabolism and yield in maize (Sharma 2020). The influxes of nickel, chromium, cadmium, lead, zinc form soil declines nitrogen content, chlorophyll, leghemoglobin and yield of chickpea. The movement of cadmium mitigates morphological growth and reproductive growth, also the yield of rice (Kumari 2022). The Nickel absorption depresses growth, yield, photosynthetic pigments, nutrients, ROS enhancement and membrane damage in tomato (Chandrasekaran, Boopathi and Manivannan, 2021). The influence of cadmium and lead reduces growth and yield with disruptions of morphological, physiological and biochemical in rocket (*Eruca sativa* L.). The influx of copper, cadmium, lead, zinc affects the productivity of forage crops (clover and perennial cereal grasses) (Riyazuddin 2021).

11.7 ASSESSMENT OF FOOD SECURITY WITH SPECIAL REFERENCE TO E-WASTES

E-waste pollution can be improved by implementing potential policies and management to secure crop production and loss. The recycling amendment is practiced in developed and developing countries to control waste mass and conserve the environment (Monika & Kishore, 2010). The amendment of urban mining mitigates global carbon concentrations, natural resource excavation, and energy consumption. RepresentativesAof 186 countries attended the Bamako Convention in Africa, where the amelioration treaty of e-waste was signed. The Organization for Economic Co-operation and Development (OECD) introduced the Extended Producer Responsibility (EPR) policy, which mentioned that producer bears the cost of waste of products such as electrical equipment, restrict waste generation, and develop eco-friendly product designs (Pouikli, 2020). The Environment, Forest, and Climate Change Ministry revised the E-Waste Rules 2011. It introduced the Extended Producer Responsibility (EPR) policy targeting commercial and residential premises, manufacturers, transport stations for goods and passengers, and worship spots (Borthakur, 2022). The developing countries China, India, and South Africa are hubs of the electronic manufacturing industry. The following challenges are countered in these countries: the high volume of e-waste consigned to developing countries without legal treaty (Monika & Kishore, 2010). The irregular awareness and regulation of policy are prevalent in the public circle. The lack of mobilization of hazardous e-waste prevails in the public zone. The physical targets and implementation are deficient for the retransformation of e-waste. The legal legitimacy is absent for restricting inter transboundary movement of e-waste (Mishra et al., 2017). Hence, managing e-waste effectively (Agarwal et al., 2023) is becoming crucial to any country's food safety and security.

REFERENCES

Agarwal, S., Darbar, S., Saha, S., Choudhury, M., & Singh, R. P. (2023). E-waste management using different cost-effective, eco-friendly biological techniques: an overview In Waste Management and Resource Recycling in the Developing World, 205–235. https://doi.org/10.1016/B978-0-323-90463-6.00026-9

Alengebawy, A., Abdelkhalek, S. T., Qureshi, S. R., & Wang, M. Q. (2021). Heavy metals and pesticides toxicity in agricultural soil and plants: Ecological risks and human health implications. *Toxics*, *9*, 1.

Anderson, K. (2010). Globalization's effects on world agricultural trade, 1960-2050. Philosophical transactions of the Royal Society of London. *Series B, Biological Sciences*, *365*, 3007.

Arora, P. K. (2020). Bacilli-Mediated Degradation of Xenobiotic Compounds and Heavy Metals. *Frontiers in Bioengineering and Biotechnology*, *8*, 1.

Bitto, A., Pizzino, G., Irrera, N., Galfo, F., & Squadrito, F. (2014). Epigenetic modifications due to heavy metals exposure in children living in polluted areas. *Current Genomics*, *154*, 464.

Bodie, A. R., Micciche, A. C., Atungulu, G. G., Rothrock, M. J., & Ricke, S. C. (2019). Current trends of rice milling byproducts for agricultural applications and alternative food production systems. *Frontiers in Sustainable Food Systems*, *3*, 1.

Borthakur, A. (2022). Design, adoption and implementation of electronic waste policies in India. *Environmental Science and Pollution Research*, *4*, 1.

Briffa, J., Sinagra, E., & Blundell, R. (2020). Heavy metal pollution in the environment and their toxicological effects on humans. *Heliyon*, *6*, 1.

Cadet, J., & Wagner, J. R. (2013). DNA base damage by reactive oxygen species, oxidizing agents, and UV radiation. *Cold Spring Harbor Perspectives in Biology*, *5*, 1.

Chandrasekaran, M., Boopathi, T., & Manivannan, P. (2021). Comprehensive Assessment of Ameliorative Effects of AMF in Alleviating Abiotic Stress in Tomato Plants. *Journal of Fungi*, *7*, 1.

Choudhury, M., Jyethi, D. S., Dutta, J., Purkayastha, S. P., Deb, D., Das, R., … & Bhattacharyya, K. G. (2022). Investigation of groundwater and soil quality near to a municipal waste disposal site in Silchar, Assam, India. *International Journal of Energy and Water Resources*, 6(1), 37–47. https://doi.org/10.1007/s42108-021-00117-5

Dastagiri, M. B., Prasad, G. M. N. V., & Ganeshagouda, I. P. (2014). World and Indian agriculture: Revolutions & multi speed strategies for future. *Science Discovery*, *2*, 14.

Dutta, S., Mitra, M., Agarwal, P., Mahapatra, K., De, S., Sett, U., & Roy, S. (2018). Oxidative and genotoxic damages in plants in response to heavy metal stress and maintenance of genome stability. *Plant Signaling & Behavior*, *13*, 1.

Ferragut Cardoso, A. P., Udoh, K. T., & States, J. C. (2020). Arsenic-induced changes in miRNA expression in cancer and other diseases. *Toxicology and Applied Pharmacology*, *409*, 1.

Gogoi, N., Sarma, A., Choudhury, M., Samanta, P., Faizan, M., & Sadhak, S. (2024). Soil heavy metal pollution and ecological risk assessment in disturbed and undisturbed soil of Morigaon, Assam. *Discover Environment*, 2(1), 13. https://doi.org/10.1007/s44274-024-00039-z

Grek, C. L., Zhang, J., Manevich, Y., Townsend, D. M., & Tew, K. D. (2013). Causes and consequences of cysteine S-glutathionylation. *The Journal of Biological Chemistry*, *288*, 26497.

Gulati, A., Terway, P., & Hussain, S. (2021). Performance of Agriculture in Uttar Pradesh. *Revitalizing Indian Agriculture and Boosting Farmer Incomes*. Edited by Gulati, A., Roy, R., & Saini, S., Springer, Singapore, 1.

Haider, F. U., Liqun, C., Coulter, J. A., Cheema, S. A., Wu, J., Zhang, R., Wenjun, M., & Farooq, M. (2021). Cadmium toxicity in plants: Impacts and remediation strategies. *Ecotoxicology and Environmental Safety*, *211*, 1–19.

Hossain, Z., & Komatsu, S. (2013). Contribution of proteomic studies towards understanding plant heavy metal stress response. *Frontiers in Plant Science*, *3*, 1.

Ittis, Y., Folberth, C., Bundle, S. C., & Obersteiner M. (2021). Restoring Nature at Lower Food Production Costs. *Frontiers in Environmental Science*, *9*, 1.

Jaishankar, M., Tseten, T., Anbalagan, N., Mathew, B. B., & Beeregowda, K. N. (2014). Toxicity, mechanism and health effects of some heavy metals. *Interdisciplinary Toxicology*, *7*, 60.

Juan, C. A., Pérez, J. M., Plou, F. J., & Pérez-Lebeña, E. (2021). The Chemistry of Reactive Oxygen Species (ROS) Revisited: Outlining Their Role in Biological Macromolecules (DNA, Lipids and Proteins) and Induced Pathologies. *International Journal of Molecular Sciences*, *2*, 1.

Karolina, P., & Małgorzata, K. (2020). The Role of Agriculture in Ensuring Food Security in Developing Countries: Considerations in the Context of the Problem of Sustainable Food Production. *Sustainability*, *12*, 1.

Kumari, V. V., Banerjee, P., Verma, V. C., Sukumaran, S., Chandran, M. A. S., Gopinath, K. A., Venkatesh G., Yadav S. K., Singh V. K., & Awasthi N. K. (2022). Plant Nutrition: An Effective Way to Alleviate Abiotic Stress in Agricultural Crops. *International Journal of Molecular Sciences*, *23*, 1.

Manisalidis, I., Stavropoulou, E., Stavropoulos, A., & Bezirtzoglou, E. (2020). Environmental and Health Impacts of Air Pollution: A Review. *Froniers in Public Health*, *8*, 1.

Meyer, T., Vigouroux, A., Aumont-Nicaise, M., Comte, G., Vial, L., Lavire, C., & Moréra, S. (2018). The plant defense signal galactinol is specifically used as a nutrient by the bacterial pathogen Agrobacterium fabrum. *The Journal of Biological Chemistry*, *293*, 7930.

Mishra, S., Shamanna, B. R., & Kannan, S. (2017). Exploring the awareness regarding E-waste and its health hazards among the informal handlers in Musheerabad area of Hyderabad. *Indian Journal of Occupational and Environmental Medicine*, *21*, 143.

Monika., & Kishore, J. (2010). E-waste management: as a challenge to public health in India. *Indian Journal of Community Medicine*, *35*, 382.

Moretton, A., & Loizou, J. I. (2020). Interplay between Cellular Metabolism and the DNA Damage Response in Cancer. *Cancers*, *12*, 1.

Nguyen, D. T. C., Tran, T. V., & Kumar, P. S. (2022). Invasive plants as biosorbents for environmental remediation: a review. *Environmental Chemistry Letters*, *20*, 1421.

Nita, M., & Grzybowski, A. (2016). The Role of the Reactive Oxygen Species and Oxidative Stress in the Pathomechanism of the Age-Related Ocular Diseases and Other Pathologies of the Anterior and Posterior Eye Segments in Adults. *Oxidative Medicine and Cellular Longevity*, 1, https:.//doi.org/10.1155/2016/3164734.

Okereafor, U., Makhatha, M., Mekuto, L., Uche-Okereafor, N., Sebola, T., & Mavumengwana, V. (2020). Toxic Metal Implications on Agricultural Soils, Plants, Animals, Aquatic life and Human Health. *International Journal of Environmental Research and Public Health*, *17*, 1.

Olaniran, A. O., Balgobind, A., & Pillay, B. (2013). Bioavailability of heavy metals in soil: impact on microbial biodegradation of organic compounds and possible improvement strategies. *International Journal of Molecular Sciences*, *14*, 10197.

Pande, V., Pandey, S. C., Sati, D., Bhatt, P., & Samant, M. (2022). Microbial Interventions in Bioremediation of Heavy Metal Contaminants in Agroecosystem. *Frontiers in Microbiology*, *13*, 1.

Parthasarathy, A., Cross, P. J., Dobson, R., Adams, L. E., Savka, M. A., & Hudson, A. O. (2018). A Three-Ring Circus: Metabolism of the Three Proteogenic Aromatic Amino Acids and Their Role in the Health of Plants and Animals. *Frontiers in Molecular Biosciences*, *5*, 1.

Pellaud, S., & Mene-Saffrane, L. (2017). Metabolic Origins and Transport of Vitamin E Biosynthetic Precursors. *Frontiers in Plant Science*, *8*, 1.

Pouikli, K. (2020). Concretising the role of extended producer responsibility in European Union waste law and policy through the lens of the circular economy. *ERA Forum*, *20*, 491.

Qin, R., Wang, C., Chen, D., Björn, LO., & Li, S. (2015). Copper-induced root growth inhibition of Allium cepa var. agrogarum L. involves disturbances in cell division and DNA damage. *Environmental Toxicology and Chemistry*, *34*, 1045.

Raklami, A., Meddich, A., Oufdou, K., Baslam, M. (2022). Plants-microorganisms-based bioremediation for heavy metal cleanup: Recent developments, phytoremediation techniques, regulation mechanisms, and molecular responses. *International Journal of Molecular Sciences*, *23*, 1.

Riyazuddin, R., Nisha, N., Ejaz, B., Khan, M., Kumar, M., Ramteke, P. W., & Gupta, R. (2021). A Comprehensive review on the heavy metal toxicity and sequestration in plants. *Biomolecules*, *12*, 1.

Roy, A., Sharma, A., Yadav, S., Jule, L. T., & Krishnaraj, R. (2021). Nanomaterials for remediation of environmental pollutants. *Bioinorganic Chemistry and Applications*, 1, https://doi.org/10.1155/2021/1764647

Sathya, K., Nagarajan, K., Carlin Geor Malar, G., Rajalakshmi, S., & Raja Lakshmi, P. (2022). A comprehensive review on comparison among effluent treatment methods and modern methods of treatment of industrial wastewater effluent from different sources. *Applied Water Science*, *12*, 1.

Sharma, A., Kapoor, D., Wang, J., Shahzad, B., Kumar, V., Bali, A. S., Jasrotia, S., Zheng, B., Yuan, H., & Yan, D. (2020). Chromium bioaccumulation and its impacts on plants: An overview. *Plants (Basel, Switzerland)*, *9*, 1.

Shashi A., & Sunil, K. (2020). E-waste in India at a glance: Current trends, regulations, challenges and management strategies. *Journal of Cleaner Production*, *271*, 1.

Singh, N., Gupta, V. K., Kumar, A., & Sharma, B. (2017). Synergistic Effects of Heavy Metals and Pesticides in Living Systems. *Frontiers in Chemistry*, *5*, 1.

Singh, S., Singh, P., Sharma, A., & Choudhury, M. (Eds.). (2022). *Agriculture Waste Management and Bioresource: The Circular Economy Perspective*. John Wiley & Sons.

Sinha, A. K., Jaggi, M., Raghuram, B., & Tuteja, N. (2011). Mitogen-activated protein kinase signaling in plants under abiotic stress. *Plant Signaling & Behavior*, *6*, 196.
Vats, M. C., & Singh, S. K. (2014). E-Waste characteristic and its disposal. *International Journal of Ecological Science and Environmental Engineering*, *1*, 49.
Verena, S., & Navin, R. (2017). Many shades of gray—The context-dependent performance of organic agriculture. *Science Advances*, *3*, 1.
William, A. M., Amelia, B. F., & Steven, A. B. (2022). The economics of malnutrition: Dietary transition and food system transformation. *Handbook of Agricultural Economics*, *6*, 4997.
Xiong, Y., Uys, J. D., Tew, K. D., & Townsend, D. M. (2011). S-glutathionylation: from molecular mechanisms to health outcomes. *Antioxidants & Redox Signaling*, *15*, 233.

12 Electronic Wastes, Its Impact on Food Safety

Vaishali Soni
PGIMER-Post Graduate Government Institute of Medical Education & Research, Chandigarh, India

Vatsala Soni
PEC-Punjab Engineering College, Chandigarh, India

Meenakshi Sharma
IEC University, Baddi, India

Dipti Bharti
Darbhanga College of Engineering, Darbhanga, India

Charu Arora
Guru Ghasidas Vishwavidyalaya, Bilaspur, India

Richa Saxena
Invertis University, Bareilly, India

Richard Baker
University of St Andrews, St. Andrews, United Kingdom

Basem E. Kestha
Faculty of Science, Tanta University, Tanta, Egypt

12.1 FOOD, FOOD SAFETY, AND SECURITY – AN INTRODUCTION

Growing demand associated with lifestyle and health awareness changes is driving the use of packaged and convenience foods. The need to protect the environment from the waste generated is largely ignored. This ignorance is a matter of concern for food safety and security. The food chain needs to be improved quickly in order to be more effective and efficient. As the world grows more interconnected, governments, businesses, and individuals face a more complicated challenge in guaranteeing food safety and security (Lin et al., 2009; Fanzo et al., 2021). It is challenging to put a number on the risks connected to improper management of food ingredients and related services. Each year in poor nations, diseases caused by contaminated food and water claim the lives of almost 2 million people, many of whom are children. More than 200 different diseases, from infections to cancer, are brought on by contaminants in food, which might include pathogenic bacteria, parasites, prions, viruses, chemicals, or radioactive materials (Brijnath et al., 2014; Fukuda, 2015).

DOI: 10.1201/9781003582311-12

In this regard, on April 7, 2015, the World Health Organization (WHO) chose to highlight food safety in honor of World Health Day. These days, food ingredients are frequently imported from multiple nations, thousands of kilometers away from the farm, ranch, or factory. People on the other side of the globe may be impacted by pollution at one end of the food chain. For strengthening Food Security, multidisciplinary and international cooperation is needed, taking into account possible communication delays and the influence of multiple actors over long distances (INFOSAN, 2015). Policies, standards, rules, guidelines, training, and recommendations—including those pertaining to food safety—are continuously being created or revised in the context in which the global food industry operates (Dey et al., 2024). Either the efficiency and efficacy of the food chain have improved to accommodate these changes, or they are not internationally coordinated and consumers lack adequate knowledge about nutrition and food safety. It can lead to complexity and confusion (Vågsholm et al., 2020).

12.2 E-WASTES AND ASSOCIATED PRODUCTS – UNDERSTANDING THE RELATION

E-waste is a growing concern universally. As technology advances, industry moves toward automation and increases the use of electronic and electrical apparatus. Electrical and electronic products permeate the daily life of ordinary consumers and are often utilized in manufacturing industries. Simultaneously, the progress of fast, advanced and trustworthy computing and processing technologies has shortened product life cycles, causing consumers to discard older products while purchasing the latest technological products. The output of e-waste has increased exponentially as a result of these expansions (Robinson, 2009). According to research by Balde et al. (2015), 41.8 million tonnes (5.9 kg per native) of e-waste were produced worldwide in 2014. To address issues, a thorough examination of e-waste management is currently being conducted on a national and international scale (Kiddee et al., 2013). Namias (2013) suggests that e-waste contains 60 metals as well including gold, silver, copper, platinum and palladium. Recovering some metals by electronic waste can decrease the comprehensive demand for the production of new metals to some extent. Recycling e-waste also facilitates to reduction the quantity of material that ends up in landfills. But just 15% of the world's e-waste is entirely recycled, in spite of all the potential compensation (Heacock et al., 2016). It is crucial to have a thorough definition when creating standards and procedures for disposal. A global initiative called STEP (Solving the E-trash Problem) aims to provide more clarity on the global issue of electronic trash. As per Step Initiative (2017), electronic and electrical equipment (EEE) and its components that are disposed of as waste by their owners without any purpose of reuse are referred to as e-waste (Shirodkar and Terkar, 2017). The property of electronic waste pollutants i.e. organic pollutants (PAHs, PCBs, etc.) and some metals (i.e. Hg, Be, Cd, Pb, Cr) on the activity of soil enzymes, soil communities and changes in chemical and physical properties have been shown by some researches (Chaperon and Sauv, 2008). According to a recent study, recycled e-waste might include dangerous chemicals into commonplace products including toys, coat hangers, food packaging, plastic jewelry, office supplies, and storage containers. Concerns regarding trash recycling for food contact packaging have been raised by the food and drug administration (FDA) (Grigore, 2017).

12.3 E-WASTE TOXICITY AND RISING COMPLICATIONS IN FOOD

When e-waste is handled and processed improperly, it can contaminate water, air, and soil, release toxic fumes from the combustion of e-waste, and introduce toxicants into the environment that directly affect the health of living things (Chen et al., 2016; Ahirwar and Tripathi, 2021; Agarwal et al., 2023). Food systems may eventually be impacted by pharmaceutical, chemical, and other e-waste contamination sources that find their way into surrounding water and soil (Ghosh et al., 2022). Only 2–5% of the entire volume of solids is made up of e-waste, yet over 70% of it is dangerous. Lead (Pb), chromium (Cr), beryllium (Be), palladium (Pd), gallium (Ga), and other poisons contaminate the air, water, soil and food chain (Akram et al., 2019; Liu et al., 2020a). It has a number of negative consequences on the human body when consumed. E-waste is linked to several chemical contaminants such esters and phthalates, in addition to hazardous metals. In addition to directly harming people, benzene, dioxins, polyaromatic hydrocarbons, and polyvinyl chloride also negatively impact neighboring plants, soils, and microbiological organisms (Reid and Chris, 2015).

Southeast China now has hundreds of modest, open, generalized e-waste recycling shelters or sites. This electronic trash that enters the environment contains a lot of hazardous substances, including lead, cadmium, beryllium, mercury, polychlorinated biphenyls, and brominated flame retardants. These substances are extremely dangerous to human health, independent of occupational or environmental exposure. This care should be given to people who live downstream or downwind of the e-waste facility in addition to the locals who live close by. The pollution of living things by the food chain is another important issue that needs to be addressed. More research should be done because there are currently insufficient relevant data (Needhidasan et al., 2014). Typical hazardous sources of e-waste that are harmful to human health are depicted in Figure 12.1. E-waste is a significant global environmental health concern, particularly in developing nations, as numerous researchers have demonstrated. Hexabromocyclododecane (HBCDD) is frequently found in samples of living things taken from e-waste recycling sites, including fish and birds (Heeb et al., 2008; Sun et al., 2012). Bioaccumulation and biomagnification in insect-dominated aquatic and terrestrial food webs are described in a study by Liu et al. (2020b).

This study collected information on the presence of hexabromocyclododecane (HBCDD) in insects and their predators from ponds that had previously been contaminated with e-waste and the surroundings around them. This review demonstrates how HBCDD exhibits opposing trophic transfer trends in aquatic and terrestrial environments. For α-HBCDD, terrestrial birds had a greater biomagnification factor (BMF) of 2.03 compared to frogs (0.29), toads (0.85), and lizards (0.63). The distinctions between poikilotherms and homeotherms in terrestrial ecosystems could be the cause of this. Electronic garbage contains heavy metals that can build up in the soil to dangerously high amounts and have detrimental impacts on both the environment and human health. Because of their superior metal-binding abilities, bacteria and fungus have been the main focus of most bioremediation studies when it comes to the bioaccumulation of heavy metals. The goal of the current study was to separate the most promising lead, copper, and zinc-resistant bacteria from polluted soil

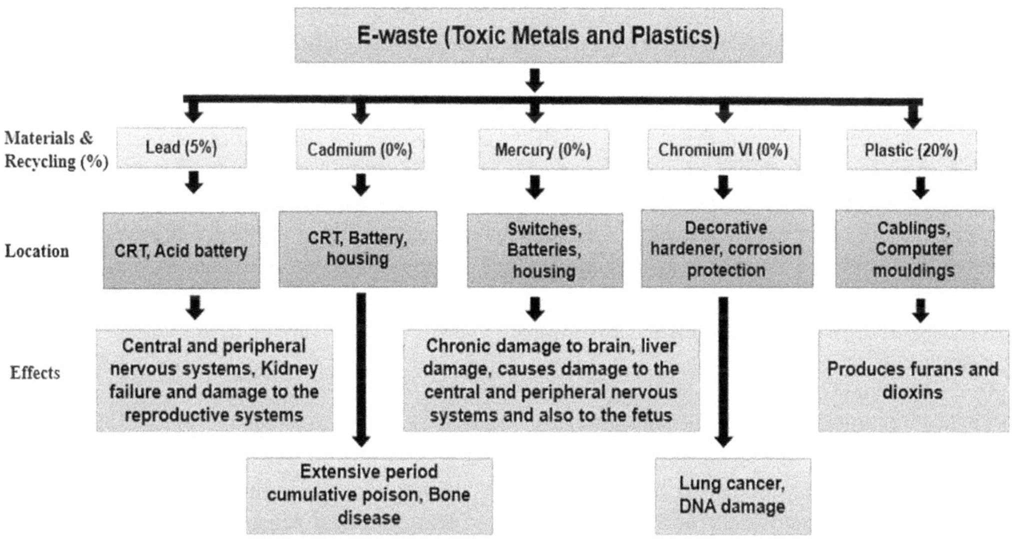

FIGURE 12.1 The effect of E-waste (toxic metals and plastics) on human health (Needhidasan et al., 2014).

and assess their ability to accumulate metals. Based on their established resistance to heavy metals, *Bacillus licheniformis, B. polymyxa, Pseudomonas aeruginosa, Micrococcus roseus, Aspergillus niger*, and *A. flavus* were chosen for bioaccumulation experiments (Akintokun et al., 2017).

12.4 RECYCLED E-WASTES, BLACK PLASTICS, AND FOOD SAFETY

More than a thousand different chemicals, many of them valuable and dangerous, are found in e-waste. Recycling releases harmful compounds into the air, water, and soil when it burns electronic waste and separates plastic from metal. For example, high concentrations of lead, aluminum, zinc, copper, cadmium, hydrochloric acid, polycyclic aromatic hydrocarbons (PAHs), and polybrominated diphenyl ether (PBDEs) were identified in dust, ash, and soil at the Abogbloshie e-waste site in Ghana, according to a recent research. According to Fujimori et al. (2016), breast milk from nearby women who work or reside in the vicinity of the site contained significant amounts of PCBs and PBDEs.

Developed nations with the capacity and infrastructure to recycle e-waste include North America, the European Union, Japan, and others. Nonetheless, substantial volumes of e-waste are shipped, whether legally or illegally, to developing nations with lax or nonexistent laws governing e-waste. Strict e-waste regulations and the greater expense of recycling hazardous items in affluent nations compared to shipping them to underdeveloped nations are the reasons behind this trash transportation. According to the 2020 Global E-Waste Monitor, the amount of e-waste that is formally collected and recycled has grown by 1.8 million tonnes since 2014, reaching 17.4% in 2019, or it is disposed of in landfills (Abalansa et al., 2021; Paul et al., 2019).

According to a recent study, hazardous chemicals can seep into commonplace goods including food packaging, plastic jewelry, toys, clothing hangers, storage containers, and office equipment from recycled e-waste. Toxic chemicals are being dumped in recycling or recycling factories as a result of inadequate e-waste disposal and rising demand for black plastics, which are utilized in the design of electronic devices, according to research by scientists at the University of Plymouth in the UK. Electronic equipment like music systems and laptops contain flame retardants made of toxic compounds including lead, antimony, and bromine. Black plastic is making its way into food contact items and other products since the need for it in consumer goods is largely satisfied by sourcing from e-waste (Tanskanen, 2013). Black plastics are generally more hazardous than plastics, according to a study published in Environmental International, which suggests that the black pigments used in traditional plastic sorting facilities are less sensitive to infrared radiation, making them difficult to recycle properly. This affects not only human health but also the marine and coastal environment as litter and microplastics (Hopewell et al., 2009). With this awareness, participants suggested either a complete ban on single-use plastics as a solution or raising awareness and encouraging people to reuse single-use plastics before recycling, multiple times if possible. In addition, the introduction of stricter measures by local governments was supported by a majority of 81.9%. The overall feeling from their comments is that more public awareness is needed, that all stakeholders should be held accountable and that people recycle effectively. Incentives based on rent or local government tax reductions should be created to motivate people. For plastic trading, the issue of potential food safety threats remains. There is also a sense that the level of awareness may be related to the level of education, as most of the respondents have attained higher education, although this cannot be proven statistically (Cox et al., 2019).

12.5 E-WASTES IN VEGETABLES, EGGS, AND MEAT

Heavy metal contamination contributes significantly to environmental pollution and poses a serious global hazard to human health and the environment. Eating polluted vegetables can cause diseases in both humans and animals. Ecosystems in water and soil are harmed worldwide. Heavy metal concentrations, including Pb, Cu, Cd, Zn, and Ni, were measured in food products, household dust, ground/drinking water, and soil from the Southern China e-waste region. Health risks associated with food and dust consumption as well as exposure to metals in drinking water were evaluated in local communities.

The main non-carcinogenic risks for an ordinary e-waste inhabitant are from vegetables (hazard index = 2.2), rice (hazard index = 3.3), and house dust (1.9). For small children, the predominant risk factor is house dust (15). Zheng et al. (2013) because of their detrimental effects on human health and the sustainability of food, heavy metals in soil, air, and water are a serious concern. Foodborne heavy metal pollution can lead to heavy metal deposition in the kidney and liver of humans and interfere with a number of biochemical processes, which can be the cause of the onset of neurological, renal, cardiovascular, and bone illnesses (Briffa et al., 2020). Human health, food safety, and environmental contaminants are all intimately related. The sources of heavy metals in vegetables contrast amongst developing and developed countries. The primary causes of soil and plant system pollution in developed nations are the application of sewage sludge and industrial wastewater as fertilizers, as well as PM deposition on food crops. Several types of research have proved that the heavy metals contained in this kind of landfill cause soil and air pollution on a large scale. Breaking down the discarded electronic apparatuses and burning plastics to recuperate metals releases pollutants this can cause harm to animals. For example, eggs from chickens in the Agbogbloshie slum contain dangerous levels of contaminants such as dioxins and polychlorinated biphenyls (Fujimori et al., 2016; Mitra et al., 2022; Mondal et al., 2023). Dioxins are "exceptionally noxious in extremely trivial amounts", JindrichPetrlikas a researcher clarifies. Later, the confirmation to this sentence was given by Marie-Noel BruneDrisse, the main author of the WHO reports Children and digital dumpsites: "A child intake a single chicken egg from Agbogbloshie will consume 220 times the bearable everyday consumption of chlorinated dioxins set by the European Food Safety Authority" (Weber et al., 2015; Pajurek et al., 2022). The major entry into the food chain of e-waste particularly heavy metals is found to be rivers and oceans. An investigation that is been published in the journal Science of the Total Environment found a fish of African oceans named Tuna largely contains high levels of various metals, that includes lead and mercury. The concluding remarks of the researchers were intake of contaminated fish poses health risks to native inhabitants and customers in other nations (Liu et al., 2012; Pandey and Madhuri, 2014; Houessionon et al., 2021).

12.6 HOW TO DEAL WITH E-WASTE FOR FOOD SAFETY?

The problems related to e-waste are rising continuously at an alarming rate in India and even globally. While several developed countries are equipped with established well-defined e-waste disposal systems, Indian organizations face obstructions due to socio-economic, infrastructural and legitimate causes. Recycling and recovery of E-waste, opportunities proficient in India are highly obsolete and dangerous, posing serious ecological and professional jeopardies. There is a basic differentiation between the recent European e-waste disposal model and the reality of e-waste recycling in India. Additional focus should be on Tier 2 and Tier 3 cure systems, including recycling and revitalization options for better disposal and organization of e-waste. The responsibility of public-private partnerships plays a crucial part in emerging and establishing all-encompassing e-waste management strategies basically in India. In some cities such as Bangalore, Chennai and Noida few private companies have set up e-waste disposal amenities that process and dispose of e-waste in a further systematic and environmentally friendly way, while in other cities. More such companies should be established, likewise. In what manner, have experience and knowledge of existing e-waste management models from abroad been inherited into India's present e-waste system and how well they are combined to form India's official and well-tuned e-waste management system.

12.7 CONCLUSION

E-waste can be found everywhere and is notable for its unfamiliar composition of chemicals and difficulty in determining its mass and flux locally and globally. Pollution related to e-waste has already been causing substantial ecological damage to developed and economically weaker countries, adversely affecting the health of ordinary citizens who inhabit those countries. It is not feasible to control e-waste contamination, as they are heavily contaminated, and many have not been well studied. However, the adverse effects of contaminants at these sites can be optimized with the help of standard remediation techniques. Some e-waste pollutants, for instance, Sb and Li, are not typically conservational, so knowledge of their ecological impacts, human health risks, and remediation options is limited. Well-developed and economically more muscular countries have a strong interest in reducing the harmful and toxic environmental effects of e-waste. E-waste adversely affects the quantity and quality of food and products from economically weaker countries.

REFERENCES

Abalansa, S., El Mahrad, B., Icely, J., & Newton, A. (2021). Electronic waste, an environmental problem exported to developing countries: The GOOD, the BAD and the UGLY. *Sustainability*, *13*(9), 5302.

Agarwal, S., Darbar, S., Saha, S., Choudhury, M., & Singh, R. P. (2023). E-waste management using different cost-effective, eco-friendly biological techniques: An overview In *Waste Management and Resource Recycling in the Developing World*, 205–235. https://doi.org/10.1016/B978-0-323-90463-6.00026-9

Ahirwar, R., & Tripathi, A. K. (2021). E-waste management: A review of recycling process, environmental and occupational health hazards, and potential solutions. *Environmental Nanotechnology, Monitoring & Management*, *15*, 100409.

Akintokun A.K., Onatunde, O.O., Shittu O.B., Okeyode I.C., Taiwo, M.O., 2017, Bioaccumulation of Heavy Metals using Selected Organisms Isolated from Electronic Waste Dumpsite of two South-Western States in Nigeria, *Applied Environmental Research 39*(2): 29–40.

Akram, R., Fahad, S., Hashmi, M. Z., Wahid, A., Adnan, M., Mubeen, M., ... & Nasim, W. (2019). Trends of electronic waste pollution and its impact on the global environment and ecosystem. *Environmental Science and Pollution Research*, *26*, 16923–16938.

Baldé, C. P., Wang, F., Kuehr, R., & Huisman, J. (2015). *The global e-waste monitor–2014, United Nations University, IAS–SCYCLE*, Bonn, Germany. *ISBN Print*, 978–992.

Brett H. Robinson, 2009. E-waste: An assessment of global production and environmental impacts, *Science of The Total Environment*, *408*(2), 183–191, ISSN 0048-9697, https://doi.org/10.1016/j.scitotenv.2009.09.044.

Briffa, J., Sinagra, E., & Blundell, R. (2020). Heavy metal pollution in the environment and their toxicological effects on humans. *Heliyon*, *6*(9), e04691.

Brijnath, B., Butler, C. D., & McMichael, A. J. (2014). In an interconnected world: joint research priorities for the environment, agriculture and infectious disease. *Infectious Diseases of Poverty*, *3*(1), 1–3.

Chaperon, S., & Sauvé, S. (2008). Toxicity interactions of cadmium, copper, and lead on soil urease and dehydrogenase activity in relation to chemical speciation. *Ecotoxicology and Environmental Safety*, *70*(1), 1–9.

Chen, M., Ogunseitan, O. A., Wang, J., Chen, H., Wang, B., & Chen, S. (2016). Evolution of electronic waste toxicity: Trends in innovation and regulation. *Environment international*, *89*, 147–154.

Cox, K. D., Covernton, G. A., Davies, H. L., Dower, J. F., Juanes, F., & Dudas, S. E. (2019). Human consumption of microplastics. *Environmental Science & Technology*, *53*(12), 7068–7074.

Dey, S., Santra, M., Choudhury, M., Ghosh, A. R., & Samanta, P. (2024). Food waste generation and its industrial utilization: An overview. *Environmental Science and Pollution Research*, 1–20. https://doi.org/10.1007/s11356-024-34252-3

Fanzo, J., Bellows, A. L., Spiker, M. L., Thorne-Lyman, A. L., & Bloem, M. W. (2021). The importance of food systems and the environment for nutrition. *The American Journal of Clinical Nutrition*, *113*(1), 7–16.

Fujimori, T., Itai, T., Goto, A., Asante, K. A., Otsuka, M., Takahashi, S., & Tanabe, S. (2016). Interplay of metals and bromine with dioxin-related compounds concentrated in e-waste open burning soil from Agbogbloshie in Accra, Ghana. *Environmental Pollution*, *209*, 155–163.

Fukuda, K. (2015). Food safety in a globalized world. *Bulletin of the World Health Organization*, *93*, 212–212.

Ghosh, B. K., Mekhilef, S., Ahmad, S., & Ghosh, S. K. (2022). A review on global emissions by e-products based waste: Technical management for reduced effects and achieving sustainable development goals. *Sustainability*, *14*(7), 4036.

Grigore, M. E. (2017). Methods of recycling, properties and applications of recycled thermoplastic polymers. *Recycling*, *2*(4), 24.

Heacock, M., Kelly, C. B., Asante, K. A., Birnbaum, L. S., Bergman, Å. L., Bruné, M. N., ... & Suk, W. A. (2016). E-waste and harm to vulnerable populations: a growing global problem. *Environmental Health Perspectives*, *124*(5), 550–555.

Heeb, N. V., Schweizer, W. B., Mattrel, P., Haag, R., Gerecke, A. C., Schmid, P., ... & Vonmont, H. (2008). Regio-and stereoselective isomerization of hexabromocyclododecanes (HBCDs): Kinetics and mechanism of γ-to α-HBCD isomerization. *Chemosphere*, *73*(8), 1201–1210.

Hopewell, J., Dvorak, R., & Kosior, E. (2009). Plastics recycling: challenges and opportunities. *Philosophical Transactions of the Royal Society B: Biological Sciences*, *364*(1526), 2115–2126.

Houessionon, M. K., Ouendo, E. M. D., Bouland, C., Takyi, S. A., Kedote, N. M., Fayomi, B., ... & Basu, N. (2021). Environmental heavy metal contamination from Electronic Waste (e-waste) recycling activities worldwide: A systematic review from 2005 to 2017. *International Journal of Environmental Research and Public Health*, *18*(7), 3517.

INFOSAN. (2015). In action in the Americas. Geneva: World Health Organization; Geneva. Available from: http://www.who.int/foodsafety/areas_work/infosan/en/#story-02 [cited 2015 Mar 3].

Kiddee, P., Naidu, R., & Wong, M. H. (2013). Electronic waste management approaches: An overview. *Waste Management*, *33*(5), 1237–1250.

Lin, A. Y. C., Huang, S. T. Y., & Wahlqvist, M. L. (2009). Waste management to improve food safety and security for health advancement. *Asia Pacific Journal Of Clinical Nutrition*, *18*(4), 538–545.

Liu, L., Hu, L., Tang, J., Li, Y., Zhang, Q., & Chen, X. (2012). Food safety assessment of planting patterns of four vegetable-type crops grown in soil contaminated by electronic waste activities. *Journal of Environmental Management*, *93*(1), 22–30.

Liu, Y., Luo, X., Zeng, Y., Deng, M., Tu, W., Wu, Y., & Mai, B. (2020a). Bioaccumulation and biomagnification of hexabromocyclododecane (HBCDD) in insect-dominated food webs from a former e-waste recycling site in South China. *Chemosphere*, *240*, 124813.

Liu, Y., Luo, X., Zeng, Y., Tu, W., Deng, M., Wu, Y., & Mai, B. (2020b). Species-specific biomagnification and habitat-dependent trophic transfer of halogenated organic pollutants in insect-dominated food webs from an e-waste recycling site. *Environment International*, *138*, 105674.

Mitra, S., Chakraborty, A. J., Tareq, A. M., Emran, T. B., Nainu, F., Khusro, A., ... & Simal-Gandara, J. (2022). Impact of heavy metals on the environment and human health: Novel therapeutic insights to counter the toxicity. *Journal of King Saud University-Science*, *34*(3), 101865.

Mondal, T., Choudhury, M., Kundu, D., Dutta, D., & Samanta, P. (2023). Landfill: An eclectic review on structure, reactions and remediation approach. *Waste Management*, *164*, 127–142.

Namias, J. (2013). *The future of electronic waste recycling in the United States: obstacles and domestic solutions*. Columbia University.

Needhidasan, S., Samuel, M., & Chidambaram, R. (2014). Electronic waste–an emerging threat to the environment of urban India. *Journal of Environmental Health Science and Engineering*, *12*, 1–9.

Pajurek, M., Warenik-Bany, M., & Mikolajczyk, S. (2022). Feed as a source of dioxins and PCBs. *Chemosphere*, *308*, 136243.

Pandey, G., & Madhuri, S. (2014). Heavy metals causing toxicity in animals and fishes. *Research Journal of Animal, Veterinary and Fishery Sciences*, *2*(2), 17–23.

Paul, S., Choudhury, M., Deb, U., Pegu, R., Das, S., & Bhattacharya, S. S. (2019). Assessing the ecological impacts of ageing on hazard potential of solid waste landfills: A green approach through vermitechnology. *Journal of Cleaner Production*, *236*, 117643. https://doi.org/10.1016/j.jclepro.2019.117643

Reid, S., & Chris, X. (2015). Consumption of rice and fish in an electronic waste recycling area contributes significantly to total daily intake of mercury. *Journal of Environmental Science*, (12), 83–86. https://doi.org/10.1016/j.jes.2015.10.003

Shirodkar, N., & Terkar, R. (2017). Stepped recycling: the solution for E-waste management and sustainable manufacturing in India. *Materials Today: Proceedings*, *4*(8), 8911–8917.

Sun, Y. X., Luo, X. J., Mo, L., He, M. J., Zhang, Q., Chen, S. J., ... & Mai, B. X. (2012). Hexabromocyclododecane in terrestrial passerine birds from e-waste, urban and rural locations in the Pearl River Delta, South China: Levels, biomagnification, diastereoisomer-and enantiomer-specific accumulation. *Environmental Pollution*, *171*, 191–198.

Tanskanen, P. (2013). Management and recycling of electronic waste. *Acta Materialia*, *61*(3), 1001–1011.
Vågsholm, I., Arzoomand, N. S., & Boqvist, S. (2020). Food security, safety, and sustainability—getting the trade-offs right. *Frontiers in Sustainable Food Systems*, *4*, 16. https://doi.org/10.3389/fsufs.2020.00016
Weber, R., Watson, A., Petrlik, J., Fernandez, J., Winski, A., Schwedler, O., … & Behnisch, P. (2015). PCDD/F, PBDD/F and PCB contamination in eggs as sensitive indicator for soil contamination around pollution sources. In *13th International HCH & Pesticides Forum*, Zaragoza, 3-6 November 2015 (pp. 204–208). International HCH & Pesticides Association.
Zheng, J., Chen, K. H., Yan, X., Chen, S. J., Hu, G. C., Peng, X. W., … & Yang, Z. Y. (2013). Heavy metals in food, house dust, and water from an e-waste recycling area in South China and the potential risk to human health. *Ecotoxicology and Environmental Safety*, *96*, 205–212.

13 Electronic Wastes, Its Impact on Animal Farming Practices and Health Concerns

Poonam Singhal
St. Joseph Degree and PG College for Women, Hyderabad, India

Ananda Sahithi Taduru and Heeba Fatima
St. Ann's College for Women, Hyderabad, India

Geetanjali Kaushik
Hi Tech Institute of Technology, Aurangabad, India

Chaudhery Mustansar Hussain
New Jersey Institute of Technology, Newark, New Jersey, United States

Arvind Laxmikant Chel
MGM University, Aurangabad, India

13.1 INTRODUCTION

Electronic waste or (e-waste) is considered the world's briskly accumulating waste. It mainly comprises of a vast range of compounds and elements which includes both precious and dangerous ones. E-waste can defile the environment and risk health because of no proper recycling and disposal of e-waste (Kiddee et al., 2020). The electronic industry is considered the fastest-growing, and most innovative industry in the last century which changed people's lifestyles. Daily used electronic items are also contemplated e-waste which includes cameras, mobiles digital calculators, clocks, ovens, fridges, grinders, and other laboratory electronics, etc. (Verma et al., 2020; Agarwal et al. 2023). Electronic items are composed of dangerous metals such as beryllium, cadmium, lead, and brominated flame-retardants such as polybrominated biphenyls and polybrominated diphenyl ethers. Approximately 50 million tons of e-waste is generated every year in the world. E-waste is usually sent to recyclers, dumped, or burnt. During this dismantling, processes like crushing, ripping, and blazing, the smoke and dirt particles are removed. The fumes and dust produced have carcinogens and many dangerous chemicals which lead to inflammations and sores including many skin and respiratory disorders (Verma et al., 2020). Destroyed circuits are having metals such as platinum, gold, and cadmium which creates harmful ashes and toxic fragments from the toners considered as carcinogens and eventually lung and skin cancers are diagnosed (Verma et al., 2020).

DOI: 10.1201/9781003582311-13

13.2 IMPACT ON SOIL

The soil quality is of utmost essential as a resource that helps in providing facilities for food produce and also the well-being of the human race and wildlife (He et al., 2019). The soil composition is altered chaotically when the e-waste materials reach the soil. The presence of acids, dioxins, PBDEs, metals like cadmium, lead, etc., and other harmful chemicals in e-waste impact development of plants, damage the plant cells as well as hamper the growth of critical microorganisms present in soil (Forti et al., 2020). Due to a number of toxins being leached into the soil and groundwater, farming and drinking water is unsafe (Li, 2020). Excess heavy metals accumulate in the soil which is lethal to humans and other animals (Gogoi et al., 2024). PBDEs are a group of flame retardants that are widely utilized in numerous products such as electronic equipment, plastics, electrical and furniture, textiles and foam-based products. Semi-urban places are specifically more prone to e-waste pollution loads due to the depositions of effluents and waste. The impact is majorly seen in agriculture as the areas take up heavy metals which pass on to the food crops and then contaminate the whole food chain (Sinha, 2014). A classification of contaminated is given on the concentrations of PAHs in four categories as given as for clean soil/non-contaminated the PAHs should be less than 200 $\mu g\ kg^{-1}$, heavily contaminated soil > 1000 $\mu g\ kg^{-1}$, for weakly and moderately contaminated soil 200–600 $\mu g\ kg^{-1}$ and 600–1000 $\mu g\ kg^{-1}$ respectively (He et al., 2019).

13.3 IMPACT ON FARMING PRACTICES

E-waste is also causing havoc among animals and also animal farmers. The produce from the animals is being contaminated along with a large number of occupational hazards being caused to animal rarer and in turn affecting their families. Aquatic life is affected because of air pollution which occurs due to informal method of recycling of e-waste such as burning openly (Okunola and Adekunle, 2021). Cattle and poultry are usually not selective eaters. The landfills where the e-waste is being processed contaminate the soil and during grazing, it leads to toxic substances entering their body and which often leads to deterioration of the quality of products they produce such as eggs, milk, etc. (Paul et al., 2019). Some of the most commonly seen occupational hazards to animal farmers are the incineration and burning of e-waste which releases of toxic gases when inhaled is determinantal to nervous system (Mondal et al., 2023). Whereas, Polyvinyl chloride (PVC) releases chlorine at high temperatures are burning and leading to reproductive and developmental problems. Mercury, chromium, and arsenic are acutely poisonous substances that are injurious to health on long-term exposure. Another occupational hazard is seen with exposure to zinc sulphide when it is combined with rare earth metals which are very toxic to human health. Asthma is one of the most commonly seen health hazards in animal farming with exposure to airborne dust after the processing of e-waste.

13.4 IMPACT ON PLANTS

Researchers have found high levels of polybrominated diphenyl ethers (PBDEs) and polychlorinated biphenyls (PCBs) in plants from Guiyu and adjacent areas. The leaves of *Erigerunannuus* L, *Rumexjaponicus* Huntt., *Sorghum bicolor* L., *Pteridiumaquilinum* L, and *Pterismultifida* showed PBDE concentrations of between 326, 278, 162, 144 and 116 ng/g (dry matter) respectively when the plants were grown in soil containing PBDE (25,479 ng/g) (Yang et al. 2008). Not only PBDE but other toxicants can translocate from soil to the plants and ultimately, they reach the consumers, affecting their health. The investigators have found that rice grown in Taizhou in eastern China, contained more than double to quadruple concentration of permissible limit of Pb and Cd in food samples (Fu et al. 2008) as is an e-waste processing town. Other researchers documented that paddy soil in the vicinity of e-waste recycling site in Zhejang showed reduced rate in germination (Zhang and Min 2009). Similar results have been obtained from other countries, high levels of PAE have been found in vegetables harvested near e-waste sites (Ma et al. 2013). Another study found that soil contaminated by e-waste in Nigeria contained high concentration of Cd, Cu, Cr, Pb, Ni, PCBs, PBDEs, and PAHs. Likewise, the results showed high concentration of Ni, Pb, Cr and Cd in the upper ground parts like stems and leaves as well as in roots of Eleusine indica and Ipomeabatatas from e-waste dumping sites in Nigeria (Ofudje et al. 2008). The e-waste leachates were responsible for genetic damage and cytotoxicity in Allium cepa (Bakare et al. 2012, Bakare et al. 2013; Babatunde & Anabuike 2015).

13.5 IMPACT ON HUMAN HEALTH

Impacts of e-waste on children are directly linked to adverse effects on respiratory, and lung functioning, damage to DNA, thyroid dysfunction, and a huge prevalence of chronic disorders such as cardiovascular diseases, cancer are diagnosed. Children who are exposed to e-waste are particularly more prone to dangerous chemicals, as children are smaller in size, have less developed organs, and have a rapid rate of growth and development hence are more susceptible to absorption of pollutants. Around the world, there are more than 18 million children and adolescents between 5 and 17 years of age engaged in industries in which waste processing is a subsector. In the E-waste recycling process, child labor is commonly seen where children as young as 5 years are found working, which is proven to be very dangerous to their physical and mental growth and performance. In the case of pregnant women, exposure to toxic e-waste can have many effects on their health as well as on the development of the foetus for the rest of their life. The health effects are stillbirth, preterm births, and low birth weight. The extreme hazard to various substances of lead generated from e-waste is causing neonatal behavioral changes along with child temperament and elevated risk of neurological problems like attention-deficit/hyperactivity disorder (ADHD), sensory difficulties, reduced cognition and language adaptability scores (Forti et al., 2020). An updated systematic review evaluated on the detrimental consequences of human health on exposure to e-waste. It was found that out of 5645 records detected, only 70 research articles met the present parameter. It has been concluded that the population residing in e-waste regions had increased degree of organic pollutants and heavy metals. The juveniles and expecting women were highly at risk which eventually had detrimental effects on their biological system and organs. It has also been recorded those elevated levels of toxic chemicals had negative effects on neonate growth and hormone level change in e-waste-exposed regions. A study was conducted where the impact is seen in informal e-waste recycling in Africa. Four electronic databases were considered for articles connected to e-waste and it has been found that increased levels of pollutants in soil, air, water, vegetable, and human matrices signify that not only e-waste employers are at increased risk but the general population and future generations will also be affected. The symptoms such as migraine, cough, chest pain, malaise, abnormal levels of thyroid, and reduction of the gonadal hormone were recorded in those people who are involved with the processing of e-waste (Orish et al., 2019).

A study was done to estimate the health effect of disposal areas and incineration in Italy, Slovakia, and England. An overall number of forty-nine incinerators in Italy, Slovakia and England were working in 2001 whereas for landfills a total of 619, 121, 232 in Italy, Slovakia, and England respectively. The study inhabitants were residing within a distance of 3 kilometers from an incinerator and 2km from a landfill. The combined data from air pollution dispersion modelling for particle pollution (PM10), Nitrogen content and epidemiological studies were used to estimate the increased risk estimates to the inhabitants. It was found that about 1,000,000 residents in Italy lived close to incinerators, 16,000 people resided in Slovakia close to incinerators and 1,200,000 residents lived close to incinerators in England. The yearly number of cancer cases due to exposure in the year 2001 to 2020 was projected to peak in 2020 and then rapidly decline by 2050. The total exposed population to landfills in Italy was 1,350,000, the subjects in Slovakia were 329,000, and 1,425,000 subjects were in England (Forastiere et al., 2011).

13.6 CONCLUSION

The increase in e-waste has shown the importance and need for safe recycling that can minimize the effects on the environment, public health, and food safety issues. Society together, including both producers and consumers, should take up the responsibility to decrease electronic waste and, in turn, diminish its hazardous effects on human beings as the development of technologies is leading to increased production of electronic devices, and this situation is causing colossal e-waste generation and increased toxic substances. Research must be done to assess the direct and indirect effects of informal e-waste recycling on health and the environment. As a result, the disposal of electronic materials is risky for humanity, and it is a great responsibility to reduce e-waste and the leakage of harmful substances during recycling. Hence, many people involved in e-waste recycling are affected by one or the other disease resulting from unsafe e-waste recycling.

REFERENCES

Agarwal, S., Darbar, S., Saha, S., Choudhury, M., & Singh, R. P. E-waste management using different cost-effective, eco-friendly biological techniques: an overview In *Waste Management and Resource Recycling in the Developing World*, (2023) 205–235. https://doi.org/10.1016/B978-0-323-90463-6.00026-9

Babatunde, B., Anabuike, F.. In Vivo cytogenotoxicity of electronic waste leachate from Iloabuchi electronic market, Diobu, Rivers state, Nigeria on Allium cepa (2015) *Challenges*. 6(1), 173–187.

Bakare, A. A., Adeyemi, A. O., Adeyemi, A., Alabi, O. A., Osibanjo, O. Cytogenotoxic effects of electronic waste leachate in Allium cepa (2012) *Caryologia*, 65(2), 94–100.

Bakare, A. A., Alabi, O. A., Gbadebo, A. M., Ogunsuyi, O. I., Alimba, C. G.. In vivo cytogenotoxicity and oxidative stress induced by electronic waste leachate and contaminated well water. (2013) *Challenges*. 4(2), 169–187.

Forastiere, F., Badaloni, C., de Hoogh, K., von Kraus, M. K., Martuzzi, M., Mitis, F., ... & Briggs, D. (2011). Health impact assessment of waste management facilities in three European countries. *Environmental Health*, *10*(1), 1–13.

Forti, V., Baldé, C. P., Kuehr, R., & Bel, G. (2020). The global e-waste monitors. *Quantities, Flows, and the Circular Economy Potential*, 1–119.

Fu, J. J., Zhou, Q. F., Liu, J. M., Liu, W., Wang, T., Zhang, Q. H., Jiang, G. B. High levels of heavy metals in rice (Oryza sativa L.) from a typical E-waste recycling area in Southeast China and its potential risk to human health. (2008), *Chemosphere*. 71, 1269–1275.

Gogoi, N., Sarma, A., Choudhury, M., Samanta, P., Faizan, M., & Sadhak, S. (2024). Soil heavy metal pollution and ecological risk assessment in disturbed and undisturbed soil of Morigaon, Assam. *Discover Environment*, 2(1), 13. https://doi.org/10.1007/s44274-024-00039-z

He, M., Yang, S., Zhao, J., Collins, C., Xu, J., & Liu, X. (2019). Reduction in the exposure risk of farmer from e-waste recycling site following environmental policy adjustment: A regional scale view of PAHs in paddy fields. *Environment International*, *133*, 105136.

Kiddee, P., Pradhan, J. K., Mandal, S., Biswas, J. K., & Sarkar, B. (2020). An overview of treatment technologies of e-waste. *Handbook of Electronic Waste Management*, 1–18.

Li, W., & Achal, V. (2020). Environmental and health impacts due to e-waste disposal in China–A review. *Science of the Total Environment*, 737, 139745.

Ma, T. T., Christie, P., Luo, Y. M., Teng, Y.. Phthalate esters contamination in soil and plants on agricultural land near an electronic waste recycling site (2013) *Environ Geochem Health*, 35, 465–476.

Mondal, T., Choudhury, M., Kundu, D., Dutta, D., & Samanta, P., Landfill: An eclectic review on structure, reactions and remediation approach. (2023) *Waste Management*, 164, 127–142.

Ofudje, E. A., Akinwunmi, F., Ogundiran, A. A.. Uptake of heavy metals by plants obtained from electronic-waste dumpsites (2008), *Nigerian Research Journal of Chemical Sciences* 4, 134–148.

Okunola, A. Alabi, Adekunle, A. Bakare, Environmental contamination and public health effects of electronic waste: an overview (2021), *Journal of Environmental Health Science and Engineering*, 19(1)1209–1227.

Orish, Ebere, Orisakwe, Chiara, Frazzoli, Cajetan, Elochukwu, Ilo, Benjamin, Oritsemuelebi, Public Health Burden of E-waste in Africa (2019), *Journal of Health and Pollution*. 9(22), 190610.

Paul, S., Choudhury, M., Deb, U., Pegu, R., Das, S., & Bhattacharya, S. S. Assessing the ecological impacts of ageing on hazard potential of solid waste landfills: A green approach through vermitechnology. (2019), *Journal of cleaner production*, 236, 117643. https://doi.org/10.1016/j.jclepro.2019.117643

Sinha, S., Mittal, A., Rajankar, P., & Sharma, V. (2014). *Impact of E-Waste Recycling on Water and Soil. Toxics Link*: New Delhi, India.

Verma, A. K. E-wastes and their impact on environment and public health. (2020), *International Journal of Applied Research* 6(9): 164–168.

Yang, Z. Z., Zhao, X. R., Zhao, Q., Qin, Z. F., Qin, X. F., Xu, X. B.. Polybrominated diphenyl ethers in leaves and soil from typical electronic waste polluted area in South China. (2008), *Bulletin of Environmental Contamination and Toxicology* 80, 340–344.

Zhang, J., Min, H.. Eco-toxicity and metal contamination of paddy soil in an e-wastes recycling area. (2009) *Journal of Hazardous Materials*, 165, 744–750.

14 Electronic Wastes and Its Impact on Public Health

Sucheta Mandal and Sabina Pradhan
Banwarilal Bhalotia College, Asansol, India

14.1 INTRODUCTION

Life in the present scenario is unimaginable without electronic devices. Our very survival depends on those. Everyday a new equipment comes in the market and lures us into buying them with the bait of making our lives simpler, better. There is no denying the fact that they do exactly that. Taking into account the demand from the buyers there has been an exponential growth in the electrical and electronic industries. Along with that comes the darker side of the story: generation of an equally insurmountable amount of waste, known as the electronic waste or e-waste. The electronic and electrical devices such as mobile phones, computers, television sets, office electronic equipment, refrigerators, etc., which are no longer useful to the end-users are termed as e-waste (Needhidasan et al., 2014). Although E-waste is a global concern, the main problem lies in the fact that the developed countries dump their e-waste to developing or less developed countries where it is recycled using primitive methods causing harm to environment and linked ecosystem. This transboundary movement of e-waste is a matter of greatest concern (Sthiannopkao et al., 2013).

The technological advancement has undoubtedly helped mankind but it has also created the problem of e-waste which is growing at a menacing pace. The multitude of harmful substances such as heavy metals, lithium, barium, dioxin leaking from polyvinylchloride and polybrominated flame retardants are responsible for toxicity in humans (Pramila et al., 2012). The toxins significantly impact the reproductive system, the skeletal system as well as the nervous system leading to heart, drain, kidney, liver damage and birth defects in humans (Kishore, 2010).

As per the report of Global Futures Foundation, though the U.S. municipal solid waste constitutes only 2–5% e-waste but it accounts for almost 70% of hazardous waste found in landfills. These toxicants bioaccumulate and are persistent creating health and environmental hazards when they openly dumped or melted down (Mondal et al., 2023). The leachate from the e-waste present in landfill affect the underground water reservoir impacting the local community. If the e-waste is burnt in open, it releases carcinogen like dioxins affecting the air we breathe.

In most of the developing countries, e-waste are recycled by traditional methods using chisel, screwdrivers, hammer and spanners for disintegrating and separating various components (Wang et al., 2010) also a long metal rod for flipping the items which are burnt (Acquah et al., 2019; Gullett et al., 2007). Such traditional methods are releasing plethora of pollutants in the environment which harm the surrounding population and the recyclers (Issah et al., 2023). Some of the chemicals, such as heavy metals and Polycyclic Aromatic Hydrocarbons (PAHs), are carcinogens (IARC, 2012). Such a level of exposure of toxicants in the e-waste recycling sites, is associated with higher cancer risks amongst the residents compared to general population (Vaccari et al., 2019; Zheng et al., 2016; Huang et al., 2016). Exposure to genotoxic agents, which may be occupational or environmental, through inhalation, ingestion, and dermal contact may damage DNA in a cell, which may be linked to the development of diseases (Dreval and Pogribny, 2019; Rodgers et al., 2018; da Silva, 2016). The researchers opine that the electronic devices need to be made with environmentally friendly and safe to use materials after they found that the when the computers, mobile devices and other

DOI: 10.1201/9781003582311-14

electronic devices are dismantled they release pollutants which can affect the lungs, heart and also cause cancer (Institute of Physics. 2011, May 31). In developing countries where most of the e-wastes are dumped and discarded, unsafe handling results in harm to human health and the environment. Problems like open-air burning and acid baths used to recover valuable materials from electronic components expose workers to harmful substances. Leaching of toxic materials into the environment is another serious threat. Heavy metals like cadmium, mercury, arsenic and lead are implicated in irreparable health effects including neurological damage, miscarriages, reduced IQ level and cancer. In 2009, 2.37 million tons of e-waste which included computers, discarded televisions, cell phones, scanners, printers, etc., was discarded by US consumers, out of which only one-fourth were recycled and rest were disposed in landfills with no recovery of precious metals.

14.2 IMPACT ON HUMAN HEALTH

E-waste cannot be treated like a common waste as it contains hazardous and carcinogenic compounds like heavy metals or dioxins and furans (when e-waste is incinerated). Electronic equipments contain approximately 36 chemical compounds among which some shows toxigenic effect if not handled properly during recycling and disposal process. TV, computers and some other electronic devices contains a significant amount of lead, cadmium, mercury, polyvinyl chloride, brominated flame retardants, chromium, and beryllium in their monitor. Constant exposure to such heavy metals damages nervous system, endocrine systems and reproductive systems. Lead, when enter into biological system via water, soil and air causes lead poising to infants as well as adults get nerve problems. Moreover, improper handling during recycling of this waste shows increased concentration of lead in blood in children in China. Dumping of E-waste on open ground contaminate the environment. People residing nearby these areas had a high incidence of skin damage, headaches, vertigo, nausea, chronic gastritis, and gastric and duodenal ulcers. A study from China states that peoples exposed to heavy metals have severe hair loss. Metal like Pb, Cu, Mn, and Ba were found in their scalp which will develop skin cancer. Similar scenario was found from rest of the developing countries. These findings are enough to alarm and strongly suggest to be replicating in occupational settings in India. Effective management (Agarwal et al. 2023) is the need of the hour to save the ecosystem from exposure to e-wastes (Table 14.1).

TABLE 14.1
Hazardous Electronic Components and Their Potential Health Effects

E-Waste Component	Sources	Impact on Health
Polychlorinated Biphenyls (PCBs)	Television sets, Computer Monitors, and Other Electronic Devices.	• Birth defects. • Damage Neuro-Endocrine System. • Damage Reproductive System. • Damage Musculoskeletal System.
Pb	Glass panels Computer monitors,	• Damage central and peripheral nervous systems, kidney, blood systems • Impaired development of children's brain and reproductive systems • Cause anorexia, muscle pain, malaise and headache
Hg	Sensors, thermostats, PCB, relays, flat panel displays	• Damages lungs, liver and kidneys • Weakening of the muscles, sleeplessness, reduced emotional stability, sleeplessness, headaches • Effects the CNS, and endocrine systems
Cd	Plastics, batteries, toys, semiconductor chips, Infra-red detectors,	• Impacts heart, kidney disfunction, hypertension, and bone toxicity

(Continued)

TABLE 14.1 (Continued)
Hazardous Electronic Components and Their Potential Health Effects

E-Waste Component	Sources	Impact on Health
As	Integrated circuit and semiconductors	• Cause lung cancer, skin cancer • Damage DNA repair mechanisms, perturbation of signaling cascades
Cr IV	Plugs, switches, screws, antenna rivets, and other accessories	• Cause of chromosomal aberration and DNA damage
Be	Electrical connectors and battery contacts	• Damage to lungs and skin
Sb	Semiconductors and Batteries,	• Stomach pain and ulcers, eye and skin irritation, vomiting and diarrhea
Dioxins & Furans	Capacitors, electrical devices, transformers, fluorescent lighting fixtures,	• Impairment of immune system and hormonal balance

14.3 CADMIUM (Cd)

Cadmium, is used in soldering, semiconductors, chip resistors, infrared detectors, contact plates, switches to prevent corrosion. Old monitors usually contain five-ten grams of Cd; even some batteries are made by mixing nickel with This commonly used compound is known as human carcinogen, damages lungs and leads to death following inhalation or ingestion by food. Exposure to cadmium can impact heart, kidneys, liver and bones leading to hypertension, kidney failure, liver damage and fragile bones. Report of epidemiological study shows that cadmium promote the development of musculoskeletal diseases, such as osteoporosis, rheumatoid arthritis (RA), and osteoarthritis (OA) (Reyes-Hinojosa et al., 2019). This heavy metal when enters to plant body through root from contaminated soil and inhibit plant growth and other metabolic activity (Shiyu et al., 2020; Gogoi et al., 2024).

Mode of action: This extremely toxic metal become more toxic when inhaled as cadmium metal or cadmium oxide. Cd present in food is rarely absorbed (5–8%) in our gut system so oral ingestion is not mainly responsible for the lethal diseases. In contrast more that 50% of Cd is present in cigarettes which after inhalation bound to alpha-2-macroglobulin and albumin, of red blood cells and make a toxigenic effect on respiratory and renal system (Wise et al., 2008). In maximum cases it was found that presence of Cd is been detected in urine after kidney failure. Prolong inhalation results into progressive fibrosis lower air causing emphysema. Such condition initiates necrosis of alveolar macrophage which release degradative enzymes and ultimately dame the membrane of alveoli. Increased proline and hydroxyproline in urine induce bone damage. Cd is a pro-oxidative agent so the reactive oxygen species (ROS) plays a significant role in development of bone abnormalities (Brzóska et al., 2011) and eventually affects immune system.

14.4 LEAD (Pb)

Lead, mainly used for soldering of circuit boards, motherboards, wiring plastic and other device components. A major amount of Pb is obtained from CRT of monitor. First exposure to Pb cause anorexia, muscle pain, malaise and headache but prolonged exposure hampers brain development and nervous system of fetuses and young children (Yohannes et al., 2022). It also affects the reproductive system in both male and female hence considered as carcinogen. Many researches also indicate that Pb damages both central and peripheral nervous systems, cause seizures, retardation, high blood pressure, damage kidneys & liver and adversely affect child development.

Mode of action: Pb toxicity is more lethal as it can mimic calcium and hamper many fundamental cellular processes. It moves through erythrocyte membrane using calcium channel. After entering into cytoplasm it completely occupy the calcium binding site and binds with calcium depending proteins. Such condition alters the functioning of enzyme Protein kinase C which is a crucial enzyme for hormone and neurotransmitter. Pb also binds with the thio- group of different enzymes which inhibit heme biosynthesis. Moreover ROS generated by Pb creates renal problem.

14.5 MERCURY (Hg)

Mercury, a common ingredient of LCD screen, fluorescent light, computer or monitor switches, which enable them to shut down while idle, and finally in batteries, circuit boards, etc., is a potent-neurotoxin (Calabrese et al., 2018). Mercury effects the liver, kidney, CNS, endocrine system and lungs, chronic exposure leads to headaches, sleeplessness, weakening of the muscles and bones, brings emotional imbalance, also degenerates the digestive system. Exposure of mercury is fatal to the neurological development of unborn fetus and it causes harm to teeth and gums. Short term exposure causes nausea, vomiting, diarrhea, high pressure, and, skin and eye irritation. Hg even got accumulated in fish and enters into food chain (Abbott and Nigussie, 2021).

Mode of action: It has an affinity with sulfhydryl (thiol; -SH) groups of low- and high-molecular-weight proteins and inhibit sulphydryl dependent enzymes like manganese-superoxide dismutase, arginase I, sorbitol dehydrogenase, δ-aminolevulinatedehydratase, lactate dehydrogenase and fatty acid synthetase. Accumulation of Hg in lysosome increase lysosomal acid phosphatase which results into cellular toxicity. In case of blood Hg binds with albumin and form CH_3Hg^+, $C_2H_5Hg^+$, $C_6H_5H^+$. Moreover Hg-SH also plays a role in apoptosis by interfering with Akt/CREB, Keap1/Nrf2, NF-κB, and mitochondrial pathways (Ajsuvakova et al., 2020). Hg induced oxidative stress shows renal toxicity.

14.6 ARSENIC

Arsenic is a poison and is used in semiconductors and some integrated circuits, exposure to arsenic can cause damage to the digestive system, affecting the liver, kidney and also causing skin cancer, pigmentation and thickening of the skin leading to keratosis. Acute toxicity of arsenic results into cardiac arrhythmias, anemia, leucopoenia, sensory loss, nausea, gastrointestinal distress and diarrhea (Hu et al., 2020).

Exposure to arsenic induces ROS which alter signaling pathway and cause direct oxidative damage. Methylated form of this compound is carcinogenic, it induces oxidative damage, decrease DNA repair mechanisms, perturbation of signaling cascades, disrupt translational and transcriptional activities, change microRNA expression, resulting from iAS interaction into changes in expression of proteins and genes (Tchounwou, et al., 2019).

14.7 CHROMIUM (Cr)

Chromium is a known as toxin in plant system and carcinogenic for human and animal system, causing toxicity even at low concentrations. Exposure to cadmium can damage liver, kidney, skin and lungs. Cr binds with DNA and shows chromosomal abnormalities, oxidative stress, chromosome breaks, and DNA adduct. Such genomic instability induces cancer by producing ROS.

14.8 BROMINATED FLAME RETARDANTS (BFRs)

BFRs can accumulate in animal and human tissues and chronic exposure leads delayed puberty, metabolic abnormalities, and hampers normal development of skeletal and CNS especially in the fetus.

14.9 ANTIMONY

The production of batteries, diodes and semiconductors all use antimony and it is toxicity is quite similar to arsenic in humans. It is fatal in large doses and interferes in DNA repair process or alteration in gene expression due to excess production of ROS (Boreiko et al., 2020). Long term exposure leads to stomach ulcers, diarrhea, pneumoconiosis, and modified electrocardiogram. The presence of antimony in air can cause irritation in the lungs, eyes and skin.

14.10 PHTHALATES

Phthalates are also known as endocrine disruptor and are used commonly in consumer products such as cosmetics, fragrances, food packaging and clothing industries. They effect the reproductive health in males and females. Exposure to phthalates has shown to cause allergy and asthma in children. It is related to fertility disorders in females and males, testicular dysgenesis syndrome, cancer and modify the release of peripheral, pituitary and hypothalamic hormones (Deng et al., 2021).

14.11 POLYVINYL CHLORIDE (PVC)

Polyvinyl chloride is commonly used toxicant and on burning it discharges dangerous chemicals. It obstructs development, multiple growth factors, immune and pulmonary system. It is carcinogen and is harmful to animals and humans.

14.12 BERYLLIUM

Beryllium is a well-known carcinogen and used in battery contacts and electrical connectors. On inhalation it can damage the lungs and chronic exposure can cause Acute Beryllium Disease. It can also be detrimental to skin and is harmful to both animals and man.

14.13 CONCLUSION

Suppose we carefully observe the trend, the continuous chain of demand and supply, the gimmicks of the market, and the exponential rise of the electrical and electronic industries. In that case, all the items and equipment are designed for the dump. Newer products keep flooding the market, and consumers lap it up without being concerned about the waste they are generating and their impact on the environment and health. To safeguard our health and environment, consumers should dispose of waste properly in e-waste bins, give it to certified recyclers, or trade it for further usage. The new products should, therefore, be designed not for the dump but to last.

REFERENCES

Abbott, L. C., & Nigussie, F. (2021). Mercury toxicity and neurogenesis in the mammalian brain. *International Journal of Molecular Sciences*, *22*(14), 7520.

Acquah, A. A., D'Souza, C., Martin, B., Arko-Mensah, J., Nti, A. A., Kwarteng, L., ..& Fobil, J. N. (2019). Processes and challenges associated with informal electronic waste recycling at Agbogbloshie, a suburb of Accra, Ghana. In *Proceedings of the Human Factors and Ergonomics Society Annual Meeting* (Vol. *63*, No. 1, pp. 938–942). Sage CA: Los Angeles, CA: Sage Publications.

Agarwal, S., Darbar, S., Saha, S., Choudhury, M., & Singh, R. P. (2023). E-waste management using different cost-effective, eco-friendly biological techniques: an overview In *Waste Management and Resource Recycling in the Developing World*, 205–235. https://doi.org/10.1016/B978-0-323-90463-6.00026-9

Ajsuvakova, O. P., Tinkov, A. A., Aschner, M., Rocha, J. B., Michalke, B., Skalnaya, M. G., & Bjørklund, G. (2020). Sulfhydryl groups as targets of mercury toxicity. *Coordination Chemistry Reviews*, *417*, 213343.

Boreiko, C. J., & Rossman, T. G. (2020). Antimony and its compounds: Health impacts related to pulmonary toxicity, cancer, and genotoxicity. *Toxicology and Applied Pharmacology*, *403*, 115156.

Brzóska, M. M., Rogalska, J., & Kupraszewicz, E. (2011). The involvement of oxidative stress in the mechanisms of damaging cadmium action in bone tissue: a study in a rat model of moderate and relatively high human exposure. *Toxicology and Applied Pharmacology*, *250*(3), 327–335.

Calabrese, E. J., Iavicoli, I., Calabrese, V., Cory-Slechta, D. A., & Giordano, J. (2018). Elemental mercury neurotoxicity and clinical recovery of function: A review of findings, and implications for occupational health. *Environmental Research*, *163*, 134–148.

da Silva, J. (2016). DNA damage induced by occupational and environmental exposure to miscellaneous chemicals. *Mutation Research/Reviews in Mutation Research*, *770*, 170–182.

Deng, Y., Yan, Z., Shen, R., Huang, Y., Ren, H., & Zhang, Y. (2021). Enhanced reproductive toxicities induced by phthalates contaminated microplastics in male mice (Mus musculus). *Journal of Hazardous Materials*, *406*, 124644.

Dreval, K., & Pogribny, I. P. (2019). Chemical carcinogen exposure, DNA damage, and epigenetic alterations. In *Carcinogens, DNA damage and cancer risk: mechanisms of chemical carcinogenesis* (pp. 309–340).

Gogoi, N., Sarma, A., Choudhury, M., Samanta, P., Faizan, M., & Sadhak, S. (2024). Soil heavy metal pollution and ecological risk assessment in disturbed and undisturbed soil of Morigaon, Assam. *Discover Environment*, *2*(1), 13. https://doi.org/10.1007/s44274-024-00039-z

Gullett, B. K., Linak, W. P., Touati, A., Wasson, S. J., Gatica, S., & King, C. J. (2007). Characterization of air emissions and residual ash from open burning of electronic wastes during simulated rudimentary recycling operations. *Journal of Material Cycles and Waste Management*, *9*, 69–79.

Hu, Y., Li, J., Lou, B., Wu, R., Wang, G., Lu, C., … & Xu, Y. (2020). The role of reactive oxygen species in arsenic toxicity. *Biomolecules*, *10*(2), 240.

Huang, C. L., Bao, L. J., Luo, P., Wang, Z. Y., Li, S. M., & Zeng, E. Y. (2016). Potential health risk for residents around a typical e-waste recycling zone via inhalation of size-fractionated particle-bound heavy metals. *Journal of Hazardous Materials*, *317*, 449–456.

Institute of Physics. (2011, May 31). 'E-waste pollution' a threat to human health, new research suggests. *ScienceDaily*. www.sciencedaily.com/releases/2011/05/110531084958.htm)

International Agency for Research on Cancer. (2012). *A Review of Human Carcinogens. F*. Chemical Agents and Related Occupations: IARC Monographs on the Evaluation of Carcinogenic Risks to Humans.

Issah, I., Arko-Mensah, J., Agyekum, T. P., Dwomoh, D., & Fobil, J. N. (2023). Electronic waste exposure and DNA damage: A systematic review and meta-analysis. *Reviews on Environmental Health*, *38*(1), 15–31.

Kishore, J. (2010). E-waste management: as a challenge to public health in India. *Indian Journal of Community Medicine: Official Publication of Indian Association of Preventive & Social Medicine*, *35*(3), 382.

Mondal, T., Choudhury, M., Kundu, D., Dutta, D., & Samanta, P. (2023). Landfill: An eclectic review on structure, reactions and remediation approach. *Waste Management*, *164*, 127–142.

Needhidasan, S., Samuel, M., & Chidambaram, R. (2014). Electronic waste–an emerging threat to the environment of urban India. *Journal of Environmental Health Science and Engineering*, *12*, 1–9.

Pramila, S., Fulekar, M., & Bhawana, P. (2012). E-waste-A challenge for tomorrow. *Research Journal of Recent Sciences*, *1*(3), 86–93, ISSN, 2277-2502.

Reyes-Hinojosa, D., Lozada-Pérez, C. A., Cuevas, Y. Z., López-Reyes, A., Martínez-Nava, G., Fernández-Torres, J., … & Martínez-Flores, K. (2019). Toxicity of cadmium in musculoskeletal diseases. *Environmental Toxicology and Pharmacology*, *72*, 103219.

Rodgers, K. M., Udesky, J. O., Rudel, R. A., & Brody, J. G. (2018). Environmental chemicals and breast cancer: An updated review of epidemiological literature informed by biological mechanisms. *Environmental Research*, *160*, 152–182.

Shiyu, Q. I. N., Hongen, L. I. U., Zhaojun, N. I. E., Rengel, Z., Wei, G. A. O., Chang, L. I., & Peng, Z. H. A. O. (2020). Toxicity of cadmium and its competition with mineral nutrients for uptake by plants: A review. *Pedosphere*, *30*(2), 168–180.

Sthiannopkao, S., & Wong, M. H. (2013). Handling e-waste in developed and developing countries: Initiatives, practices, and consequences. *Science of the Total Environment*, *463*, 1147–1153.

Tchounwou, P. B., Yedjou, C. G., Udensi, U. K., Pacurari, M., Stevens, J. J., Patlolla, A. K., … & Kumar, S. (2019). State of the science review of the health effects of inorganic arsenic: perspectives for future research. *Environmental Toxicology*, *34*(2), 188–202.

Vaccari, M., Vinti, G., Cesaro, A., Belgiorno, V., Salhofer, S., Dias, M. I., & Jandric, A. (2019). WEEE treatment in developing countries: Environmental pollution and health consequences—An overview. *International Journal of Environmental Research and Public Health*, *16*(9), 1595.

Wang, H., Lv, S., Li, F., Liu, Q., & Ke, S. (2010). Study on the changes of urinary 8-hydroxydeoxyguanosine levels and burden of heavy metal around e-waste dismantling site. *Science of the Total Environment, 408*(24), 6092–6099.

Wise, S. S., Holmes, A. L., & Wise, Sr., J. P. (2008). Hexavalent chromium-induced DNA damage and repair mechanisms. *Reviews on Environmental Health, 23*(1), 39–58.

Yohannes, Y. B., Nakayama, S. M., Yabe, J., Toyomaki, H., Kataba, A., Nakata, H., … & Ishizuka, M. (2022). Glutathione S-transferase gene polymorphisms in association with susceptibility to lead toxicity in lead- and cadmium-exposed children near an abandoned lead-zinc mining area in Kabwe, Zambia. *Environmental Science and Pollution Research, 29*(5), 6622–6632.

Zheng, X., Xu, X., Yekeen, T. A., Zhang, Y., Chen, A., Kim, S. S., … & Huo, X. (2016). Ambient air heavy metals in PM2. 5 and potential human health risk assessment in an informal electronic-waste recycling site of China. *Aerosol and Air Quality Research, 16*(2), 388–397.

15 Endocrine Disrupting Chemicals (EDCs) from Electronic Wastes

Raghunath Satpathy
Gangadhar Meher University, Sambalpur, India

15.1 INTRODUCTION

The human endocrine system is an important and complex system that plays a crucial role in regulating various body processes. The chemical messengers are the hormones which are produced by the endocrine glands, help maintain homeostasis by coordinating the functions of different organs and tissues in the body. Examples of endocrine glands and location in the body are, the pituitary gland (base of the brain), hypothalamus (lower central part of the brain), pineal gland (middle part of the brain) adrenal glands, testes (in males), ovaries (in females), thyroid gland (at the lower neck position), parathyroid glands (attached to the thyroid), and pancreas. The endocrine system receives the specific stimulus and releases the hormones to the bloodstream and subsequently reached the target organ. Upon reaching the target, the hormones bind to the receptor molecules, usually located on the surface of the target cell. The hormone molecule and the receptor of the target cell interacts and that triggers a series of biochemical reaction (cellular signaling) in the target cell ultimately lead to alteration of its function (homeostasis, growth and development). An uninterrupted route from the endocrine gland and target cell receptor is essential for the proper functioning of the hormones (Martyniuk et al., 2022; Ghosh et al., 2022).

Endocrine-disrupting chemicals (EDCs) are a group of chemical of a compound that are released in the form of environmental pollutant and interrupts hormone function and causes diseases and metabolic disorders. For example, the metabolic regulation of carbohydrates, lipids and proteins are controlled by the endocrine system. So, any kind of interference in the metabolism controlled by the hormones may lead to the excess deposit of fat in the body hence *obesity* condition arises. As per the report framed by U.S. Environmental Protection Agency in 2020, about 86,000 chemicals have been categorized to have the endocrine disruption activity. However, the detailed study about many of these compounds in the disruption of endocrine function is not fully understood. Considering the mode of action, the EDCs, have the structural similarity with the normal hormones and binds to the receptor, hence the hormone cannot bind to perform the normal metabolic function. Also some of the EDCs directly interrupts to the hormone signaling activity and generates a wide range of metabolic disorder and disease such as obesity, reduced reproduction ability, cancer, cardiovascular disease, and neural disorders (Francis et al., 2021; Satpathy, 2020; Varticovski et al., 2022). The human being irrespective of age, and sex, are exposed to several of the EDCs in their lifetime by the environmental samples (soil, air, water, milk, direct contact with these substances). Also, the toxic effects of the EDCs are observed to increases in a proportional manner with the duration of exposure and the dosage (Tanner et al., 2020; Ohayi et al., 2022). The main source of EDCs is xenobiotics which are largely produced in the form of waste and by-products from various industrial sectors.

DOI: 10.1201/9781003582311-15

15.2 E-WASTE-ASSOCIATED ENDOCRINE DISRUPTING CHEMICALS AND METABOLIC DISORDER

Among different types of waste material generated currently by human beings, the electronic waste amount is gradually increasing progressively. Electronic waste (e-waste) consists of damaged, outdated electronic product like radios, televisions, personal computers (both CPUs and monitors), scanners and printers, cell phones, cameras and so on. Due to the wide range of use of these substances in day-to-day life, a surplus amount of e-wastes is produced. The hazardous chemicals from e-waste are generated either directly from the component of electronic devices and also during their recycling process. Several categories of pollutants are available and generated from the e-wastes. Many are the persistent organic pollutants (POPs), flame retardants (FRs), polybrominated di-phenyl ethers (PBDAs), polychlorinated biphenyls (PCB), poly-brominated diphenyls, di-brominated diphenyl ethers, poly-chlorinated dioxins.

These hydrocarbons are usually entering into the environment by the combustion process of e-waste materials (Gupta et al., 2014; Ma et al., 2022, Saha et al., 2021). The continuous management of huge amount of e- wastes (recycling and disposal) creates a major source of these hazardous substances like EDCs (Jinhui et al., 2011). As compared to other conventional solid wastes the e- wastes contain many toxic substances such as EDCs and then enter into the environment badly affecting human health (Figure 15.1). For example, the toxic metallic substances such as lead, mercury, and cadmium are present in several parts of television and computer monitors. Similarly, the electronic circuit boards contain other substances such as nickel, beryllium, and zinc. A mobile phone contains more than 40 elements, including the toxic heavy metals. Other components of the personal computers (PCs) contain toxic organic compounds such as chlorinated and brominated hydrocarbons and other biologically active chemicals. Due to the presence of varieties of these compounds the process of recycling as well as disposal of e-waste is a challenging task. Improperly monitored landfills can cause environmental hazards (Otsuka et al., 2012; Rajesh et al., 2022; Mondal et al., 2023; Awasthi et al., 2022; Abalansa et al., 2021).

Another major hurdle associated with e-waste processing is its ever increasing quantity. The annual global e-waste generation amount is about 20–25 million tons, mostly produced by European countries. The annual e-waste generation for the USA, is the largest (~3 million tons) followed by China (~2.3 million tons) and Brazil (de Oliveira et al., 2012). As the EDCs are associated with these e-wastes, so it is essential to trace and analyze these chemicals in the environmental sample. For example, several analytical methods have been developed to determine the concentration of the EDCs (in the level of ng/L) in the contaminated water (Kasonga et al., 2021). Although documentation regarding the toxic effect of several individual substances are from the e-waste source available but little is known about the toxic profile combinations are of substances (Saha et al., 2021; Tsydenova and Bengtsson, 2011; Igharo et al., 2018). The adverse health effect due to e-waste exposure may include a range of activity such as alteration in the function of lung, thyroid dysfunction, higher/lower amount of hormone secretion, abnormal birth, cytotoxicity, and genotoxicity, carcinogenic effects, dysfunction in the nervous system and reproductive development. For

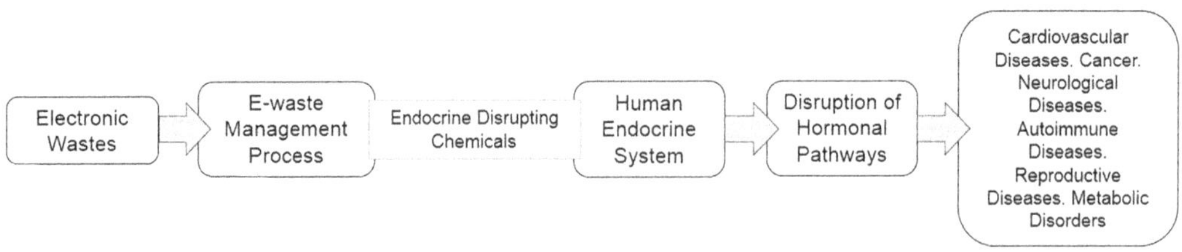

FIGURE 15.1 Harmful interference of the endocrine-disrupting chemicals on the endocrine system.

TABLE 15.1
Harmful Effects of Some Important EDCs from E-waste and Their Effect

Compound	Major Source of E-waste	Endocrine Disruption Effect	References
PCB (polychlorinated biphenyls)	Condensers and Transformers	Negative impact on Thyroid and Parathyroid Glands.	(Cao et al., 2011)
PVC (polyvinyl chloride)	Cable insulation and computer housing	Estrogenic effector.	(Verma and Prakash, 2020)
Lead (PB)	CRT monitors, dead batteries, panels of the circuit boards	Altered activity of uterine estrogen receptors also affects the ovarian steroid and gonadotropin secretion.	(Yoshida et al., 2016)
Cadmium (Cd)	Semiconductors, rechargeable batteries, switches, coating as fluorescent layer in the computer monitor	Inhibition of progesterone synthesis also having estrogenic effect.	(Awuchi et al., 2020)
Mercury (Hg)	Electronic switches and relays, circuit boards, used alkaline batteries	Interferes with thyroid metabolism	(Uddin et al., 2021)
Arsenic (As)	Light-emitting diode (LED)	Estrogenic effect Inhibition of spermatogenesis.	(Tian et al., 2022)
Lithium (Li)	Batteries	Interfere with thyroid gland functioning.	(Krysiak, and Okopien, 2014)
Nickel (Ni)	CRT	Affects hypothalamus and pituitary gland.	(Yang, and Ma, 2021)
Hexavalemnt chromium (Cr VI)	Coating (anti-corrosive) on steel plates, audio and video tapes, floppy disks.	Reproductive effects in women.	(Wuri et al., 2022)

example, the exposure to PBDEs from e-waste interfere with the thyroid hormone functioning system causes disorders. Due to the heterogeneous nature of these substances makes it difficult to study the health effect causes by the e-waste exposure. The potential health impact of these EDCs has been documented by several researchers (Verma and Prakash, 2020; Monika and Kishore, 2010; Gravel et al., 2020).

The mechanism of endocrine-disrupting activity has been explored for many of the substances (Table 15.1). For example, the EDC activity of Bisphenol A is due to its structural similarity to thyroid hormones thyroxine and triiodothyronine. The plasticizers in the PVC tubes have shown enhanced binding affinities for the human estrogen receptor alpha, causing endocrine disruption. Additionally, the effects of heavy metals are directly linked to the altered activity of hormone receptors. The experiment has shown that the exposure of Cadmium (Cd) mimics the estrogen hormone action and causes an abnormal expression of the estrogen receptor (ER) progesterone receptor (PR) gene. Hg interferes with thyroid metabolism by interacting negatively with type I iodothyronine deiodinase enzyme, which plays a vital role in the activation of the thyroid T4 to T3 form. The endocrine disruption effect of Arsenic revealed that the presence of sodium arsenite decreases the synthesis of key enzymes like delta 5, 3 beta-HSD, 17 beta-HSD, and sorbitol dehydrogenase required for androgen production. Lithium was found to inhibit iodine uptake and change thyroglobulin structure by interfering with iodotyrosine residues leading to inhibition of the release of thyroid hormone. Nickel also shows endocrine disruption activity by damaging the targeted endocrine organs by inducing apoptosis. Cr (VI) alters the cytoskeleton dynamics in oocytes leading to defective oocyte production.

15.3 EFFECTIVE MANAGEMENT STRATEGIES FOR E-WASTE-ASSOCIATED EDCs

Several approaches have been suggested by researchers to minimize the release of e-waste into the ecological system (Agarwal et al., 2023). The various approaches are the reuse of e-waste, recycling (~20%), landfilling (~80%), and chemical and biological leaching processes (Needhidasan et al., 2014). Since the increase in the amount of e-waste causes a great concern, hence further improvement in research and development is necessarily emphasizing the proper treatment methods leading to reduce the concentration EDCs generated from the e-waste materials. Some of the specific aspects are described below:

- Much research is required to access the adverse health impacts of e-waste workers due to the generation of specific EDCs. Hence, this will lead to establishing the links between occupational exposure to hazardous e-wastes and the establishment of disease related to it.
- Through study is required to understand the relationship between biological effects and the secondary chemical compounds that are generated as a combination of EDCs from the e-wastes during their processing.
- Development of a potentially sensitive instrument to measure the accurate concentration of the EDCs generated from the e-waste.

15.4 CONCLUSION

Electronic waste (e-waste) contains several hazardous chemical substances, metals, and semiconductors. Many of these compounds have endocrine disruption activity, hence called Endocrine disrupting Chemicals (EDCs). Research suggests that the contamination due to e-waste is causing severe environmental degradation and also affected living organisms drastically. Further, research on the effect of EDCs from e-waste exposure on human beings is desirable at local, regional and international levels. The results from the research can be used to propose the new e-waste recycling policy and help the society at large. Hence, the need for proper e-waste management is important to protect the health of individuals and environment.

REFERENCES

Abalansa, S., El Mahrad, B., Icely, J., & Newton, A. (2021). Electronic waste, an environmental problem exported to developing countries: The GOOD, the BAD and the UGLY. *Sustainability*, *13*(9), 5302.

Agarwal, S., Darbar, S., Saha, S., Choudhury, M., & Singh, R. P. (2023). E-waste management using different cost-effective, eco-friendly biological techniques: an overview In Waste Management and Resource Recycling in the Developing World, 205–235. https://doi.org/10.1016/B978-0-323-90463-6.00026-9

Awasthi, A. K., Awasthi, M. K., Mishra, S., Sarsaiya, S., & Pandey, A. K. (2022). Evaluation of E-waste materials linked potential consequences to environment in India. *Environmental Technology & Innovation*, *28*, 102477.

Awuchi, C. G., Hannington, T., Awuchi, C. G., Igwe, V. S., & Amagwula, I. O. (2020). Industrial waste management, treatment, and health issues: wastewater, solid, and electronic wastes. *European Academic Research*, *8*(2), 1081–1119.

Cao, J., Guo, L. H., Wan, B., & Wei, Y. (2011). In vitro fluorescence displacement investigation of thyroxine transport disruption by bisphenol A. *Journal of Environmental Sciences*, 23(2), 315–321.

de Oliveira, C. R., Bernardes, A. M., & Gerbase, A. E. (2012). Collection and recycling of electronic scrap: A worldwide overview and comparison with the Brazilian situation. *Waste Management*, 32(8), 1592–1610.

Francis, C. E., Allee, L., Nguyen, H., Grindstaff, R. D., Miller, C. N., & Rayalam, S. (2021). Endocrine disrupting chemicals: Friend or foe to brown and beige adipose tissue? *Toxicology*, 463, 152972.

Ghosh, A., Tripathy, A., & Ghosh, D. (2022, March). Impact of endocrine disrupting chemicals (EDCs) on reproductive health of human. In *Proceedings of the Zoological Society* (Vol. 75, No. 1, pp. 16–30). New Delhi: Springer India.

Gravel, S., Lavoué, J., Bakhiyi, B., Lavoie, J., Roberge, B., Patry, L., Bouchard, M. F., Verner, M. A., Zayed, J., & Labrèche, F. (2020). Multi-exposures to suspected endocrine disruptors in electronic waste recycling workers: Associations with thyroid and reproductive hormones. *International Journal of Hygiene and Environmental Health*, 225, 113445.

Gupta, S., Modi, G., Saini, R., & Agarwala, V. (2014). A review on various electronic waste recycling techniques and hazards due to its improper handling. *International Refereed Journal of Engineering and Science (IRJES)*, 3(5), 05–17.

Igharo, O. G., Anetor, J. I., Osibanjo, O., Osadolor, H. B., Odazie, E. C., & Uche, Z. C. (2018). Endocrine disrupting metals lead to alteration in the gonadal hormone levels in Nigerian e-waste workers. *Universa Medicina*, 37(1), 65–74.

Jinhui, Li, Huabo, Duan, & Pixing, Shi. (2011). Heavy metal contamination of surface soil in electronic waste dismantling area: Site investigation and source-apportionment analysis. *Waste Management and Research*, 29(7), 727–738.

Kasonga, T. K., Coetzee, M. A. A., Kamika, I., Ngole-Jeme, V. M., & Benteke Momba, M. N. B. (2021). Endocrine-disruptive chemicals as contaminants of emerging concern in wastewater and surface water: A review. *Journal of Environmental Management*, 277, 111485.

Krysiak, R., & Okopien, B. (2014). Potentiation of endocrine adverse effects of lithium by enalapril and verapamil. *West Indian Medical Journal*, 63(7), 803–806.

Ma, S., Lin, M., Tang, J., Liu, R., Yang, Y., Yu, Y., ... & An, T. (2022). Occurrence and fate of polycyclic aromatic hydrocarbons from electronic waste dismantling activities: A critical review from environmental pollution to human health. *Journal of Hazardous Materials*, *424*, 127683.

Martyniuk, C. J., Martínez, R., Navarro-Martín, L., Kamstra, J. H., Schwendt, A., Reynaud, S., & Chalifour, L. (2022). Emerging concepts and opportunities for endocrine disruptor screening of the non-EATS modalities. *Environmental Research*, *204*, 111904.

Mondal, T., Choudhury, M., Kundu, D., Dutta, D., & Samanta, P. (2023). Landfill: An eclectic review on structure, reactions and remediation approach. *Waste Management*, 164, 127–142.

Monika, J., & Kishore, J. (2010). E-waste management: As a challenge to public health in India. *Indian Journal of Community Medicine: Official Publication of Indian Association of Preventive and Social Medicine*, 35(3), 382–385.

Needhidasan, S., Samuel, M., & Chidambaram, R. (2014). Electronic waste–an emerging threat to the environment of urban India. *Journal of Environmental Health Science and Engineering*, 12(1), 1.

Ohayi, R. S., Chime, O. H., & Ndu, I. K. (2022). Endocrine-disrupting chemicals and their adverse health effects: A review of current knowledge and the Nigerian situation. *Medical Science and Discovery*, 9(5), 267–271.

Otsuka, M., Itai, T., Asante, K. A., Muto, M., & Tanabe, S. (2012). Trace element contamination around the e-waste recycling site at Agbogbloshie, Accra City, Ghana. *Interdisciplinary Studies on Environmental Chemistry*, 6(6), 161–167.

Rajesh, R., Kanakadhurga, D., & Prabaharan, N. (2022). Electronic waste: A critical assessment on the unimaginable growing pollutant, legislations and environmental impacts. *Environmental Challenges*, 7, 100507.

Saha, L., Kumar, V., Tiwari, J., Rawat, S., Singh, J., & Bauddh, K. (2021). Electronic waste and their leachates impact on human health and environment: Global ecological threat and management. *Environmental Technology & Innovation*, 24, 102049.

Satpathy, R. (2020). Application of molecular docking methods on endocrine disrupting chemicals: A review. *Journal of Applied Biotechnology Reports*, 7(2), 74–80.

Tanner, E. M., Hallerbäck, M. U., Wikström, S., Lindh, C., Kiviranta, H., Gennings, C., & Bornehag, C. G. (2020). Early prenatal exposure to suspected endocrine disruptor mixtures is associated with lower IQ at age seven. *Environment International*, 134, 105185.

Tian, X., Xie, J., Hu, L., Xiao, H., & Liu, Y. (2022). Waste LEDs in China: Generation estimation and potential recycling benefits. *Resources, Conservation and Recycling*, 187, 106640.

Tsydenova, O., & Bengtsson, M. (2011). Chemical hazards associated with treatment of waste electrical and electronic equipment. *Waste Management*, 31(1), 45–58.

Uddin, M. N., Arifa, K., & Asmatulu, E. (2021). Methodologies of e-waste recycling and its major impacts on human health and the environment. *International Journal of Environment and Waste Management*, 27(2), 159–182.

Varticovski, L., Stavreva, D. A., McGowan, A., Raziuddin, R., & Hager, G. L. (2022). Endocrine disruptors of sex hormone activities. *Molecular and Cellular Endocrinology*, 539, 111415.

Verma, A. K., & Prakash, S. (2020). E-wastes and their impact on environment and public health. *International Journal of Applied Research*, 6(9), 164–168.

Wuri, L., Arosh, J. A., Wu, J. Z., & Banu, S. K. (2022). Exposure to hexavalent chromium causes infertility by disrupting cytoskeletal machinery and mitochondrial function of the metaphase II oocytes in superovulated rats. *Toxicology Reports*, 9, 219–229.

Yang, J., & Ma, Z. (2021). Research progress on the effects of nickel on hormone secretion in the endocrine axis and on target organs. *Ecotoxic*, 213, 1–10.

Yoshida, A., Terazono, A., Ballesteros Jr, F. C., Nguyen, D. Q., Sukandar, S., Kojima, M., & Sakata, S. (2016). E-waste recycling processes in Indonesia, the Philippines, and Vietnam: A case study of cathode ray tube TVs and monitors. *Resources, Conservation and Recycling*, 106, 48–58.

16 Electronic Wastes and Impact on Nervous System

Shivani Narwal and Rajesh Dhankhar
Maharshi Dayanand University Rohtak, Rohtak, India

Mansi Chawla
The Energy and Resources Institute (TERI), New Delhi, India

16.1 INTRODUCTION

Technological advances led to the continuous upgrading of electronic equipment, resulting in the generation of electronic waste (E-waste), whose components are primary contaminants. E-waste encompasses the end-of-life of different electronic equipment, such as television, computers, printers, mobile phones, washing machines, refrigerators, telecommunication, and medical equipment (Grant et al., 2013). E-waste developed into a briskly spreading waste stream and accumulated in the surroundings, significantly impacting the environment. As per the UNU report named "Global E-waste Monitor", around 53.6 Mt of e-waste was produced worldwide in 2019, and it was approximately 44.75 Mt in 2016 (Balde et al., 2017). The e-waste generation rate depends on time, population, socio-economic status, and on lifestyle choices. Asia comes on number one in the list with 24.9 Mt, America follows it at number two with 13.1 Mt, and approximate E-waste generation per capita in the year 2019 was 7.3 Kg, an average America with 13.1 Kg, Africa with 2.9 Kg and Asia with 5.6 Kg (Forti et al., 2020). The composition of E-waste has discrete metals and non-metals, which makes E-waste exclusive and diversified. In general, E-waste is composed of 30% of organics (glass fibres, flame retardants), 40% inorganics (ferrous and non-ferrous metals) and 30% of ceramics (mica and silica). The inorganic part of E-waste contains noble metals, base metals, rare earth metals and heavy metals (Kaya et al., 2016). The existence of hazardous substances such as mercury, cadmium, lead, hexavalent chromium and brominated flame retardants, these materials inflate the complexity of the waste treatment process (recovery). Thus the disposal of this lethal waste is a dominating objection in its management. Electronic equipment subsists a considerable quantity of chemical compounds and elements, even a mobile phone relatively owes more than 40 elements from the periodic table. Different metals found in E-waste are known to induce neurotoxicity. Neuroscience is conjointly acknowledged as neural science, the study of nervous system development, how it is structured, and what functions it does. Neuroscience focuses on the brain and its attendant effects on an individual's cognitive functions and behavior. Neuroscience is an emerging area consisting of numerous multifarious investigations that pursue the perceptive understanding of the relationship between the brain and body (Agarwal et al., 2023). In the early and initial stages of life, the central nervous system (CNS) is susceptible to neurotoxic substances. Brain development occurs during different phases of childhood and foetal stages at vacillating rates encompassing pruning, migration, cell division and synaptogenesis (Caito and Aschner, 2017). Neurogenesis is an absolutely composed process that starts at a specific time and betides in particular structures, all the while neurodevelopment. Maintaining the right timing and control of neurogenesis is essential for functions and development of the nervous system, as interruptions can proceed to neuro diseases and altered functions. Exposure to numerous chemicals during this susceptible neurogenesis period can rattle the development of the brain and endures continuing functional deterioration (Kim et al., 2016).

DOI: 10.1201/9781003582311-16

TABLE 16.1
Potential Affected Functions and Exposure Routes of Neurotoxicants in E-waste

Neurotoxicants	Possible Neuropsychological Function	Exposure Routes
Hg	Cognition and motor functions, catastrophically damages brain and Central Nervous System	Vapor form, Air, Seafood
Cr	Motor functions	Dust, Air, Water
Pb	Delinquent and tardy behavior, memory and academic achievements, Cognition and motor skills intellectual impairment, prolonged neurotoxicity	Lead paint, Dust, Air, Water, Soil
As	Cognition, Neurological Disorders	Water, Dust, Air

16.2 HARMFUL SUBSTANCES IN E-WASTE

E-waste includes perilous metals (Pb, Cd, As, Hg) and dangerous chemicals (brominated flame retardants and polycyclic aromatic hydrocarbons) (Lecler et al., 2018). Scrapped E-waste composition will progressively differ owing to the entrance of short life period electronic equipment such as smartphones. Illustratively a mobile phone is enclosed with 500–1000 components and much more pernicious heavy metals (Song and Li, 2014). A typical cathode ray tube (CRT) used in televisions consists of 2–3 kg of Pb (Chen et al., 2011), although CRTs are being replaced by much light flat screens which contain a lesser amount of Pb (Bigum et al., 2017) a vast quantity of Pb is approaching the waste streams due to previously used CRTs. According to estimates that in 2014 about 2.2 million tonnes was found to be Pb glass in the total volume of E-waste (41.8 million tonnes). Monitors, batteries and lamps have Liquid crystal displays (LCD) which can be exceedingly dangerous if mobilized. A major problem in fluorescent lamp recycling is the exposure to mercury and relative occupational hazards. Although several metallic contaminants are not toxic in their metal form but become dangerous only when they are ingested above particular concentrations and amounts in gaseous, dust and liquid form. During waste recycling, hazardous substances might be formed/released continuously into the environment. Several investigators present and discuss various toxicological studies to assess the likely possible hazardous effects (Tsydenova and Bengtsson, 2011). Exposure routes significantly depend on the recycling process and the substance being recycled; humans are primarily exposed via ingestion, dermal contact and inhalation (Grant et al., 2013). Additionally, exposure to pollutants such as Pb, Cd and polychlorinated biphenyls occurs via indirect trails through substantial contact with contaminated objects such as dust, soil, air, water and food (Table 16.1).

16.3 NEUROTOXICANTS EXPOSURE DURING RECYCLING AND SMELTING

Children who live in the surroundings where mining, E-waste recycling, smelting and scrapping occur are on the verge of exposure to menacing neurotoxic metals and related pollutants. Families residing nearby to mineral deposits will be primarily exposed to higher levels of As, Pb, Hg and Cd (Castro et al., 2014) (Table 16.2). Owing to the concurrent co-occurrence of neurotoxicants in contaminated sites, it is essential to discern the biomarkers representing exposure and an early life sensitivity to neurotoxic effects (Dórea, 2021). The metallic form of Hg is utilized to amalgamate gold during extraction procedures but infects the space used during mining and processing (Gracia et al., 2015). Pb is the most perilous material in gold mining, and it becomes bioavailable, thus more toxic than arsenic and cadmium. Nervous system vulnerability to neurotoxicants is extensively acknowledged during foetal and postnatal stages (Grandjean et al., 1994). Certainly, maternal transmission to offspring continues due to breastfeeding; in this situation, further exposure to

TABLE 16.2
Elevated Concentration of Toxicants in Human Blood Owing to Informal Recycling of Electronic Wastes

Toxicant	Site	Concentration (ug/L) (in blood)	References
Arsenic	E-waste dismantling industry	15.02	(Li et al., 2020)
Cadmium	Vulnerable exposure site	21	(Chen et al., 2019)
Lead	E-waste dismantling industry	45.3	(Li et al., 2020)
Mercury	Recycling site	1.8	(Srigboh et al., 2016)
Zinc	Recycling site	5574	(Srigboh et al., 2016)
Nickel	Dismantling industry	4.68	(Li et al., 2020)
Copper	Recycling site	1063	(Srigboh et al., 2016)

neurotoxicants expands the danger to still-sensitive nervous systems. In poor developing countries, recycling of E-waste is generally done informally, and recovery of materials remains a home-based activity using indigenous methods. The procedure of E-waste recycling and recovery requires washing with hazardous solvents, acid leaching, burning and dismantling (Tansel, 2017; Orisakwe et al., 2019) described recycling and recovery practices in Africa; proclaimed reports showed the presence of neurotoxic metals (As, Cd, Cr, Hg, Li, Pb, Al) associated with recycling activities. A few primary neurotoxicants and their effects on the nervous system are briefly discussed in the next section.

16.3.1 Arsenic (As)

Arsenic contamination can be depicted in E-waste; it is a metalloid produced by the erosion of rocks and through numerous agricultural and industrial activities. It is understandable as a toxicant and elicits its relative toxicity through disparate mechanisms. It impinges different organs and tissues present in the body through its menacing activities. It induces and persuades cognitive dysfunctions in the central nervous system (Garza-Lombo et al., 2019). Additionally, it is associated with numerous neurodevelopmental changes and neurological disorders and has been investigated as a perilous factor for various neurodegenerative diseases. In behavioral and cognitive functions testing, results of demonstrated experiments displayed that arsenic has functional accoutrements on CNS (Tolins et al., 2014). In a reported study, the spatial learning ability of weanling rats was examined after exposing them to arsenic using several tests, such as visible and hidden platform acquisition trials. Terminal results exhibited impairment and wreckage in spatial memory deceived by impoverished performance on hidden platform tests. (Luo et al., 2009). Transgenerational effects of meagre doses of arsenic on the nervous system have been studied, and established results showed the transmittal of cognitive and motor deficits through ancestral exposure (Valles et al., 2020). Additionally, exposure to arsenic reduces the intelligence quotient (IQ). In exposure to arsenic, children of the US (Wasserman et al., 2014), Bangladesh (Parvez et al., 2011), and Mexico (Calderon et al., 2001) displayed modifications in verbal abilities and alteration in cognitive functions, motor skills, IQ and prolonged memory. However, some studies failed to reveal that arsenic solitarily can cause neurodegeneration, but neurotoxicity of arsenic can correlate and coactions with molecular processes related to neurodegeneration (Garza lombo et al., 2019). Arsenic induces lethal effects on neurotransmitters that work during cell signaling in the brain. Arsenic appears to induce an increment in serotonin and dopamine levels at specific regions and, on the contrary, cutbacks the norepinephrine levels in particular brain regions (Mejia et al., 1997). Arsenic and its related methylated metabolites trigger apoptosis in neuroblastoma and neurons, and calcium (Ca^{2+}) can provoke this induced apoptosis in the cerebral cortex (Yen et al., 2011).

16.3.2 Lead (Pb)

Pb is among the primary pollutant exhibited by E-waste; it is released while burning, dumping and roasting electronic equipment and adversely impacts the environment. Children are much more sensitive to its exposure due to several differences in demeanor and physiology. Pb is a potential neurotoxicant that causes fatal diseases and disorders and modifies morphological features of the brain, which in turn affects cognitive functions, cell signaling and neurotransmission (Verstraeten et al., 2008). Pb induces neurotoxicity by disturbing the calcium flux; it outcompetes calcium and mimics calcium cation, and associates itself with the calcium transport mechanisms in the nervous system. Accordingly, exposure to Pb triggers calcium homeostasis alterations and further disturbs neurotransmission (Westerink and Vijverberg, 2002). It inhibits cholinesterase activity; Cholinesterase catabolizes acetylcholine and plays a peculiar role in neurotransmission. In severe cases, it can lead to death, and this inhibition is advised as a fair criterion for the assessment of neurotoxicity (Nunes et al., 2014). There is credible documentation that exposure to Pb has a detrimental effect on the nervous system, and environmental aspects amplify Pb susceptibility; exposure in initial phases can induce neurodegeneration in the later stages. Pb reduces N-acetylaspartate-to-creatine and phosphocreatine ratios in the frontal gray matter of relatively exposed children revealed in Magnetic resonance spectroscopy (MRS) (Bellinger 2011). Pb-related deficits have been well elaborated in several fields, including performance IQ, verbal ability, intelligence quotient, gross and fine motor skills and language capability. Determined meta-analyses indicate that child's IQ score decreases as blood Pb level increases by ten ug/dl (Bellinger, 2011).

16.3.3 Cadmium (Cd)

According to Agency for Toxic Substance and Disease Registry (ATDSR), cadmium ranks seven in extremely toxic heavy metals. Cd got enters in the human body via various mechanisms, particles of Cd are carried out along olfactory neurons and it gets accumulates there in the olfactory bulb. Cd penetrates into neurons through voltage-gated calcium channels. Several in vitro and in vivo investigations displayed that Cd exposure substantially affects CNS and PNS (Miura et al., 2013) with a broad range of ailments such as peripheral neuropathy, mental adjournment, learning infirmity, behavioral and activity impairments in adults as well as in children (Wang and Du., 2013). Furthermore, Cd neurotoxicity is associated with neurodegenerative diseases, for example, Alzheimer's (AD) and Parkinson's diseases (PD), along with amyotrophic lateral sclerosis and multiple sclerosis (Sheykhansari et al., 2018).

16.3.4 Mercury (Hg)

Prolonged exposure to Hg outcomes well defined clinical term known as "mercurialism". Ailments in mercurialism comprise behavioral changes, memory loss, tremors, insomnia and irritability (Magos, 1997). Studies on the person exposed to Hg exhibit symptoms that might last for numerous years due to retention in the brain (Steckling et al., 2014). Hg causes neurotoxicity by disrupting glutamate transport, which is the primary excitatory transmitter (fonfria et al., 2005). It alters glutamate-mediated signals, enhancing release from the pre-synaptic terminal by blocking their uptake by astrocytes and creating neuronal dysfunction (Yuan and Atchison, 2003). Analysis of the methylmercury poisoning timeline showed significant confirmation on the array of maladies associated relatively to methylmercury neurotoxicity (Clarkson and Magos, 2006). Catastrophic numbness and coordination loss are the early symptoms reflecting cerebellar activity and function deterioration. With time when exposure advances, sensory and mobility capability gets extremely afflicted, speech gets slurred and death may befall (Weiss et al., 2002). Correspondingly several studies with native people of Amazon displayed weakened performance in neurobehavioral assessment by adults as well as by children with extremely higher hair mercury (Yokoo et al., 2003). Exposure to Hg alters social behavior, and changes the neurochemistry and histology of the brain (Olczak et al., 2009).

16.4 MANAGEMENT AND FUTURE PROSPECTS

The existence of hazardous substances in E-waste escalates its complexity and causes adversities in E-waste treatment. Thus, the primary objection is the disposal and management of this toxic waste, and with the arrival of smartphones and other electronic equipment, the concentration of these lethal elements is continuously increasing with time. Because of the precarious nature of E-waste, its resource recovery and recycling are compelling issues (Mmereki et al., 2016). Moreover, the percentage of E-waste collected formally is only 17.4%, while the rest of the waste is traded illegally in markets or dumped in landfills (Forti et al., 2020; Mondal et al., 2023). To minimize waste generation, numerous technological interventions should be made for effective production. Highly updated instruments and equipment use raw materials proficiently and generate a low amount of waste. Even though resource recovery might be a good source for livelihood and business ventures, its management can deteriorate due to the inadequacy of infrastructure and lack of competent scientific methodologies (Lakshmi & Raj, 2017). To minimize the lethal effects of these menacing metals, several technological interventions have been developed and adopted, such as:

1. Contemporarily replacing CRTs with LCD screens to phase out Pb.
2. Usage of rechargeable batteries typically reduces nickel (Ni) and cadmium (Cd) but increases Lithium (Li).
3. Introducing "halogen-free" equipment to eliminate PCBs and dioxins.
4. Legislative rules and restraints.

The omnipresence of E-waste and its unusual composition creates a harmful situation in the surroundings. Pollutants and neurotoxicants associated with E-waste contaminate the environment and deteriorate the public health of those who live near E-waste facilities. Reclaiming precious metals from e-waste is becoming a booming business in developing countries, but associated neurotoxicity comes with stringent hidden costs for the people involved. One of the primary ways to overcome these toxic effects is to manage e-waste. While significant countries put effort into managing solid waste, E-waste is somehow ignored. One reason is that e-waste management facilities require high capital costs, but only some countries have allocated a budget for these e-waste recycling units. Mainly, e-waste is managed informally and in unauthorized ways; the extensive population is still unaware of the rules and relevant regulations associated with e-waste due to the dearth of enforcement of current legislation. A coupled system of E-waste management and sustainable development might play a critical role in condensing death and health problems due to E-waste's toxic and hazardous composition. Transformation of informal stakeholders into micro-entrepreneurs can promote well-being, economic growth, and financial opportunities, and viable guidelines to manage neurotoxicants in E-waste can be a sustainable roadmap to make the management system resilient.

REFERENCES

Agarwal, S., Darbar, S., Saha, S., Choudhury, M., & Singh, R. P. (2023). E-waste management using different cost-effective, eco-friendly biological techniques: an overview. *Waste Management and Resource Recycling in the Developing World*, 205–235. https://doi.org/10.1016/B978-0-323-90463-6.00026-9

Baldé, C. P., Forti, V., Gray, V., Kuehr, R., & Stegmann, P. (2017). *The global e-waste monitor 2017: Quantities, flows and resources*. United Nations University, International Telecommunication Union, and International Solid Waste Association.

Bellinger, D. C. (2011). The protean toxicities of lead: new chapters in a familiar story. *International Journal of Environmental Research and Public Health*, *8*(7), 2593–2628.

Bigum, M., Damgaard, A., Scheutz, C., & Christensen, T. H. (2017). Environmental impacts and resource losses of incinerating misplaced household special wastes (WEEE, batteries, ink cartridges and cables). *Resources, Conservation and Recycling*, *122*, 251–260.

Caito, S., & Aschner, M. (2017). Developmental neurotoxicity of lead. *Neurotoxicity of Metals*, 18, Springer Cham, 3–12, https://doi.org/10.1007/978-3-319-60189-2.

Calderon, J., Navarro, M. E., Jimenez-Capdeville, M. E., Santos-Diaz, M. A., Golden, A., Rodriguez-Leyva, I., … & Dıaz-Barriga, F. (2001). Exposure to arsenic and lead and neuropsychological development in Mexican children. *Environmental Research, 85*(2), 69–76.
Castro, F., Harari, F., Llanos, M., Vahter, M., & Ronco, A. M. (2014). Maternal–Child Transfer of Essential and Toxic Elements through Breast Milk in a Mine-Waste Polluted Area. *American Journal of Perinatology, 31*(11), 993–1002.
Chen, M., Zhang, F. S., & Zhu, J. (2011, February). Leaching toxicity of Pb and Ba containing in cathode ray tube glasses by SEP–TCLP. In *Recycling of Electronic Waste II: Proceedings of the Second Symposium Wiley Online Library*.
Chen, Y., Xu, X., Zeng, Z., Lin, X., Qin, Q., & Huo, X. (2019). Blood lead and cadmium levels associated with hematological and hepatic functions in patients from an e-waste-polluted area. *Chemosphere, 220*, 531–538.
Clarkson, T. W., & Magos, L. (2006). The toxicology of mercury and its chemical compounds. *Critical Reviews in Toxicology, 36*(8), 609–662.
Dórea, J. G. (2021). Neurodevelopment in mining environments entails different types of exposure and non-essential element interactions: Broadening the significance of the Nyanza et al study in Tanzania. *Environment International, 149*, 106407.
Fonfría, E., Vilaró, M. T., Babot, Z., Rodríguez-Farré, E., & Sunol, C. (2005). Mercury compounds disrupt neuronal glutamate transport in cultured mouse cerebellar granule cells. *Journal of Neuroscience Research, 79*(4), 545–553.
Forti, V., Balde, C. P., Kuehr, R., & Bel, G. (2020). *The Global E-waste Monitor 2020: Quantities, flows and the circular economy potential*, UNU Collections. https://collections.unu.edu/view/UNU:7737#viewMetadata
García, O., Veiga, M. M., Cordy, P., Suescún, O. E., Molina, J. M., & Roeser, M. (2015). Artisanal gold mining in Antioquia, Colombia: a successful case of mercury reduction. *Journal of Cleaner Production, 90*, 244–252.
Garza-Lombó, C., Pappa, A., Panayiotidis, M. I., Gonsebatt, M. E., & Franco, R. (2019). Arsenic-induced neurotoxicity: a mechanistic appraisal. *JBIC Journal of Biological Inorganic Chemistry, 24*, 1305–1316.
Grandjean, P., Weihe, P., & Nielsen, J. B. (1994). Methylmercury: significance of intrauterine and postnatal exposures. *Clinical Chemistry, 40*(7), 1395–1400.
Grant, K., Goldizen, F. C., Sly, P. D., Brune, M. N., Neira, M., van den Berg, M., & Norman, R. E. (2013). Health consequences of exposure to e-waste: a systematic review. *The Lancet Global Health, 1*(6), e350–e361.
Kaya, M. (2016). Recovery of metals from electronic waste by physical and chemical recycling processes. *International Journal of Chemical and Molecular Engineering, 10*(2), 259–270.
Kim, H. Y., Wegner, S. H., Van Ness, K. P., Park, J. J., Pacheco, S. E., Workman, T., … & Faustman, E. M. (2016). Differential epigenetic effects of chlorpyrifos and arsenic in proliferating and differentiating human neural progenitor cells. *Reproductive Toxicology, 65*, 212–223.
Lakshmi, S., & Raj, A. (2017). A Review study of E-waste management in India. *Asian Journal of Applied Science and Technology (AJAST) 1*, 33–36.
Lecler, M. T., Zimmermann, F., Silvente, E., Masson, A., Morèle, Y., Remy, A., & Chollot, A. (2018). Improving the work environment in the fluorescent lamp recycling sector by optimizing mercury elimination. *Waste Management, 76*, 250–260.
Li, Z., Guo, C., Li, X., Wang, Z., Wu, J., Qian, Y., & Wei, Y. (2020). Associations between metal exposure and global DNA methylation in potentially affected people in E-Waste recycling sites in Taizhou City, China. *Science of the Total Environment, 711*, 135100.
Luo, J. H., Qiu, Z. Q., Shu, W. Q., Zhang, Y. Y., Zhang, L., & Chen, J. A. (2009). Effects of arsenic exposure from drinking water on spatial memory, ultra-structures and NMDAR gene expression of hippocampus in rats. *Toxicology Letters, 184*(2), 121–125.
Magos, L. (1997). Physiology and toxicology of mercury. *Metal Ions in Biological Systems, 34*, 321–370.
Mejia, J. J., Diaz-Barriga, F., Calderon, J., Rios, C., & Jiménez-Capdeville, M. E. (1997). Effects of lead–arsenic combined exposure on central monoaminergic systems. *Neurotoxicology and Teratology, 19*(6), 489–497.
Miura, S., Takahashi, K., Imagawa, T., Uchida, K., Saito, S., Tominaga, M., & Ohta, T. (2013). Involvement of TRPA1 activation in acute pain induced by cadmium in mice. *Molecular Pain, 9*, 1744–8069.
Mmereki, D., Li, B., Baldwin, A., & Hong, L. (2016). The generation, composition, collection, treatment and disposal system, and impact of E-waste. In *InTech eBooks*, https://doi.org/10.5772/61332.

Mondal, T., Choudhury, M., Kundu, D., Dutta, D., & Samanta, P. (2023). Landfill: An eclectic review on structure, reactions and remediation approach. *Waste Management*, *164*, 127–142.

Nunes, B., Brandão, F., Sérgio, T., Rodrigues, S., Gonçalves, F., & Correia, A. T. (2014). Effects of environmentally relevant concentrations of metallic compounds on the flatfish Scophthalmus maximus: biomarkers of neurotoxicity, oxidative stress and metabolism. *Environmental Science and Pollution Research*, *21*, 7501–7511.

Olczak, M., Duszczyk, M., Mierzejewski, P., & Majewska, M. D. (2009). Neonatal administration of a vaccine preservative, thimerosal, produces lasting impairment of nociception and apparent activation of opioid system in rats. *Brain Research*, *1301*, 143–151.

Orisakwe, O. E., Frazzoli, C., Ilo, C. E., & Oritsemuelebi, B. (2019). Public health burden of e-waste in Africa. *Journal of Health and Pollution*, *9*(22), 190610. https://doi.org/10.5696/2156-9614-9.22.190610.

Parvez, F., Wasserman, G. A., Factor-Litvak, P., Liu, X., Slavkovich, V., Siddique, A. B., ... & Graziano, J. H. (2011). Arsenic exposure and motor function among children in Bangladesh. *Environmental Health Perspectives*, *119*(11), 1665–1670.

Sheykhansari, S., Kozielski, K., Bill, J., Sitti, M., Gemmati, D., Zamboni, P., & Singh, A. V. (2018). Redox metals homeostasis in multiple sclerosis and amyotrophic lateral sclerosis: a review. *Cell Death & Disease*, *9*(3), 348.

Song, Q., & Li, J. (2014). Environmental effects of heavy metals derived from the e-waste recycling activities in China: a systematic review. *Waste Management*, *34*(12), 2587–2594.

Srigboh, R. K., Basu, N., Stephens, J., Asampong, E., Perkins, M., Neitzel, R. L., & Fobil, J. (2016). Multiple elemental exposures amongst workers at the Agbogbloshie electronic waste (e-waste) site in Ghana. *Chemosphere*, *164*, 68–74.

Steckling, N., Bose-O'Reilly, S., Pinheiro, P., Plass, D., Shoko, D., Drasch, G., ... & Hornberg, C. (2014). The burden of chronic mercury intoxication in artisanal small-scale gold mining in Zimbabwe: data availability and preliminary estimates. *Environmental Health*, *13*(1), 1–20.

Tansel, B. (2017). From electronic consumer products to e-wastes: Global outlook, waste quantities, recycling challenges. *Environment International*, *98*, 35–45.

Tolins, M., Ruchirawat, M., & Landrigan, P. (2014). The developmental neurotoxicity of arsenic: cognitive and behavioral consequences of early life exposure. *Annals of Global Health*, *80*(4), 303–314.

Tsydenova, O., & Bengtsson, M. (2011). Chemical hazards associated with treatment of waste electrical and electronic equipment. *Waste Management*, *31*(1), 45–58.

Valles, S., Hernández-Sánchez, J., Dipp, V. R., Huerta-González, D., Olivares-Bañuelos, T. N., González-Fraga, J., & Bardullas, U. (2020). Exposure to low doses of inorganic arsenic induces transgenerational changes on behavioral and epigenetic markers in zebrafish (Danio rerio). *Toxicology and Applied Pharmacology*, *396*, 115002.

Verstraeten, S. V., Aimo, L., & Oteiza, P. I. (2008). Aluminium and lead: molecular mechanisms of brain toxicity. *Archives of Toxicology*, *82*, 789–802.

Wang, B., & Du, Y. (2013). Cadmium and its neurotoxic effects. *Oxidative Medicine and Cellular Longevity*, *2013*, 898034. https://doi.org/10.1155/2013/898034.

Wasserman, G. A., Liu, X., LoIacono, N. J., Kline, J., Factor-Litvak, P., van Geen, A., ... & Graziano, J. H. (2014). A cross-sectional study of well water arsenic and child IQ in Maine schoolchildren. *Environmental Health*, *13*(1), 1–10.

Weiss, B., Clarkson, T. W., & Simon, W. (2002). Silent latency periods in methylmercury poisoning and in neurodegenerative disease. *Environmental Health Perspectives*, *110*(suppl 5), 851–854.

Westerink, R. H., & Vijverberg, H. P. (2002). Ca^{2+}-independent vesicular catecholamine release in PC12 cells by nanomolar concentrations of Pb^{2+}. *Journal of Neurochemistry*, *80*(5), 861–873.

Yen, C. C., Ho, T. J., Wu, C. C., Chang, C. F., Su, C. C., Chen, Y. W., ... & Huang, C. F. (2011). Inorganic arsenic causes cell apoptosis in mouse cerebrum through an oxidative stress-regulated signaling pathway. *Archives of Toxicology*, *85*, 565–575.

Yokoo, E. M., Valente, J. G., Grattan, L., Schmidt, S. L., Platt, I., & Silbergeld, E. K. (2003). Low level methylmercury exposure affects neuropsychological function in adults. *Environmental Health*, *2*(1), 1–11.

Yuan, Y., & Atchison, W. D. (2003). Methylmercury differentially affects GABAA receptor-mediated spontaneous IPSCs in Purkinje and Granule cells of rat cerebellar slices. *The Journal of Physiology*, *550*(1), 191–204.

17 Electronic Wastes, It's Impact on Maternal and Child Health

Sangita Agarwal
RCC Institute of Information Technology, Kolkata, India

Dibya Das
JIS University, Kolkata, India

Sudipta Saha
Health Research Board, Grattan House, Dublin, Ireland

Preeta Bose, Gourav Samajdar, and Himangshu Sekhar Maji
JIS University, Kolkata, India

Anupa Ruwali
St. Joseph's Health Centre, Toronto, Ontario, Canada

17.1 INTRODUCTION

The phrase "electronic garbage," often referred to as "e-waste," refers to discarded consumer goods with an electrical or electronic component. The United Nations Environment Programme (UNEP) estimates that the great majority of it is illegally traded or abandoned, making it the world's fastest-growing solid waste stream at the moment. Even though there are safe and efficient ways to recycle e-waste, most of it ends up in developing countries with outdated recycling practices. Recycling of e-waste in an informal way involves burning to extract metals or get rid of unwanted plastic parts, as well as physically sorting and dismantling products and recovering valuable metals in acid baths. Because a number of hazardous substances, including as lead (Pb), cadmium (Cd), hexavalent chromium (Cr [VI]), manganese (Mn), mercury (Hg), and brominated flame retardants, are present in consumer electronics and may be harmful to both human health and the environment. During informal recycling processes, burning, for instance, generates more chemical byproducts. Workers and neighboring residents may experience a range of detrimental health impacts from exposure to these substances from informal e-waste recycling, especially vulnerable populations like expectant mothers and their growing fetuses. Developed countries, particularly the United States, frequently export their electronic waste (e-waste) to India and other Asian nations. A recent study has uncovered that a significant portion of the electronics collected for recycling in the United States is ultimately sent to Asia. There, these items are either discarded improperly or recycled under conditions that often compromise worker health and safety. The primary motivations behind this export trend are the availability of cheap labor and the absence of stringent environmental and occupational regulations in these Asian countries. Consequently, the hazardous by-products of e-waste from developed nations are disproportionately impacting some of the world's poorest communities. The full extent

 DOI: 10.1201/9781003582311-17

of the resulting environmental and health issues remains largely undocumented (Verma 2020). This study's main focus is on the health risks that e-waste poses to children's and mothers' health.

17.2 MATERNAL AND CHILD HEALTH

Maternal and child health deals (MCH) with the well-being of pregnant mother and her child. When addressing MCH it is important to understand that the basis of such important agenda is equitability in all environments such as workplace, home and school as well as removing the disparities associated to health. Ascertaining health and fostering well-being of all, is one of the main objectives of United Nations sustainable development goals and to ensure this goal it is pertinent to improve the maternal and child health. Pregnant women when exposed to e-waste can have serious negative impact on the health, upbringing of the unborn child and also throughout his life.

17.3 E-WASTE EXPOSURE IN MOTHER AND CHILD

In many primitive e-waste recycling unit's children are involved in manual dismantling and burning activities related to e-waste, hence gets exposed to such toxicants. Those children who are involved in choosing and selecting of recyclable materials from waste which are mixed together are also at risk. Another route of exposure in case of children is through their parents, who are working with e-waste recycling and this brings them in contact with contaminated dust. The routes of exposure to children are through various pathways such as contamination of surrounding areas which can be soil, water and food as well drinking water. Mothers' intake is transferred over the placenta and by breast milk. Inhalation represents a significant route of exposure to e-waste contaminants for newborn infants. As their bodily systems—such as the central nervous system, immune system, reproductive system, and digestive system—are still in the developmental stage, the intake of these harmful substances can severely disrupt their growth. This exposure may result in irreversible damage to their health and development (Meem et al. 2021). Pregnant women growing up in such sites would have higher history of exposure. Pregnant women exposed to a variety of persistent organic pollutants (POPs) such as polybrominated diphenyl ethers (PBDEs), polychlorinated biphenyls (PCBs), perfluorooctanoic acid (PFOA), and polycyclic aromatic hydrocarbons (PAHs) may experience several adverse outcomes. These include an increased risk of spontaneous abortion, premature delivery, and fetal loss. Additionally, these exposures can interfere with thyroid development, leading to abnormal thyroid function. There are also potential neurobehavioral effects, genetic toxicity, low birth weight, and congenital malformations that may arise from such exposures(Meem et al. 2021).Children's e waste exposure depends upon number of factors such as duration of recycling activities, types of e-waste, location of workshop, day-to-day activities of the children, and involvement of parents in recycling processes (Figure 17.1). Many recycling sites are present for a long time causing greater damage to inhabitants living in the vicinity too (Cabrera-Rodríguez et al. 2019).

17.4 E-WASTE TOXICITY AND EMBRYONIC DEVELOPMENT

The probable harmful effects of e-waste exposure in pregnant women are low birth weight, reduced height, premature or still birth. E-waste recycling activities release lead and its contact results in high rates of attention deficit hyperactivity disorder, lowering of neonatal behavioral neurological assessment score, diminished language and cognitive functions, behavioral challenges, radical alteration in temperament of child, and difficulties in integration at sensory levels (Figure 17.2). Other determinantal health impacts are DNA damage, respiratory effects, lung function, augmented risk of developing cardiovascular disease and cancer.

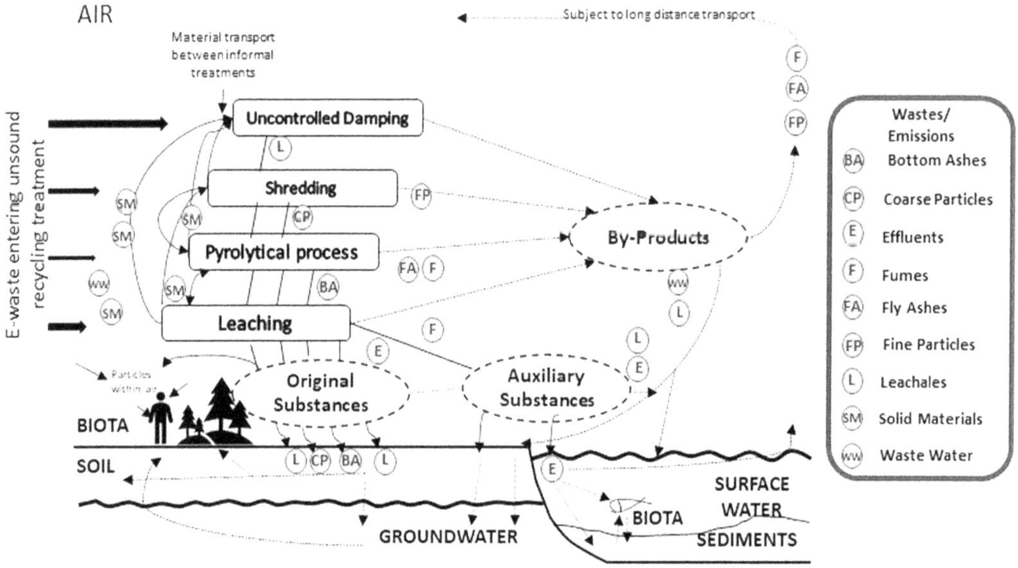

FIGURE 17.1 Pathway of Substances Released in e-waste treatment A review of the environmental fate and effects of hazardous substances released from electrical and electronic equipment during recycling: Examples from China and India 28–41, 2010, with permission from Elsevier.

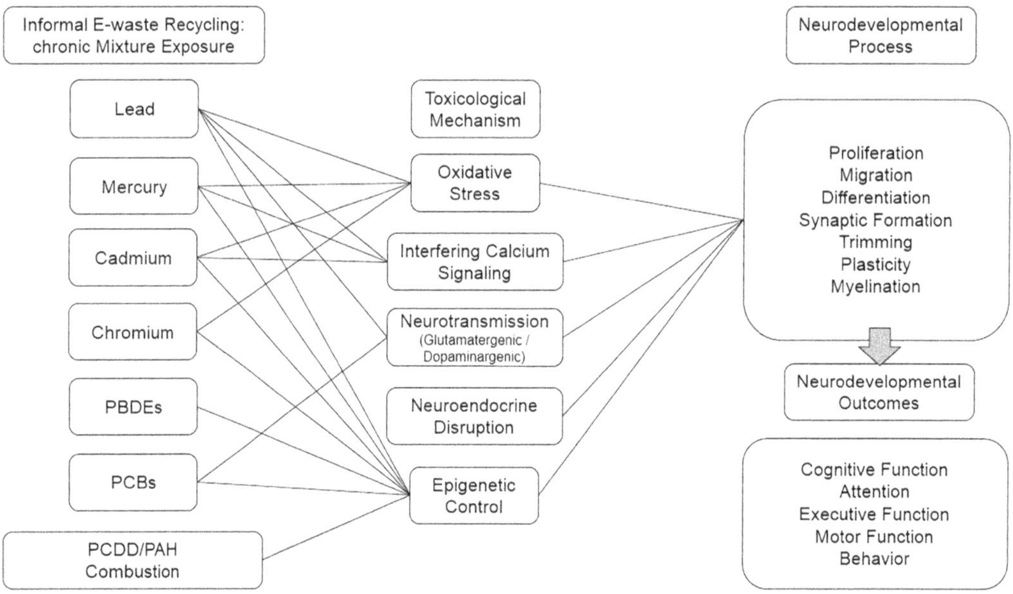

FIGURE 17.2 E-waste and Neurotoxicity (Adapted from Developmental Neurotoxicants in E-Waste: An Emerging Health Concern, 119(4): 431–438., 2010, with free to use permission from, Environmental Health Perspectives.)

17.5 E-WASTE TOXICITY, PRENATAL EXPOSURE, AND NEONATES' HEALTH

It has been shown that being exposed to e-waste during pregnancy has a deleterious impact on the pollutant load of the infant. Around eighty percent of residents in the town of Guiya, China, take part in various recycling activities, making it home to one of the most advanced electronic waste recycling facilities in the world. It has also been shown that parental involvement in the processes of e-waste recycling and the length of time the women spent in Guiyu and in e-waste recycling companies or workshops during pregnancy were linked to higher cord blood lead concentrations in newborns. This was the finding from a study that was published in the journal Environmental Health Perspectives. (Grant et al., 2013) Additionally, neonates from Guiyu had higher placental metallothionein levels owing to Cd exposure, and their parents' involvement in recycling e-waste was connected to their higher Cd levels. This was shown to be a correlation between the two (a tiny protein that indicates exposure to harmful metals). (Li et al., 2011) Mothers' high PFOA exposure in Guiyu is associated with a negative impact on their newborns' development and prepotency there. Prenatal exposure to illegal e-waste recycling may cause a number of undesirable birth outcomes, such as stillbirths, low birth weights, low Apgar scores, and others, as well as long-term consequences including behavioral and learning issues for the newborns in the future.

17.6 E-WASTES TOXICITY AND PLACENTAL RESPONSE

Most of the research whose findings were included in the literature suggested a possible connection between exposure to e-waste during pregnancy and unfavorable consequences for the health of the unborn child. These results include variations in baby development according to sex, the placenta's ability to transport toxicants, oxidative damage, thyroid hormones, DNA methylation disruption, ALAD genotypes, carcinogenic hazards, and changes in sex hormones in both pregnant women and the growing fetus. The findings (Chen et al., 2018) indicate that exposure to e-waste is linked to poorer outcomes for the newborn's and mother's health. However, further study in South Asia, Africa, and Latin America is required to support these links, in addition to advancements in the design of epidemiological research studies (Ben et al. 2014, Grant et al. 2013, Heacock et al., 2015).

17.7 E-WASTE TOXICITY, HIGH RISK PREGNANCY, AND FETAL DISORDERS

E-hazardous waste's components have been associated with a number of harmful health effects. The impact of e-waste exposure on the health of the fetus and the environment. Associated with the health of newborns, pregnancy outcomes, and e-waste. The majority of studies suggested a connection between e-waste exposure and pregnancy and harmful neonatal health outcomes, such as differences in infant sex-specific growth, toxicant transfer through the placenta, disruption of thyroid hormones, etc., cancer risks, and disruption of sex hormones (Wu et al. 2012; Xu et al. 2012). The findings are consistent with the growing body of research showing that maternal and newborn health outcomes are adversely impacted by exposure to e-waste (Figure 17.3).

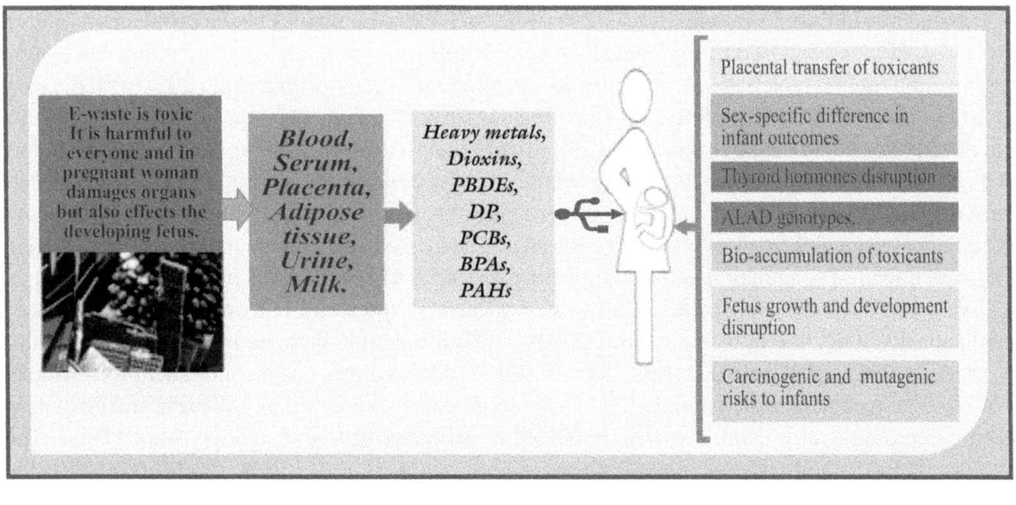

FIGURE 17.3 Effect of e-waste on the fetal and maternal health (Heacock et al., 2018).

17.8 E-WASTE TOXICITY AND BREAST FEEDING

Researchers have found that various e-waste recycling methods, particularly open burning and acid leaching, are significant sources of dioxin emissions (Luo et al., 2021). Among the various biological matrices available for monitoring dioxin levels, breast milk is regarded as the most effective indicator (Luo et al., 2021). Although polybrominated dibenzo-p-dioxins/dibenzofurans and polychlorinated are both common at unofficial e-waste recycling facilities, human exposure to the latter has not been well researched. Researchers have found evidence of dioxin-related chemical accumulation levels and profiles in breast milk samples from mothers who lived near two Vietnamese e-waste recycling facilities. (Asamoah et al., 2018) Results from screening using the Dioxin-Responsive Chemically Activateddioxin-like ed LUciferase gene eXpression assay revealed that samples from the Bui Dau (site 1) had higher dioxin-like activities than samples from the Trang Minh (site 2) and a reference site (Wu et al. 2012; Xu et al. 2012; Asamoah et al., 2019) found that women who participated in recycling acquired greater levels of PCDFs and PBDFs. Dust ingestion was estimated to contribute the majority of the intake for tetra BDF, 37.0% – 55.0% for penta-octa CDFs, but < 20% for DL-PCBs and, PCDDs and 26% for total WHO-TEQs by comparing The Dioxin-Responsive Chemical profile in these women's milk matches the profile in house dust that has been reported from the same location. (Huo et al., 2007; Bakhiyi et al., 2018) (Figure 17.3).

17.9 E-WASTE TOXICITY AND DEVELOPMENTAL COMPLICATIONS IN CHILDREN

Children are more susceptible to the negative health effects of being exposed to electronic waste for a number of reasons, including their quicker metabolism, less size, higher surface area in contrast to their weight, and varied exposure pathways (for example, take-home exposure, skin and, hand-to-mouth) (Bakhiyi et al., 2018). They were proven to represent an 8-times larger risk to their health when compared to the adult workers who recycled electronic garbage, according to the findings. According to research, children who live in a location where electronic trash is recycled have significantly higher blood lead levels and blood cadmium levels than children who live in a control zone. This is the case for both lead and cadmium (Zheng et al., 2008). Example, According to one study, Guiyu's BLL on average was over 1.5 times higher than it was in the control site.This finding was based on comparisons between the two locations. While the CDC has decided that a maximum of 5 ug/dL should be considered acceptable for lead levels in blood samples taken in the United States (Wang et al., 2011). The children of parents who recycled plastic had the lowest lead levels, whereas the children of parents who worked with circuit boards in their factory had the highest lead levels. It is possible that children who are exposed to e-waste may develop serious health problems. Children are at risk of being exposed to neurodevelopmental toxins such lead, mercury, cadmium, chromium, arsenic, and nickel, which are all components of electronic trash. (Lebbie et al., 2021) PBDEs may also raise the risk of having a lower IQ, impaired cognitive function, and exposure to chemicals that are already known to cause cancer in humans, in addition to the risk of other harmful effects (Chen et al., 2011). It has been shown that young people working in recycling facilities for electronic debris have decreased lung function in certain age groups are some of the research that have shown connections between children's exposure to e-waste and decreased vaccination antibody tilters, hearing loss, and impaired coagulation in the e-waste recycling area (Parvez et al., 2021). For instance, when 8–9-year-old boys are exposed to nickel at an e-waste site, the result is a decrease in their forced vital capacity, a decline in their catalase activity, and a discernible increase in the amount of malondialdehyde and superoxide dismutase activity. All of these changes occur simultaneously. Workers at e-waste recycling facilities were found to have significant concentrations of Sb, Pb, and Fe in their urine, according to a research that was conducted in the scrap market of Agbogbloshie, Ghana. The lead levels in the soil samples taken from these locations were found to be much higher than the levels that are considered safe by the US EPA. In addition, increased amounts of Zn, Cu, Fe, Pb, and

Al were found in personal air samples as well as ambient air samples that were obtained from employees (Asante et al., 2012; Chen et al., 2011).

17.10 CONCLUSION

The growing problem of electronic waste or e-waste poses significant environmental and health challenges are especially in developing countries with prevalent outdated and informal recycling practices. The rapid increase in e-waste is primarily driven by the continuous innovation and consumption of electronic devices, resulting in a substantial amount of discarded consumer goods with electrical or electronic components. Despite the availability of safe and efficient recycling methods, much of this e-waste ends up in countries with insufficient regulations and outdated recycling techniques, leading to severe environmental contamination and health risks.

Informal recycling methods, such as open burning and acid leaching, are significant sources of hazardous emissions, including dioxins, lead, cadmium, hexavalent chromium, and brominated flame retardants. These toxic substances pose substantial threats to human health, particularly affecting vulnerable populations such as pregnant women and children. Exposure to these harmful chemicals can lead to a wide range of adverse health outcomes, including neurodevelopmental disorders, respiratory issues, cardiovascular diseases, and various forms of cancer. The situation is exacerbated by the fact that developed countries often export their e-waste to developing nations, exploiting the availability of cheap labor and the lack of stringent environmental and occupational regulations.

The impact of e-waste exposure on maternal and child health is particularly concerning. Pregnant women exposed to e-waste contaminants can experience adverse pregnancy outcomes such as spontaneous abortions, premature deliveries, and fetal loss. Furthermore, the exposure can interfere with thyroid development, lead to genetic toxicity, and cause congenital malformations. Children in e-waste recycling areas are at high risk due to their smaller size, faster metabolism, and various exposure pathways, including contaminated soil, water, food, and breast milk. The intake of harmful substances can severely disrupt their development, leading to long-term health issues.

The evidence from studies conducted in e-waste recycling hotspots like Guiyu, China, highlights the severe health impacts on mothers and their newborns. Increased levels of toxic metals such as lead and cadmium in cord blood and higher placental metallothionein levels have been documented, correlating with adverse developmental outcomes. These findings underscore the urgent need for improved e-waste management practices and stringent regulations to protect vulnerable populations.

Effective monitoring and mitigation strategies are essential to address the health risks associated with e-waste exposure. Breast milk has been identified as an effective matrix for monitoring dioxin levels, providing valuable insights into the extent of chemical accumulation in human populations. Additionally, enhancing public awareness and implementing stricter environmental and occupational safety standards are crucial steps toward mitigating the adverse effects of e-waste.

In conclusion, the rapidly growing e-waste problem requires a concerted global effort to implement safer recycling practices, enforce stricter regulations, and protect the health of vulnerable populations. Continued research and advocacy are essential to address the environmental and health challenges posed by e-waste, ensuring a safer and more sustainable future for all.

REFERENCES

Asamoah, A., Essumang, D. K., Muff, J., Kucheryavskiy, S., & Søgaard, E. G. (2018). Assessment of PCBs and exposure risk to infants in breast milk of primiparae and multiparae mothers in an electronic waste hot spot and non-hot spot areas in Ghana. *Science of the Total Environment, 612*, 1473–1479. https://doi.org/10.1016/j.scitotenv.2017.08.177

Asamoah, A., Muff, J., Essumang, D. K., & Søgaard, E. G. (2019). PAHs contamination levels in the breast milk of Ghanaian women from an e-waste recycling site and a residential area. *Science of the Total Environment, 666*, 347–354. https://doi.org/10.1016/j.scitotenv.2019.02.204

Asante, K. A., Agusa, T., Biney, C. A., Agyekum, W. A., Bello, M., Otsuka, M., Itai, T., Takahashi, S., & Tanabe, S. (2012). Multi-trace element levels and arsenic speciation in urine of e-waste recycling workers from Agbogbloshie, Accra in Ghana. *Science of the Total Environment, 424*, 63–73. https://doi.org/10.1016/j.scitotenv.2012.02.072

Bakhiyi, B., Gravel, S., Ceballos, D. M., Flynn, M. J., & Zayed, J. (2018). Has the question of e-waste opened a Pandora's box? An overview of unpredictable issues and challenges. *Environment International, 110*, 173–192. https://doi.org/10.1016/j.envint.2017.10.021

Ben, Y., Li, X., Youlin, Y., Li, L., Zheng, M., Wang, W., & Xu, X. (2014). Placental Transfer of Dechlorane Plus in Mother–Infant Pairs in an E-Waste Recycling Area (Wenling, China). *Environmental Science & Technology, 48*(9), 5187–5193. https://doi.org/10.1021/es404106b.

Cabrera-Rodríguez, R., Luzardo, O. P., Almeida-González, M., Boada, L. D., Zumbado, M., Acosta-Dacal, A., Rial-Berriel, C., & Henríquez-Hernández, L. A. (2019). Association between prenatal exposure to multiple persistent organic pollutants (POPs) and growth indicators in newborns. *Environmental Research, 171*, 285–292. https://doi.org/10.1016/j.envres.2018.12.064.

Chen, A., Dietrich, K. N., Huo, X., & Ho, S. M. (2011). Developmental Neurotoxicants in E-Waste: An Emerging Health Concern. *Environmental Health Perspectives, 119*(4), 431–438. https://doi.org/10.1289/ehp.1002452 Epub 2010 Nov 15. PMID: 21081302; PMCID: PMC3080922.

Chen, M., Ogunseitan, O. A., Duan, H., Zeng, X., & Li, J. (2018). China E-waste management: Struggling for future success. *Resources Conservation and Recycling, 139*, 48–49. https://doi.org/10.1016/j.resconrec.2018.08.006.

Grant, K., Goldizen, F. C., Sly, P. D., Brune, M. N., Neira, M., Van Den Berg, M., & Norman, R. E. (2013). Health consequences of exposure to e-waste: a systematic review. *The Lancet Global Health, 1*(6), e350–e361. https://doi.org/10.1016/s2214-109x(13)70101-3

Heacock, M., Kelly, C., Asante, K. A., Birnbaum, L. S., Bergman, Å., Brune, M. N., Buka, I., Carpenter, D. O., Chen, A., Huo, X., Kamel, M., Landrigan, P. J., Magalini, F., Díaz-Barriga, F., Neira, M., Omar, M. M., Pascale, A., Ruchirawat, M., Sly, L., … Suk, W. A. (2015). E-Waste and Harm to Vulnerable Populations: A Growing Global Problem. *Environmental Health Perspectives, 124*(5), 550–555. https://doi.org/10.1289/ehp.1509699.

Heacock, M., Trottier, B. A., Adhikary, S. P., Asante, K. A., Basu, N., Brune, M. N., Caravanos, J., Carpenter, D. O., Cazabon, D., Chakraborty, P., Chen, A., Barriga, F. J., Ericson, B., Fobil, J. N., Haryanto, R. B., Huo, X., Joshi, T., Landrigan, P. J., Lopez, A. R., … Suk, W. A. (2018). Prevention-intervention strategies to reduce exposure to e- waste. *Reviews on Environmental Health, 33*(2), 219–228. https://doi.org/10.1515/reveh- 2018-0014.

Huo, X., Peng, L., Xu, X., Zheng, L., Qiu, B., Qi, Z., Zhang, B., Han, D. H., & Piao, Z. (2007). Elevated Blood Lead Levels of Children in Guiyu, an Electronic Waste Recycling Town in China. *Environmental Health Perspectives, 115*(7), 1113–1117. https://doi.org/10.1289/ehp.9697

Lebbie, T., Moyebi, O. D., Asante, K. A., Fobil, J. N., Brune-Drisse, M. N., Suk, W. A., Sly, P. D., Gorman, J. K., & Carpenter, D. O. (2021). E-Waste in Africa: A Serious Threat to the Health of Children. *International Journal of Environmental Research and Public Health, 18*(16), 8488. https://doi.org/10.3390/ijerph18168488.

Li, Y., Huo, X., Liu, J., Peng, L., Li, W., & Xu, X. (2011). Assessment of cadmium exposure for neonates in Guiyu, an electronic waste pollution site of China. *Environmental Monitoring and Assessment, 177*(1–4),343–351. https://doi.org/10.1007/s10661-010-1638-6.

Luo T, Hang JG, Nakayama SF, Jung CR, Ma CC, Kido T, Wang FH, Wang Z, Dong JJ, Shi LL, Sun XL. Dioxins in breast milk of Chinese mothers: A survey 40 years after the e-waste recycling activities. *Science of the Total Environment* 2021 Mar 1;*758*:14362

Meem RA, Ahmed A, Hossain MS, Khan RA. A Review on the environmental and health impacts due to electronic waste disposal in Bangladesh. *GSC Advanced Research and Reviews*. 2021 Aug 30;8(2):116–125.

Parvez, S. M., Jahan, F., Brune, M. N., Gorman, J. F., Rahman, M. J., Carpenter, D., Islam, Z., Rahman, M., Aich, N., Knibbs, L. D., & Sly, P. D. (2021). Health consequences of exposure to e-waste: An updated systematic review. The Lancet Planetary Health, 5(12), e905–e920. https://doi.org/10.1016/S2542-5196(21)00263-1.

Verma, A. (2020). *E-wastes and their impact on environment and public health*. International *Journal of Applied Research* Available at SSRN: https://ssrn.com/abstract=3699964

Wang, L., Hou, M., An, J., Zhong, Y., Wang, X., Wang, Y., Wu, M., Bi, X., Sheng, G., & Fu, J. (2011). The cytotoxic and genetoxic effects of dust and soil samples from E- waste recycling area on L02 cells. *Toxicology and Industrial Health, 27*(9), 831– 839. https://doi.org/10.1177/0748233711399313

Wu, K., Xu, X., Peng, L., Liu, J., Guo, Y., & Huo, X. (2012). Association between maternal exposure to perfluorooctanoic acid (PFOA) from electronic waste recycling and neonatal health outcomes. *Environment International*, *48*, 1–8. https://doi.org/10.1016/j.envint.2012.06.018.

Xu, X., Yang, H., Chen, A., Zhou, Y., Wu, K., Liu, J., Zhang, Y., & Huo, X. (2012). Birth outcomes related to informal e-waste recycling in Guiyu, China. *Toxicology*, *33*(1), 94–98. https://doi.org/10.1016/j.reprotox.2011.12.006.

Zheng, L., Wu, K., Li, Y., Qi, Z., Han, D. H., Zhang, B., Gu, C., Chen, G., Liu, J., Chen, S., Xu, X., & Huo, X. (2008). Blood lead and cadmium levels and relevant factors among children from an e-waste recycling town in China. *Environmental Research*, *108*(1), 15– 20. https://doi.org/10.1016/j.envres.2008.04.

18 Innovative Electronic Waste Recycling Technologies, Systematic Management Strategies, and Challenges

Saufishan T A
Faculty of Engineering & Technology, AMU, Aligarh, India

Abdul Rafey
Indian Institute of Technology Delhi, New Delhi, India

Faisal Zia Siddiqui
Faculty of Engineering & Technology, AMU, Aligarh, India

18.1 INTRODUCTION

The usage of electrical and electronic equipment (EEE) is closely connected to the growth of the global economy and is crucial for improving living standards in modern societies. However, its resource-intensive nature also exemplifies the environmental challenges associated with improved living standards. After use, it is discarded, generating a waste stream known as e-waste, which contains valuable as well as hazardous materials. In 2019, the global generation of e-waste reached around 53.6 million metric tonnes (MMT) (Shittu et al., 2021). Global e-waste generation has increased by 9.2 MMT since 2014, and it is projected to reach 74.7 MMT by 2030, nearly tripling in 16 years. The growing volume of e-waste is driven by increased consumption of EEE, shorter product life cycles, and limited refurbishing options. Figure 18.1A represents the e-waste generation trend globally from 2010 to 2019 and Figure 18.1B depicts e-waste trends across continents in 2019 (Forti et al., 2020).

In Figure 18.1B, it is shown that Europe and Oceania have achieved higher rates of collection and recycling when compared with other continents. Meanwhile, the other continents have only managed to recycle a limited portion of their e-waste. Asia reported the lowest rate of e-waste collection and recycling relative to its annual revenue. Figure 18.2 illustrates the categories of EEE and their covered components, consumables, parts, and spares. The classification of e-waste can be done into two broad categories, which are further subdivided into five categories.

18.2 INNOVATIVE TECHNOLOGIES AND METHODS FOR E-WASTE RECYCLING

In this modern era, humans are reaping the benefits of technological advancements. Still, this progress has also led to significant challenges, particularly e-waste disposal, which has evolved into a significant issue nowadays. Traditional methods such as incineration, hydraulic shaking bed separation, and acid leaching are ineffective and highly detrimental to the environment (Rene et al., 2021). Figure 18.3 highlights recent advancements in e-waste recycling technologies.

DOI: 10.1201/9781003582311-18

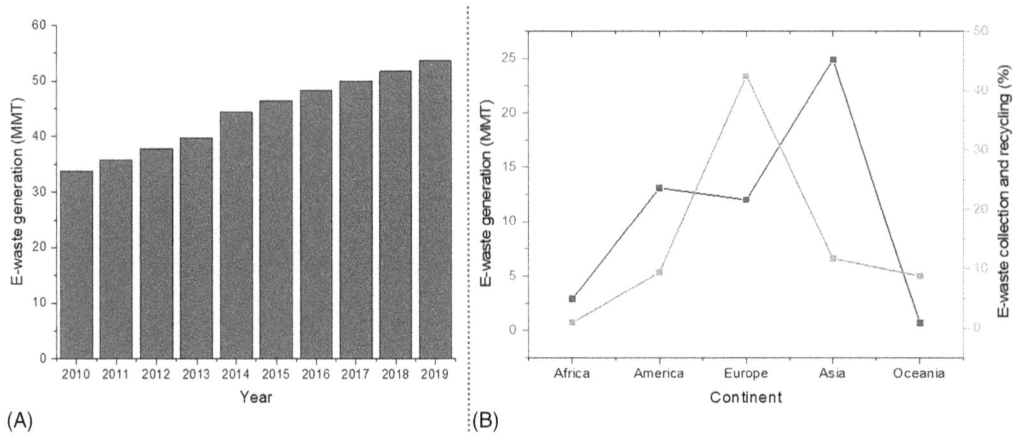

FIGURE 18.1 (A) Global e-waste generation (2010–2019); (B) E-waste generation, collection and recycling in different continents during 2019. Redrawn from Forti et al., 2020 (Open access).

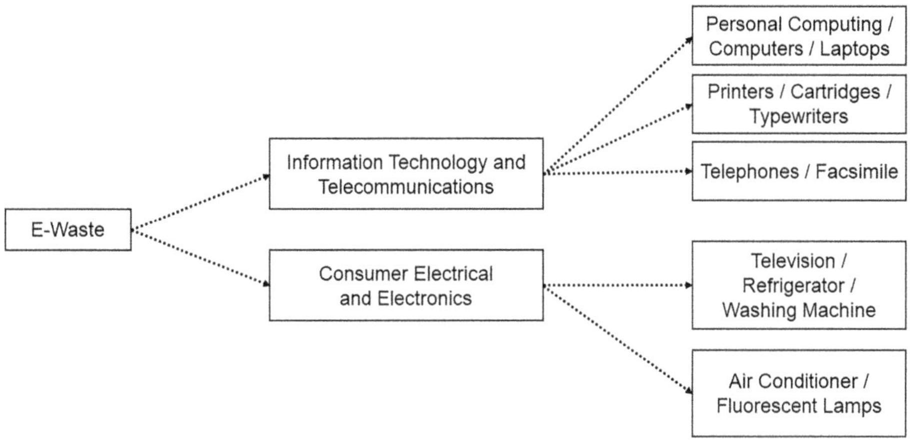

FIGURE 18.2 Categories of electrical and electronic equipment (E-waste rules 2016 Schedule I). Redrawn from MoEFCC 2022 (Open access).

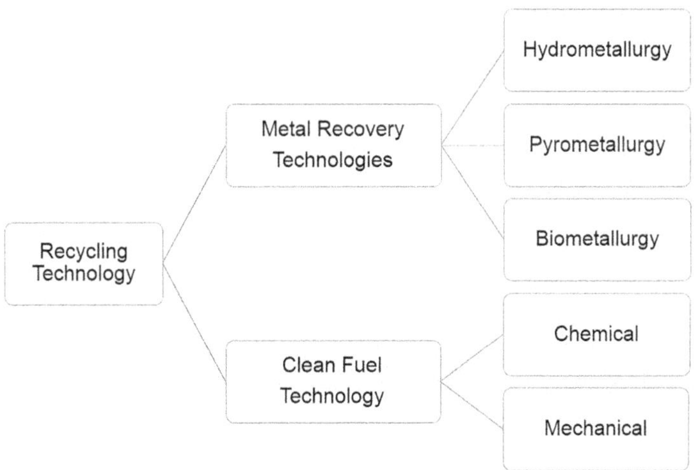

FIGURE 18.3 Recent advances in e-waste recycling technologies.

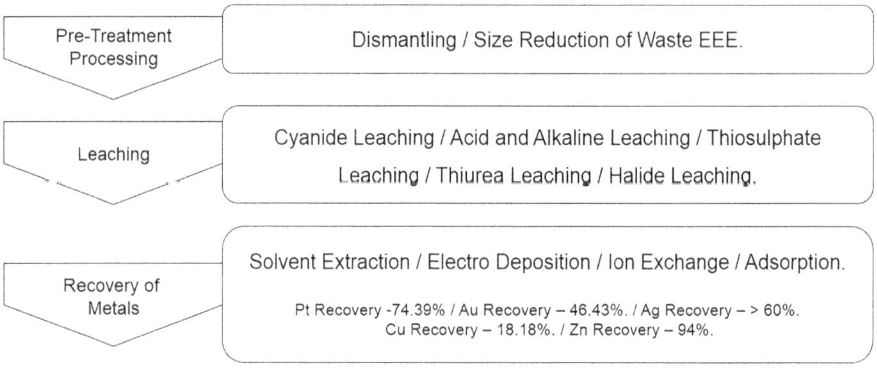

FIGURE 18.4 E-waste recycling using hydrometallurgy.

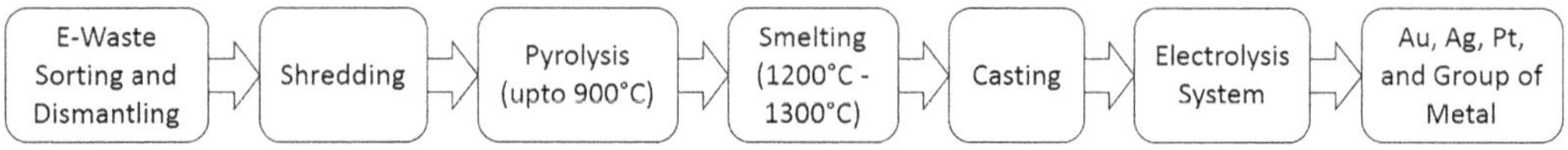

FIGURE 18.5 E-waste recycling using pyrometallurgy.

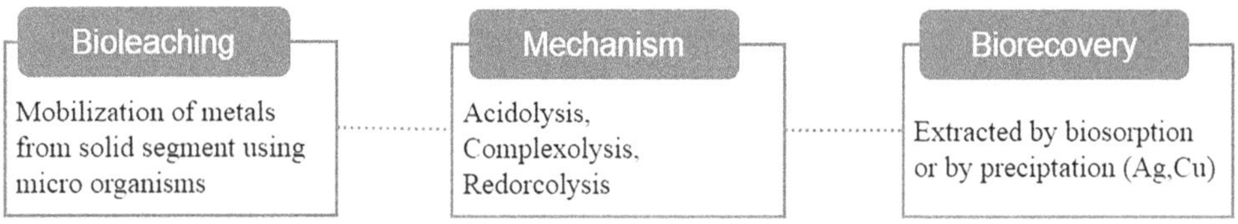

FIGURE 18.6 E-waste recycling using biometallurgy.

Figure 18.3 shows that e-waste recycling technologies are broadly classified into two categories with further sub-classification into five categories. Their metal recovery fractions are distilled and recovered in the processing stage using hydrometallurgy, pyrometallurgy, and biometallurgy.

1. **Hydrometallurgy**: The methodology for e-waste recycling using the Hydrometallurgy technique is summarised in Figure 18.4 (Rene et al., 2021). From Figure 18.4, it can be deduced that the applicability of hydrometallurgy or recovery potential of e-waste is limited to selected metals.
2. **Pyrometallurgy**: The methodology for e-waste recycling using the Pyrometallurgy technique is presented in Figure 18.5. This technology involves smelting, combustion, incineration, and pyrolysis in a blast furnace (Rene et al., 2021).
3. **Biometallugy**: The methodology for e-waste recycling using the metallurgy technique is summarised in Figure 18.6 (Habibi et al., 2020).

18.3 COMPARATIVE ANALYSIS OF E-WASTE RECYCLING TECHNOLOGIES

A comparative analysis of the e-waste recycling technologies based on selected parameters has been summarized in Table 18.1.

Table 18.1 shows the recycling using the pyrometallurgy technique is an expensive option owing to the temperature requirements and thus has limited usage in small and medium enterprises (Hsu et al., 2019). The hydrometallurgical processing technology is efficient and easy to operate, offering high selectivity for target metals and excellent metal recovery rates. Although biometallurgy has

TABLE 18.1
Comparison of E-Waste Recycling Technologies

Name of Technology	Green Technology	Sustainable Technology	Efficiency	Raw Material	Products Formed	Temperature Range (°C)	Reference
Hydrometallurgy	Low GHG emissions as compared to Pyrometallurgy	Recycling alternatives and processes for decreased energy requirements	Ag-90%, Zn-80% and Al-100%	Printed circuit boards (PCBs), photovoltaic panels, batteries, and light-emitting diodes (LEDs)	Glass-90%, semiconductors-95%, tellurium-80%	25 – 250	(Huang et al., 2015)
Pyrometallurgy	Sulphur emissions can be handled, and slag waste products can be inert.	Earth's crust occurs as metal oxides and sulfides	Metal purity > 99%.	PCBs	Black copper (70%–85%)	900 – 1200	(Ambaye et al., 2020)
Biometallurgy	It uses biodegradable compounds	It is suitable and adequate for processing ore	High leaching efficiencies ->80% of metals	PCBs	REE, Ga, In	70 – 150	(Ilyas & Lee, 2014)

TABLE 18.2
Benefits and Drawbacks of the E-waste Recycling Technologies

Advantages	Disadvantages
Hydrometallurgy	
High selectivity, easy to manage, fast reaction kinetics, high extraction efficiency	Presence of toxic materials
Leaching can be done at a meager cost in single or multi-stage reactors	Large quantity of leachate generation
Low operational temperature	Requires corrosion-resistant equipment
Low emission of pollutants	High cost of metal recovery
Pyrometallurgy	
There is no need for the pretreatment of PCBs	High energy and investment cost
Produces Cu-rich final product	Low process efficiency
Biometallurgy	
Green technology	It is difficult to maintain the purity of microorganisms and regenerate the bioreactors.
Rare metals can be reclaimed	Effect on microorganisms due to toxicity of metals
Low operational cost	Long-term process as compared to other technologies

shown great promise for leaching rare metals from urban mines owing to its cost-effectiveness and environmentally friendly nature, its full potential has yet to be fully explored.

Hydrometallurgy is limited over pyrometallurgy because mechanical pretreatment of e-waste is needed to reduce the volume. Pyrometallurgy is much more effective than hydrometallurgy because it applies to any e-waste and needs no pretreatment. Among the three techniques, bio metallurgy is more eco-friendly, and a high quantity of metals can be leached out, for example, Cu, Ni, Zn, and Al, but the drawback is the extended period required for the process (Table 18.2).

18.4 VALUE-ADDED PRODUCTS FROM E-WASTE

E-waste contains valuable metals like platinum, silver, gold, cobalt, neodymium, aluminum, and tin (Das et al., 2021; Ma et al., 2016). In 2017, the European Union (EU) identified 27 critical elements in e-waste, which was further updated to 35 elements, as shown in Table 18.3. Identifying these

TABLE 18.3
Critical Raw Materials as Identified by the European Union

• Indium	• Barytes	• Germanium
• Niobium	• Rhodium	• Holmium
• Scandium	• Diatomite	• Vanadium
• Phosphorus	• Gallium	• Natural rubber
• Hafnium	• Helium	• Ruthenium
• Iridium	• Lithium	• Phosphate rock
• Natural graphite	• Palladium	• Tantalum
• Bentonite	• Platinum group metals	• Lutetium
• Fluorspar	• Beryllium	• Antimony
• Rare earth elements (REE)	• Silicon metal	• Magnesium
• Cobalt	• Tungsten	
• Thulium	• Bismuth	

TABLE 18.4
Elemental Share of Valuable Elements in Different E-waste Sources

E-waste Source	Elements (Value-share)						
	Ag	Au	Pd	Fe	Al	Cu	Sum PM*
TV Board	7%	22%	7%	4%	10%	50%	36%
PC-Board	5%	61%	15%	0%	1%	18%	81%
Mobile phone	13%	64%	14%	0%	0%	9%	91%
Portable audio	3%	10%	2%	2%	0%	82%	15%
DVD-Player	5%	32%	5%	13%	3%	42%	42%
Calculator	7%	69%	4%	0%	5%	14%	80%

* PM (Precious Metal) – Ag, Au, Pd

critical raw materials is essential owing to their high economic impact. For instance, LCDs use indium coating, and if current usage rates continue, indium could be depleted from the Earth within the next 13 years.

Table 18.4 shows the elemental share of valuable elements in six different e-waste sources. This data can be used to identify the recycling or treatment technique that would be most suitable for recovering metal from the particular e-waste source.

High Medium Low-value boards =>80%

From Table 18.4 it can be seen that PCB and mobile phones come under the high-value board category. Copper is abundant in portable audio, whereas gold and silver are embedded in large quantities in mobile phones because of the multilayer structure in PCB. In mobile phones, more than 40 elements of the periodic table are present, making a total of 91% (Baldé et al., 2022).

Figure 18.7 shows the yield of fuel products after the thermal decomposition of mobile phones at 450°C for 1.5 hours. The yield of pyrolytic oil is approximately 57 wt.%, which is a mixture of organic compounds, 29.68% are gases, and 13.29% are solids (Anis & Prasetyo, 2019).

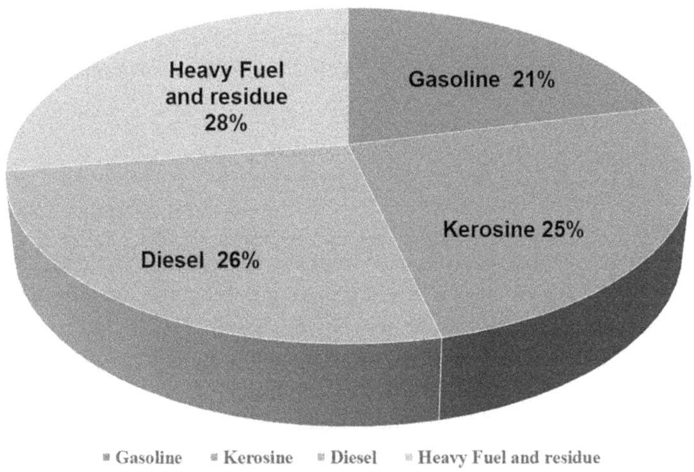

FIGURE 18.7 Product yield from pyrolysis of mobile phone at 450°C.

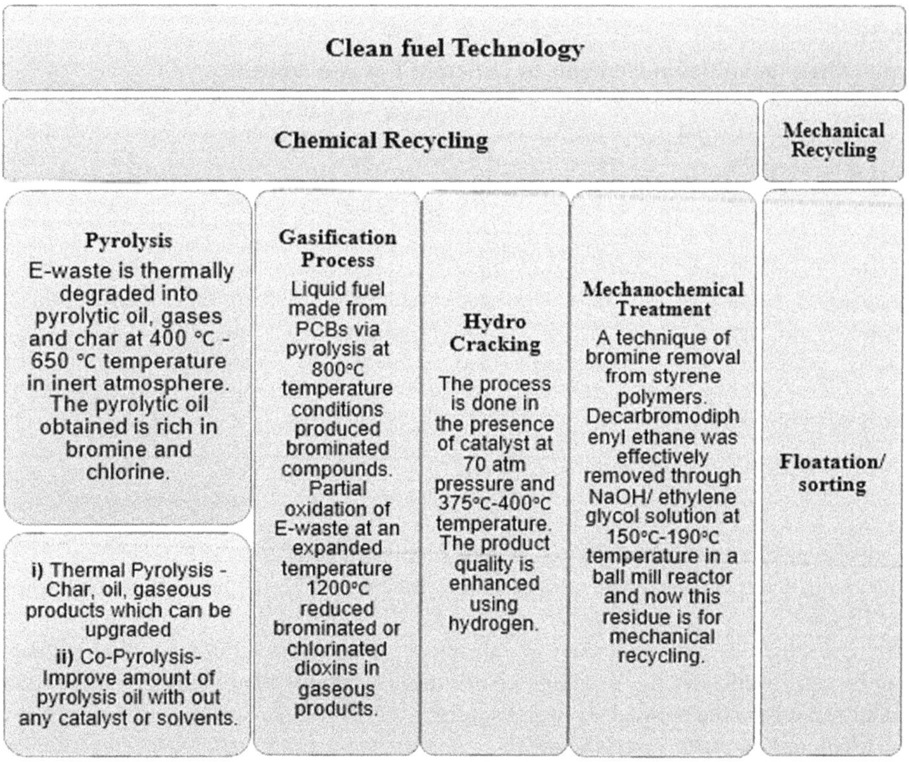

FIGURE 18.8 Classification of clean fuel technology for e-waste processing.

18.5 SYSTEMATIC MANAGEMENT STRATEGIES FOR E-WASTE

E-waste management could be a good startup enterprise platform closer to energy manufacturing and metallic restoration (Agarwal et al., 2023). Chemical recycling consists of catalytic cracking, pyrolysis, and gasification methods. E-waste pyrolysis before cracking catalytically resulted in the generation of prime oil (Shen et al., 2016). The obtained oil could be similarly transformed into clean fuel with an integrated procedure (pyrolysis and catalytic upgrading-hydro, mechano-chemical treatment), impacting significant financial and environmental benefits during e-waste processing into clean fuels. The classification of clean fuel technology for e-waste processing is presented in Figure 18.8 (Shen et al., 2018).

18.6 E-WASTE MANAGEMENT RULES OF 2016

The E-waste rule 2016 was notified by the Ministry of Environment, Forest and Climate Change (MoEFCC), Government of India, in March 2016 in the Gazette of India after considering public comments on the draft rules in June 2015. The rules have been framed considering the growing problem of e-waste disposal and utilization of valuable resources, which are generally not recognized by the common masses. The rules focus on the idea of extended producer responsibility (EPR) for discarded products disposal by producers or through financial incentives to consumers. The rules are not applicable to discarded lead-acid batteries (covered by the Batteries Waste Management Rule of 2022) or to radioactive waste (governed by the Atomic Energy Act of 1962). The roles of manufacturers, producers, consumers, bulk consumers, dismantlers, and recyclers are given in subsequent sections. The rules list down the authorities such as Central Pollution Control Board (CPCB), State Pollution Control Boards (SPCBs), Pollution Control Committees (PCCs), Urban Local

Bodies (ULBs)/Municipal Corporations, and Port authorities, highlighting their corresponding duties. A manifest system for E-waste is provided.

18.7 E-WASTE MANAGEMENT – ROLE OF MANUFACTURER

The manufacturer of WEEE shall be responsible for;

1. Proper collection of e-waste produced during the manufacturing of EEE and taking it forward for recycling or directly disposing off;
2. Applying for authorization following the procedure outlined in the rule from the respective SPCB or PCC;
3. Ensuring there is no environmental damage during the e-waste handling, storage, and transportation;
4. Maintaining database of generation, handling, and disposal of e-waste in the specified format and making these records available for audit by the relevant SPCB or PCC;
5. Submitting annual returns to the relevant SPCB or PCC on or before the 30th day of June of the applicable financial year.

18.8 E-WASTE MANAGEMENT – ROLE OF CONSUMER/BULK CONSUMER

1. Consumers and bulk consumers of EEE must ensure that the e-waste produced is sent via authorized collection centers or dealers designated by the producer, dismantler, recycler, or any assigned service provider of the producer.
2. Bulk consumers of EEE listed in Schedule I must keep a database of the generated e-waste and present them for audit by the relevant State Pollution Control Board (SPCB) or Pollution Control Committee (PCC).
3. Bulk consumers of EEE must ensure that end-of-life equipment is not mixed with e-waste containing radioactive materials, as regulated by the Atomic Energy Act of 1962.

18.9 E-WASTE MANAGEMENT – ROLE OF PRODUCER

The producer of EEE Shall be responsible for;

1. Collecting e-waste generated during the manufacturing process of EEE and sending it for recycling or disposal;
2. Ensuring e-waste collection under EPR authorization and sending it to registered recyclers.
3. Guiding in implementing the Deposit Refund Scheme to ensure the collection and delivery of end-of-life products to authorized dismantlers or recyclers.
4. Facilitating the return of end-of-life EEE from consumers or bulk consumers.
5. Maintaining a database of the e-waste handled and presenting the data for audit before CPCB or SPCB.
6. Submitting annual returns to CPCB/SPCB on or before the 30th day of June of the applicable financial year.
7. The Producer shall apply to the CPCB/SPCB for authorization, which shall subsequently grant the EPR Authorization.

18.10 E-WASTE MANAGEMENT – ROLE OF DISMANTLER

1. Obtain authorization and adhere to standards or guidelines prescribed by the CPCB/SPCB as per current requirements.
2. Ensuring any components that cannot be recycled or recovered are directed to authorized storage, treatment, and disposal facilities.

3. Maintaining a database of collected e-waste, dismantling activities, and shipments to authorized recyclers using Form-2, and ensure these records are available for inspection by the CPCB or the respective SPCB/PCC.
4. Submit returns to the relevant SPCB/PCC as required.

18.11 E-WASTE MANAGEMENT – ROLE OF RECYCLER

1. Complying with standards and guidelines as prescribed by CPCB/SPCB in a timely manner;
2. Obtaining authorization from respective SPCB/PCC following the procedure outlined in the rules.
3. Ensuring there is no damage to the environment during e-waste handling, storage, and transportation.
4. Filing annual returns to the concerned SPCB/PCC by 30th day of June of the applicable financial year.
5. WEEE or their components not included in Schedule I can be recycled, ensuring they don't possess radioactive materials, and this condition requires certification when obtaining a permit from the concerned SPCB/PCC.

Some proposed measures are as follows:

1. The list provided in Schedule–I of E-waste (Management) Rules, 2016 is dynamic, and new items would come into the market; thus, a broader definition is needed to include all the waste under the e-waste category.
2. The definition of bulk consumer should be elaborated further to cover service centers of electronic items, non-governmental organizations (NGOs), service providers, etc.
3. Developed economies hardly import E-waste from other countries. However, developing economies like India import E-waste from developed economies.
4. Developed economies have advanced online centralized and decentralized systems for managing E-waste. However, developing economies such as India do not have such facilities compared to them.
5. In developed economies, the formal sector is more active than the informal sector in developing economies like India.
6. A take-back policy exists for most electronic products in developed economies. In contrast, no such schemes exist for electronic items in developing economies, such as used refrigerators, air conditioners, or washing machines.
7. The recycling sector in developed countries has a trained workforce, which is not the case in developing economies where a limited trained workforce is used.
8. Setting up e-waste collection centers in commercial, residential, or institutional areas to encourage formal recycling.
9. Registration of collection centers with CPCB/SPCP/PCC is done to promote authorized recycling and discourage unauthorized dismantling and recycling.
10. No provision exists for imposing penalties against handlers or processors of E-waste. Rules should incorporate provisions for CPCB/SPCP/PCC to execute for violating clauses of the e-waste rules.
11. To ensure effective compliance with e-waste rules, the authority to issue penalties should be directly given to CPCB/SPCP/PCC.

18.12 CONCLUSION

The application of technologies in our day-to-day lifestyles has resulted in a significant growth in the usage of electronic gadgets by the common masses. Hence, to manage the growing e-waste problem, systematic e-waste legislation should be implemented, and continuous amendments should be done to overcome the drawbacks of systematic and periodic audits and evaluations. This study reviews four commonly used techniques: pyrometallurgy, biohydrometallurgy, hydrometallurgy, and pyrolysis. It must reevaluate e-waste management, improve recycling, and address the issue with a cost-effective, safe (bio) technological approach. Implementing the "reuse and refurbish" principle is essential for effectively managing e-waste generation and promoting reduction and recycling efforts. The techniques discussed in this chapter, from both a circular economy perspective and an environmental standpoint, could become a crucial and authoritative area shortly. To enforce this principle, the government must encourage refurbishment where consumers can be engrossed with incentives, including reduced tax prices, to purchase refurbished commodities. The current study's findings indicate that ongoing local production, interstate transportation, and importation of e-waste from foreign nations harm the ecosystem. The review and analysis reveal that stringent legislation on e-waste management is in place. However, implementation is limited. There is a need to relook into the existing policies and legislation through continued analysis and amendments. Further, Indian legislation needs to be synergized with e-waste legislation that is effective globally.

REFERENCES

Agarwal, S., Darbar, S., Saha, S., Choudhury, M., & Singh, R. P. (2023). E-waste management using different cost-effective, eco-friendly biological techniques: an overview. In *Waste Management and Resource Recycling in the Developing World*, 205–235. https://doi.org/10.1016/B978-0-323-90463-6.00026-9

Ambaye, T. G., Vaccari, M., Castro, F. D., Prasad, S., & Rtimi, S. (2020). Emerging technologies for recovering rare earth elements (REEs) from the end-of-life electronic wastes: A review on progress, challenges, and perspectives. *Environmental Science and Pollution Research*, *27*(29), 36052–36074. https://doi.org/10.1007/s11356-020-09630-2

Anis, S., & Prasetyo, E. (2019). Conversion of Hand Phone Case Waste into Liquid Fuels in a Microwave Reactor. *IOP Conference Series: Earth and Environmental Science*, *258*(1). https://doi.org/10.1088/1755-1315/258/1/012011

Baldé, C. P., Angelo, E. D., Luda, V., Deubzer, O., & Kuehr, R. (2022). Global Transboundary E-waste Flows Monitor - 2022.

Das, P., Gabriel, J. C. P., Tay, C. Y., & Lee, J. M. (2021). Value-added products from thermochemical treatments of contaminated e-waste plastics. *Chemosphere*, *269*, 129409. https://doi.org/10.1016/j.chemosphere.2020.129409

Forti, V., Baldé, C. P., Kuehr, R., & Bel, G. (2020). The Global E-waste Monitor 2020. *Quantities, flows, and the circular economy potential* (Issue July). http://ewastemonitor.info/

Habibi, A., Shamshiri Kourdestani, S., & Hadadi, M. (2020). A review of Biohydrometallurgy as an environmentally friendly approach in metals recovery from electrical waste. *Waste Management and Research*, *38*(3), 232–244. https://doi.org/10.1177/0734242X19895321

Hsu, E., Barmak, K., West, A. C., & Park, A. H. A. (2019). Advancements in the treatment and processing of electronic waste with sustainability: A review of metal extraction and recovery technologies. *Green Chemistry*, *21*(5), 919–936. https://doi.org/10.1039/c8gc03688h

Huang, X. W., Long, Z. Q., Wang, L. S., & Feng, Z. Y. (2015). Technology development for rare earth cleaner hydrometallurgy in China. *Rare Metals*, *34*(4), 215–222. https://doi.org/10.1007/s12598-015-0473-x

Ilyas, S., & Lee, J.-C. (2014). Biometallurgical Recovery of Metals from Waste Electrical and Electronic Equipment: a Review. *ChemBioEng Reviews*, *1*(4), 148–169. https://doi.org/10.1002/cben.201400001

Ma, C., Yu, J., Wang, B., Song, Z., Xiang, J., Hu, S., Su, S., & Sun, L. (2016). Chemical recycling of brominated flame retarded plastics from e-waste for clean fuels production: A review. *Renewable and Sustainable Energy Reviews*, *61*, 433–450. https://doi.org/10.1016/j.rser.2016.04.020

Rene, E. R., Sethurajan, M., Kumar Ponnusamy, V., Kumar, G., Bao Dung, T. N., Brindhadevi, K., & Pugazhendhi, A. (2021). Electronic waste generation, recycling, and resource recovery: Technological perspectives and trends. *Journal of Hazardous Materials*, *416*(March), 125664. https://doi.org/10.1016/j.jhazmat.2021.125664

Shen, Y., Chen, X., Ge, X., & Chen, M. (2018). Chemical pyrolysis of E-waste plastics: Char characterization. *Journal of Environmental Management*, *214*, 94–103. https://doi.org/10.1016/j.jenvman.2018.02.096

Shen, Y., Zhao, R., Wang, J., Chen, X., Ge, X., & Chen, M. (2016). Waste-to-energy: Dehalogenation of plastic-containing wastes. *Waste Management*, *49*, 287–303. https://doi.org/10.1016/j.wasman.2015.12.024
Shittu, O. S., Williams, I. D., & Shaw, P. J. (2021). Global E-waste management: Can WEEE make a difference? A review of e-waste trends, legislation, contemporary issues, and future challenges. *Waste Management*, *120*, 549–563. https://doi.org/10.1016/j.wasman.2020.10.016

19 Remediation Techniques for Soil Contaminated with Electronic Waste-Associated Toxic Materials

Sheetal Kumari and Manoj Chandra Garg
Amity Institute of Environmental Science (AIES), Amity University Uttar Pradesh, Noida, India

Jyoti Chowdhry
Maharishi Dayanand University, Rohtak, India

19.1 INTRODUCTION

Remediation is a novel and environment-friendly method that can restore the quality of soil contaminated by organic and inorganic pollutants (Lasat, 2002; Salt et al., 1998). Soil contamination is attracting attention as a global environmental problem due to rising focus on the safety of agricultural products (Hu et al., 2017). Contaminated soil, particularly agricultural land, is a major issue that needs addressing. The linkage between source-pathway-receptors for pollutants needs to be disrupted to prevent contaminants from reaching their target and causing harm. The "Source" refers to the areas where pollutants get generated and enter the environment. The "pathway" refers to the routes and patterns by which they enter the environment from their sources. The "receptors" refer to all the biotic and abiotic components in the environment that may get affected by the pollutants under consideration. Another important application of remediation methods is to reduce the development of unwanted expenditures in future. The effectiveness and effects of remediation strategies vary depending on the area under consideration. Several remediation methods have been created; some are widely employed while others are still in the early stages of their development. The remediation techniques for polluted soil can broadly be divided into two approaches: InSitu and ExSitu Remediation Techniques (Figure 19.1). We will be understanding each of these approaches in detail in subsequent sections.

19.2 INSITU REMEDIATION TECHNIQUES

The word "InSitu" means "in the original place", i.e., they are carried at the concerned location. Insitu remediation involves techniques and processes that aim at treating and restoring contaminated soil or groundwater in the same location. It removes the contaminants directly from soil or groundwater. It does not require any specific equipment and is not a labor-intensive process, thus making it very economical as compared to Exsitu Techniques (Tsai et al., 2008).

DOI: 10.1201/9781003582311-19

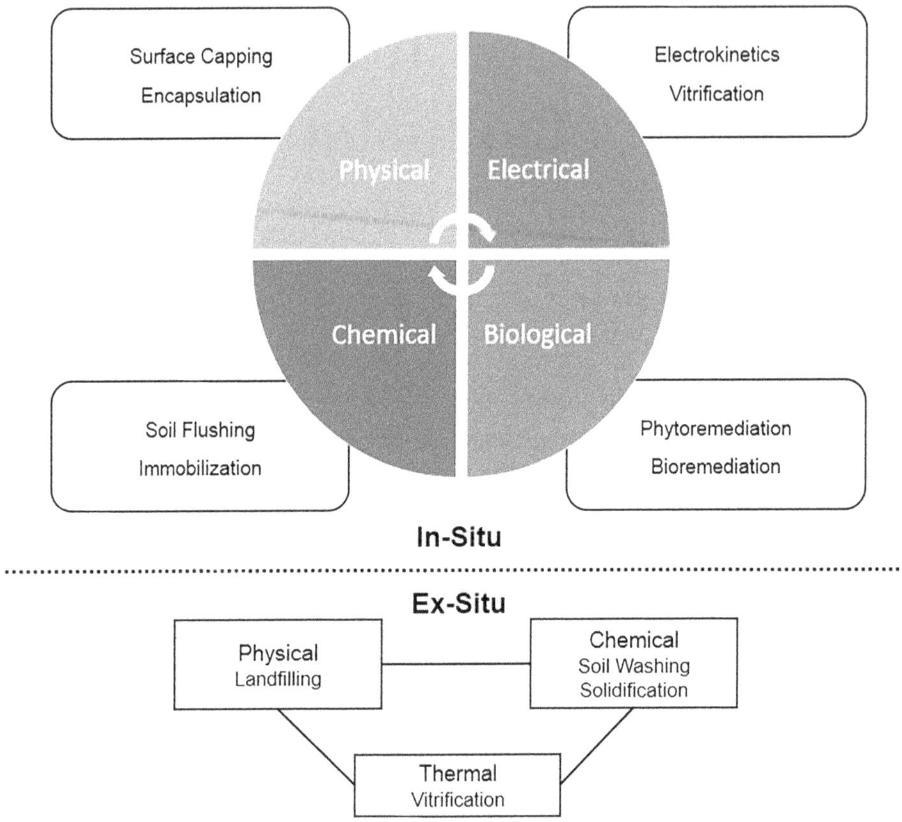

FIGURE 19.1 Insitu and exsitu remediation technique for contaminated soil.

19.2.1 Physical Approach

Insitu remediation minimizes soil disturbance, reduces worker and public exposure to toxins, and may greatly treatment costs because it does not necessitate excavating and transporting the polluted soil to facilities for off-site treatment. However, specific field conditions, such as the permeability of soil, weather, depth of contamination, and potential chemical deep leaching, must be carefully taken into account. One of the world's greatest concerns is the disposal of electronic waste. Most of them contain metals and toxic substances. The rapid development of computers and communication devices, among other electrical consumables, is largely responsible for the increase in global e-waste emissions. Metals make up about 60% of e-waste, followed by plastics at 15% and mixtures of metals and plastics at 5%. Mercury (Hg), cadmium (Cd), lead (Pb), chromium (Cr), arsenic (As), and selenium (Se), are toxic metals, and concentrations above them in the environment pose a risk to humans and environmentally permissible limits (Rautela et al., 2021).

19.2.1.1 Surface Capping

A surface cap is a waterproof coating applied over a contaminated region to establish a stable and protective surface. Due to the lack of efforts to eliminate or reduce the reactivity of pollutants in the soil, this containment-based approach cannot be considered an actual soil 'remediation method.' Nevertheless, the procedure efficiently minimizes the likelihood of direct skin contact or accidental ingestion of contaminated soil. The surface cover serves as an impenetrable barrier, preventing groundwater and surface water infiltration and thereby halting the spread of soil pollutants. The benefits of surface scraping include enhanced security, cost-effectiveness, and straightforward installation. The drawbacks encompass the reduction of land's capacity for cultivation and the limitation to certain regions and specific geographical locations (Miskewitz, 2014).

19.2.1.2 Encapsulation

Encapsulation, also known as the barrier wall, cut-off wall, or liner approach, is a corrective measure similar to surface capping. The method entails enclosing contaminated soil in an adequately constructed physical barrier system made up of enclosing underground barriers, enclosing underground barriers, and, in rare instances, barrier floors. At polluted sites, pollutants are controlled and separated to stop their spread off-site and community exposure. Constructing subsurface vertical impermeable walls at contaminated sites is the fundamental difficulty in encapsulation. Sheet pile walls, slurry walls, thin walls, and injection walls are only a few of the construction techniques that have been created (Liu et al., 2018). The advantages of encapsulation are high security and quick installation. Disadvantages of encapsulation are limited to tiny, shallow contamination zones, expensive, and losing the ability to cultivate the land.

19.2.2 Electrical Approach

19.2.2.1 Electrokinetic

To remove polluted soils, electrokinetic extraction is used. When low-density current (DC) electricity is supplied via electrodes buried in the ground, anions migrate to the anode at the attractive force of the generated electrical field while cations in the solution phase of the contaminated soil move to the cathode. Since the late 1980s, electrokinetic extraction has been studied for eliminating contaminants from soil (Hansen et al., 2016). Advantages and disadvantages of electrokinetic removal of contaminants, little soil disturbance and most effective on fine-textured soils with low permeability, time-consuming and inefficient.

19.2.2.2 Vitrification

A thermal remediation method called vitrification utilizes heat to turn polluted soil into solids that resemble glass. Since 1980, technology has been developed and put to the test. A high-temperature zone is created by intense energy being delivered to polluted soil a common method of washing soil to remove contamination (Figure 19.2). The area's soil subsequently melts into molten lava, which

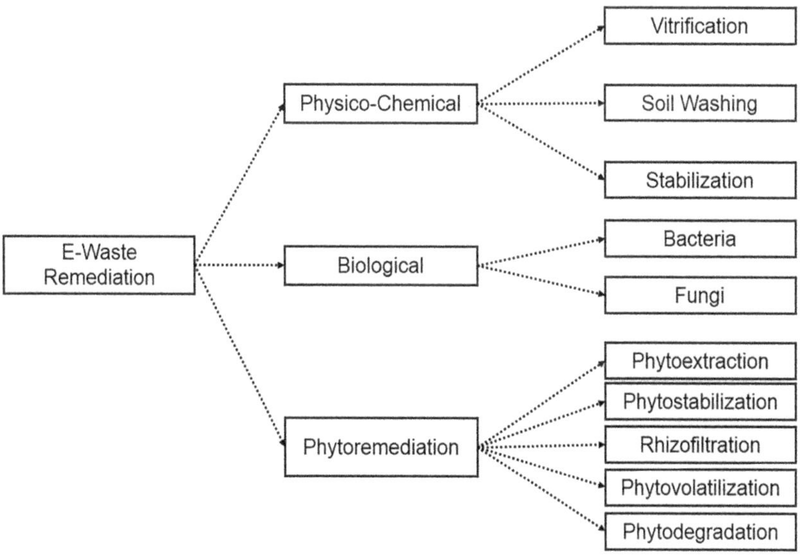

FIGURE 19.2 Available techniques to remediate contaminants from E-waste.

cools to form a glass-like substance. The resulting vitrification structure is robust, long-lasting, chemically inert, and leaching resistant. The advantages of vitrification are high performance, and the disadvantages are low soil area/volume, high price, treated land, and loss of environmental functions.

19.2.3 Chemical Approach

19.2.3.1 Soil Flushing

An insitu chemical technique for soil remediation is soil flushing. It involves injecting or infiltrating an aqueous solution into polluted soil. The saturated zone, the unsaturated zone, or both may experience this. The flushing solution makes pollutants that have been bound to the soil matrix more mobile or soluble. Surfactants, co-solvents, acids, bases, oxidants, chelates, solvents, or water may be included in this mixture. The main drawbacks are the impact of flushing solutions on the soil environment and the potential risk of spreading toxins into uncontaminated areas (Guo et al., 2018). Flushing has the following benefits: Contaminant elimination, little soil disturbance, low cost, and ease of installation.

19.2.3.2 Immobilization

Chemical immobilization is a highly successful technique for removing heavy metal contamination from agricultural soils. Nevertheless, the capacity of immobilized metals to become mobile again is a noteworthy environmental issue, especially in regions that experience acid rain. The study aimed to evaluate the stability of immobilization treatment for two cadmium (Cd) contaminated soils by testing the effectiveness of various additives under simulated soil acidification conditions (Guo et al., 2018).

19.2.4 Biological Approach

19.2.4.1 Phytoremediation

Green plants are cultivated in contaminated soils as part of phytoremediation, specifically through Phytoextraction and Phytovolatilization, to eliminate heavy metals or render them stable. This process, also known as Phyto immobilization and Phytostabilization, is illustrated in Figure 18.2. Phytoremediation refers to the use of plants in polluted areas to promote various interactions, such as biological, biochemical, physical, microbiological, and chemical, to reduce the toxicity of toxins (Godheja et al., 2019; Agarwal et al., 2022). Different methods, such as biodegradation, vaporization, filtration, and others, rely on the type of pollutant. Organic contaminants are mainly eliminated through Rhizodegradation, biodegradation, vaporization, or stabilization processes. On the other hand, elemental pollutants such as heavy metals or radioactive elements are primarily extracted, transformed, and stored (Kuiper et al., 2004).

There are various methods by that plants engage with pollutants in phytoremediation.

1. **Phytoextraction –** Hyperaccumulation of pollutants in collectable parts of the plants. This process entails removing the pollutant from the soil and causing it to accumulate in a plant's root, stem, or leaf (Sidhu et al., 2018).
2. **Phytotransformation/Phytodegradation –** The pollutant is absorbed by the plant and then transformed into a less dangerous form. A comparable transition can occur on the plant's surface (Park et al., 2011).
3. **Phytovolatilization –** Volatilization by leaves through transpiration. In this procedure, the pollutant is taken out of the soil by the plant and transformed into a volatile substance before being released into the atmosphere (Cristaldi et al., 2020).
4. **Phytostimulation –** Plant roots discharge materials (exudates) that might quicken the breakdown of some contaminants. Whenever microorganisms aid in degradation, the entire process is referred to as Phytostimulation (Borriss, 2020).

5. **Phytostabilization –** This procedure involves immobilizing contaminants in the soil to prevent erosive processes and to enable connection with lignin and humus (Shackira & Puthur, 2019).
6. **Rhizofiltration –** Rhizosphere accumulation through sorption, concentration, and precipitation. This technique uses plant roots to help pollutants in wastewater, groundwater, and surface water be absorbed, precipitated out, and removed (Awa & Hadibarata, 2020).
7. **Rhizodegradtion –** Because fungi, bacteria, and yeasts in the rhizosphere have metabolic rates that are 10 to 100 times higher than those in soil, they may efficiently break down pollutants from Energies 2020 as well as organic nutrients found in nature (Cristaldi et al., 2020).

19.2.4.2 Bioremediation

Bioremediation is a technique that employs live microorganisms to lessen or prevent natural pollution (Cristaldi et al., 2020). Remediation is the process of eliminating a problem, and the term "bioremediation" is used when it involves addressing an ecological problem, such as soil and groundwater contamination. It is a technology that, when compared to other engineering methods, is more affordable, productive, unique, eco-friendly, and solar driven (Ali et al., 2013). Ever since it was initially used in 1989 to address the environmental catastrophe caused by the Exxon Valdez, this bioprocess has emerged as one of the most promising new methods for removing oil and its byproduct (Kumpiene et al., 2019).

19.2.4.3 Bioaccumulation

When the rate of uptake exceeds the rate of excretion, a phenomenon known as bioaccumulation occurs - the accumulation of chemicals in organisms. Chemicals can enter the body through trophic transfer and through contact with food and the abiotic environment (soil, water, air) (Prokop, 2000). The increase in metal concentration in organisms over time relative to environmental levels is called bioaccumulation. If metals are absorbed and retained faster than they can be digested or excreted, they can accumulate in the body (Sharma, 2004).

19.3 EXSITU REMEDIATION TECHNIQUES

Exsitu remediation entails the removal of polluted soils or groundwater from the subsoil and either treating it on the spot or relocating it somewhere for cleanup (Tsai et al., 2008). Insitu approaches, on the other hand, are procedures that take contaminated soil or groundwater right out of the ground. Because it is less expensive than exsitu monitoring, in situ treatment is frequently chosen (Decesaro et al., 2017). The implementation of environmental policies is a serious issue for developing nations. Remediation of soil and groundwater is primarily intended to reduce contamination to a manageable level. Physical, biological and chemical methods are employed to clean soil and groundwater. Generally speaking, a number of variables, including the features and contaminants of the soil as well as the type of remediation technique selected and planned, have a considerable impact on the decision of which soil and groundwater remediation processes are best suited (Godheja et al., 2019).

Exsitu bioremediation methods won't likely be applied in some locations, like workplaces, city centers, and construction zones. Exhumation procedures used in exsitu bioremediation typically change the soil's structure, aggravating polluted areas and surrounding areas. More time and resources are required to develop the approaches because any biological restoration process carried out in situ must have a modest to wide design. These methods typically call for a broad field of action. Exsitu bioremediation technique offers an effective and versatile solution for treating a wide range of contaminants. It is more preferred method compared to insitu method because of its speed and ease of implementation. One of the key advantages of exsitu bioremediation is that it eliminates the need for a comprehensive initial assessment of the contaminated site before remediation can begin. It also allows for more control over key interaction factors such as pH, temperature, and mixing, which can be easily adjusted to optimize the bioremediation process, which enhances the overall efficiency and effectiveness of the remediation process. These techniques permit adjustments to the organic, material, physical, chemical, and other parameters necessary for convincing and successful bioremediation (Prokop, 2000).

19.3.1 Physical Approach

19.3.1.1 Landfilling

The primary method of e-waste disposal is landfilling. Mostly, the discarded electronic devices are either burnt in open or discarded along with municipality waste ending up in landfills, which releases detrimental gases in atmosphere or leachate underground (Rajpal et al., 2024; Mondal et al., 2023). The informal method of e-waste treatment used in mostly developing countries give low yield (Yu et al., 2010). The discarded electronic products contain various components and hence should be segregated before discarding in the landfill, need to be kept and then moved to the curbside or an off-site pickup location (Liu et al., 2018). Although studies have shown that curbside collection is the most practical strategy for residential waste management, off-site drop-offices are nevertheless popular. This is because curbside collections are thought to be costly, time-consuming, and difficult to design, administer and manage (Chancerel, 2010).

19.3.2 Chemical Approach

19.3.2.1 Soil Washing

Physical separation and chemical leaching by aqueous solution are the two techniques used in soil washing for removing pollutants from the soil. First step is homogenization wherein the coarse particles are segregated based on their densities. The basis of physical separation is the tendency of most inorganic and organic pollutants to adhere to silt, clay and other inorganic particles. The finer particles are separated from coarse gravel and sand. Washing operation help in concentrating the pollutants into smaller soil volume and further cleaned using bioremediation or incineration. Depending on the type of contamination that needs to be treated, different reagents and additives are added to water. The implementation of the technique typically necessitates a sequential procedure utilizing various washing solutions in soils polluted by a number of distinct chemicals, each with a particular set of properties. The technology used to treat the contaminants is applied to the contaminated water.

Pollutant extraction, degradation, or fixation is reasonably quick and easy when using chemicals for soil washing. Soil washing could be done in conjunction with oxidation or degradation. To treat soils contaminated by PAHs, PBDEs, PCBs, Pb, Ni and Cu for instance, sodium persulfate (SP), citric acid (CA), and hydroxypropyl—cyclodextrin (HPCD) were mixed. Toxic metals and organic contaminants were removed with removal efficiency of 69.4-78.1% and 93.8-96.4%, respectively, after washing for 180-240 minutes (Chen et al. 2017). Among the solvents employed, HPCD was utilized to extract POPs, CA to remove metals, and SP to oxidize POPs even further. In a different study, polyoxymethylene lauryl ether (Brij35) and TiO2 were utilized to extract and photocatalyzed the breakdown of PCBs, while HPCD was also employed to remove PCBs from soil. Different methods could be used to improve soil washing to remove various pollutants from contaminated soils. For instance, ultrasonication-assisted soil washing with tea saponin (TS, 10 g/L1) and methyl—cyclodextrin (MCD, 15 g/L1) effectively eliminated PBDEs, PAHs, PCBs, Pb, Ni and Cu by almost 90% (Chen et al., 2017). After two cycles of 60 minutes each, ultra-sonication-assisted washing mobilized 91-94% of the PBDEs, PCBs, PAHs, Pb, Cu, and Ni in the contaminated soils, but less pollutants (86-88%) were mobilized without ultra-sonication even after three cycles of 120 minutes each (Chen et al., 2017). The washing time was cut in half thanks to ultra-sonication, which also improved the mobilization of contaminants. However, in such disturbed soils, microorganisms and plants are crucial elements to restoring normal soil functioning, which will offer long-term stabilization after soil washing since it's a cost-effective process. In general, soils with between 30 and 50 percent silt, clay, or organic matter are useless for washing (Sharma, 2004). Additional disadvantage is the need for costly and time-consuming specialist treatment of the utilized wastewater, which often contain added chemicals.

19.3.2.2 Solidification

The cement solidification and high-pressure compaction methods are practical and environmentally friendly way to safely dispose of used printed wire boards (PWBs) (Niu & Li, 2007). The investigation of E-plastic waste in concrete was closely examined. According to the compressive strength tests, the inclusion of E-plastic as a percentage replacement for coarse aggregate had no significant effects on the compressive strength of concrete. When the replacement percentage is 20%, it is discovered that the compressive strength of concrete decreases by 23.6% in comparison to the control mix. Additionally, it was discovered that 12% aggregate substitution with E-plastic yields satisfactory results without any long-term negative impacts (Lakshmi, 2010).

19.3.3 Thermal Approach

19.3.3.1 Vitrification

Vitrification is an effective method for treating waste by turning the waste materials into glass or glass-like substances. It has also been shown to have positive effects on the destruction of volatile organic and semi-volatile inorganic compounds, as well as elimination of elements in glassy materials by trapping inside glass matrix (Sigmon & Skorka, 1998), Hazardous waste vitrification offers both benefits and drawbacks. The advantages of the process is that it reduces the volume of waste as well as prevents generation of dangerous byproducts. Additionally, inert glass can be addition, inert glass may be discarded in landfill as it offers strong resistance to chemical leaching (Miskewitz, 2014). Additionally, the material produced by vitrification is appropriate for use in roads and pavements as well as for inactivating the pollutants in metal slag to stop them from leaking into the environment. The drawbacks are high cost of installation cost and utilization of energy in comparison to landfill or waste stabilization utilizing the cement matrix. In addition, processing waste requires a significant amount of energy (Mmereki, 2016). On account of some of the issues as discussed, the need of the hour is creating low -cost flux agents in order to attain the desired outcomes best results for the breakdown of degradation of wastes such as metals and metalloids (Sanito et al., 2022). Only then will vitrification be able to be used to remediate waste successfully.

19.4 FUTURE PROSPECTS AND CONCLUSION

So far, we have learned that current technology has increased and developed over the years, and a considerable quantity of e-waste continues to be generated. However, given the advantages of modern technology (advanced computers, robotics, life-saving medical equipment, consumer electronics, etc.), we cannot stop developing them because they are so integral to our daily lives. Efficient recycling of electronic waste by promoting sustainable consumption, management systems, and production patterns of electronic materials and avoiding hazardous substances where feasible (e.g., using biodegradable polymers to manufacture electronic devices). Some metals used in electronic devices are highly toxic and harmful to the environment and living organisms. Complexes are formed when they enter the water and settle in the soil subsoil. The complex formation of the metal and its adverse effects on the environment and animals make it challenging to handle. Bioremediation and heat treatment procedures have the highest potential for removal and the most general acceptance of any technique used to date that is addressed in this study. Finally, exploring new techniques using different process objects based on comparable properties is essential to outperform current technology.

REFERENCES

Agarwal, S., Albeshr, M. F., Mahboobb, S., Atique, U., Pramanick, P., & Mitra, A. (2022). Bioaccumulation Factor (BAF) of heavy metals in green seaweed to assess the phytoremediation potential. *Journal of King Saud University-Science*, *34*(5), 102078.

Ali, H., Khan, E., & Sajad, M. A. (2013). Phytoremediation of heavy metals—Concepts and applications. *Chemosphere*, *91*(7), 869–881. https://doi.org/10.1016/j.chemosphere.2013.01.075

Awa, S. H., & Hadibarata, T. (2020). Removal of heavy metals in contaminated soil by phytoremediation mechanism: A review. *Water, Air, & Soil Pollution*, *231*(2), 47. https://doi.org/10.1007/s11270-020-4426-0

Borriss, R. (2020). Phytostimulation and Biocontrol by the Plant-Associated Bacillus amyloliquefaciens FZB42: An Update (pp. 1–20). https://doi.org/10.1007/978-981-15-2576-6_1

Chancerel, P. (2010). Substance flow analysis of the recycling of small waste electrical and electronic equipment. https://doi.org/10.14279/depositonce-2404

Chen, F., Luo, Z., Liu, G., Yang, Y., Zhang, S., & Ma, J. (2017). Remediation of electronic waste polluted soil using a combination of persulfate oxidation and chemical washing. *Journal of Environmental Management*, *204*, 170–178. https://doi.org/10.1016/j.jenvman.2017.08.050

Cristaldi, A., Oliveri Conti, G., Cosentino, S. L., Mauromicale, G., Copat, C., Grasso, A., Zuccarello, P., Fiore, M., Restuccia, C., & Ferrante, M. (2020). Phytoremediation potential of Arundo donax (Giant Reed) in contaminated soil by heavy metals. *Environmental Research*, *185*, 109427. https://doi.org/10.1016/j.envres.2020.109427

Decesaro, A., Machado, T. S., Cappellaro, Â. C., Reinehr, C. O., Thomé, A., & Colla, L. M. (2017). Biosurfactants during in situ bioremediation: factors that influence the production and challenges in evalution. *Environmental Science and Pollution Research*, *24*(26), 20831–20843. https://doi.org/10.1007/s11356-017-9778-7

Godheja, J., Modi, D. R., Kolla, V., Pereira, A. M., Bajpai, R., Mishra, M., Sharma, S. V., Sinha, K., & Shekhar, S. K. (2019). Environmental remediation: Microbial and nonmicrobial prospects. In *Microbial Interventions in Agriculture and Environment* (pp. 379–409). Springer Singapore. https://doi.org/10.1007/978-981-13-8383-0_13

Guo, F., Ding, C., Zhou, Z., Huang, G., & Wang, X. (2018). Stability of immobilization remediation of several amendments on cadmium contaminated soils as affected by simulated soil acidification. *Ecotoxicology and Environmental Safety*, *161*, 164–172. https://doi.org/10.1016/j.ecoenv.2018.05.088

Hansen, H. K., Ottosen, L. M., & Ribeiro, A. B. (2016). Electrokinetic soil remediation: An overview. In *Electrokinetics Across Disciplines and Continents* (pp. 3–18). Springer International Publishing. https://doi.org/10.1007/978-3-319-20179-5_1

Hu, B., Chen, S., Hu, J., Xia, F., Xu, J., Li, Y., & Shi, Z. (2017). Application of portable XRF and VNIR sensors for rapid assessment of soil heavy metal pollution. *PLoS One*, *12*(2), e0172438. https://doi.org/10.1371/journal.pone.0172438

Kuiper, I., Lagendijk, E. L., Bloemberg, G. V., & Lugtenberg, B. J. J. (2004). Rhizoremediation: A beneficial plant-microbe interaction. *Molecular Plant-Microbe Interactions®*, *17*(1), 6–15. https://doi.org/10.1094/MPMI.2004.17.1.6

Kumpiene, J., Antelo, J., Brännvall, E., Carabante, I., Ek, K., Komárek, M., Söderberg, C., & Wårell, L. (2019). In situ chemical stabilization of trace element-contaminated soil–Field demonstrations and barriers to transition from laboratory to the field–A review. *Applied Geochemistry*, *100*, 335–351. https://doi.org/10.1016/j.apgeochem.2018.12.003

Lakshmi, R., & Nagan, S. (2010). "Studies on concrete containing e plastic waste." *International Journal of Environmental Sciences*, *1*(3), 270.

Lasat, M. M. (2002). Phytoextraction of Toxic Metals. *Journal of Environmental Quality*, *31*(1), 109–120. https://doi.org/10.2134/jeq2002.1090

Liu, L., Li, W., Song, W., & Guo, M. (2018). Remediation techniques for heavy metal-contaminated soils: Principles and applicability. *Science of the Total Environment*, *633*, 206–219. https://doi.org/10.1016/j.scitotenv.2018.03.161

Miskewitz, R. B. D. (2014). Landfill Closure with dredged materials–Desktop analysis. https://rosap.ntl.bts.gov/view/dot/28427

Mmereki, D. A. B. L. H., & B. L. (2016). The management of hazardous waste in developing countries. *Management of Hazardous Wastes*. https://doi.org/10.5772/63055

Mondal, T., Choudhury, M., Kundu, D., Dutta, D., & Samanta, P. (2023). Landfill: An eclectic review on structure, reactions and remediation approach. *Waste Management*, *164*, 127–142. https://doi.org/10.1016/j.wasman.2023.03.034

Niu, X., & Li, Y. (2007). Treatment of waste printed wire boards in electronic waste for safe disposal. *Journal of Hazardous Materials*, *145*(3), 410–416. https://doi.org/10.1016/j.jhazmat.2006.11.039

Park, S., Kim, K. S., Kim, J.-T., Kang, D., & Sung, K. (2011). Effects of humic acid on phytodegradation of petroleum hydrocarbons in soil simultaneously contaminated with heavy metals. *Journal of Environmental Sciences*, *23*(12), 2034–2041. https://doi.org/10.1016/S1001-0742(10)60670-5

Prokop, Gundula. (2000). Management of Contaminated Sites in Western Europe.

Rajpal, A., Choudhury, M., Goswami, S., Chakravorty, A., & Raghavan, V. (Eds.). (2024). *Waste Management and Treatment: Advances and Innovations* (1st ed.). CRC Press. https://doi.org/10.1201/9781003258377

Rautela, R., Arya, S., Vishwakarma, S., Lee, J., Kim, K.-H., & Kumar, S. (2021). E-waste management and its effects on the environment and human health. *Science of The Total Environment*, *773*, 145623. https://doi.org/10.1016/j.scitotenv.2021.145623

Salt, D. E., Smith, R. D., & Raskin, I. (1998). Phytoremediation. *Annual Review of Plant Physiology and Plant Molecular Biology*, *49*(1), 643–668. https://doi.org/10.1146/annurev.arplant.49.1.643

Sanito, R. C., Bernuy-Zumaeta, M., You, S.-J., & Wang, Y.-F. (2022). A review on vitrification technologies of hazardous waste. *Journal of Environmental Management*, *316*, 115243. https://doi.org/10.1016/j.jenvman.2022.115243

Shackira, A. M., & Puthur, J. T. (2019). Phytostabilization of heavy metals: Understanding principles and practices. In *Plant-Metal Interactions* (pp. 263–282). Springer International Publishing. https://doi.org/10.1007/978-3-030-20732-8_13

Sharma, H. D., and K. R. Reddy. (2004). Geoenvironmental engineering: site remediation, waste containment, and emerging waste management technologies. ISBN: 0-471-21599-6

Sidhu, G. P. S., Bali, A. S., Singh, H. P., Batish, D. R., & Kohli, R. K. (2018). Ethylenediamine disuccinic acid enhanced phytoextraction of nickel from contaminated soils using Coronopus didymus (L.) Sm. *Chemosphere*, *205*, 234–243. https://doi.org/10.1016/j.chemosphere.2018.04.106

Sigmon, B., & Skorka, M. B. (1998). The Market for Waste Vitrification Technology. *Remediation Journal*, *8*(4), 47–67. https://doi.org/10.1002/rem.3440080406

Tsai, T. T., Kao, C. M., Yeh, T. Y., & Lee, M. S. (2008). Chemical oxidation of chlorinated solvents in contaminated groundwater: Review. *Practice Periodical of Hazardous, Toxic, and Radioactive Waste Management*, *12*(2), 116–126. https://doi.org/10.1061/(ASCE)1090-025X(2008)12:2(116)

Yu, J., Williams, E., Ju, M., & Shao, C. (2010). Managing e-waste in China: Policies, pilot projects and alternative approaches. *Resources, Conservation and Recycling*, *54*(11), 991–999. https://doi.org/10.1016/j.resconrec.2010.02.006

20 Electronic Wastes, Occupational Exposure, and Safety Measures

Rwitabrata Mallick
Amity University Madhya Pradesh, Gwalior, India

Sandeep Poddar
Lincoln University College, Selangor, Malaysia

20.1 INTRODUCTION

Electronic garbage, junk, e-waste refers to used electronics that have reached the end of life and have been thrown, donated, or given to a recycler. E-waste is defined as any electronic devices which have been abandoned including a battery or plug and contain poisonous and dangerous materials like mercury that can seriously endanger both human and environmental health. Computers, keyboards, monitors, printers, central processing units, refrigerators, washing machine, VCRs, stereos, copiers, fax machines, typewriters, cell phones and chargers, and other household appliances (Sengupta et al., 2022) all produce e-waste. Several household appliances, including speakers, music systems, and other appliances, contain harmful chemicals (Gangwar et al., 2019). Current management techniques for waste electrical and electronic equipment (WEEE) are different in different nations and regions (Agarwal et al. 2023). Comprehensive analysis is done on global trends in WEEE amounts and composition, and the various handling, regulating, and prevention methods used by chosen nations. The results show that there are significant and rising amounts of WEEE production. At least in terms of volume, IT and telecom equipment appear to be the predominant WEEE produced in Africa, Asia's less developed regions, and Latin and South America. The research asserts that the reported amounts of WEEE created may be significantly understated. For many nations it appears to be deficient in or lack in starting, creating, and approving WEEE rules. Waste electrical and electronic equipment handling in developing countries is characterized by an expediate rate of rescue and repair in a recycling industry that is largely unregulated. Both industrialized and developing nations still struggle with the issue of disposing of WEEE in landfills. It is common knowledge that the USA and less developed nations often hoard surplus electrical and technological goods. In addition, four global priority areas for WEEE have been highlighted, including ethical issues, health issues, resource depletion, WEEE take back programs and environmental issues.

Present study explores knowledge of handling of various e-wastes, their characteristics, fate and utilization especially during post usage stage. Studies have shown that the mobility of initiating and enacting of e-waste related acts and regulations is relatively very slow from the point of view of the whole world. Even in some countries, at present there is no at all legislations related to electronic waste materials. From various impact analysis, contamination aftereffects and umpteen health issues, it is very much evident that there is extreme need for updated database of e-wastes worldwide. Recycling technology done informally, especially in developing countries, might become a significant factor if by any chance their mode of execution is modified with stringent safety standards.

 DOI: 10.1201/9781003582311-20

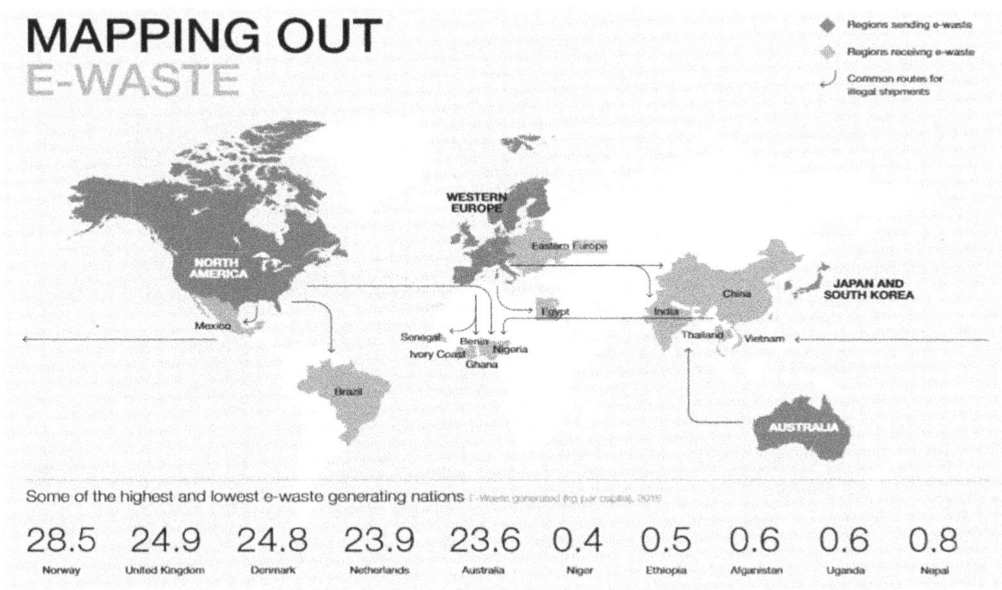

FIGURE 20.1 Worldwide E-waste generation.

Source: https://www.genevaenvironmentnetwork.org/wp-content/uploads/2020/09/mapping-out-ewaste.jpg

20.2 E-WASTE GENERATION: WORLDWIDE

The UN estimates that in 2021, e-waste production worldwide reached a staggering 57.4 million ton, or 7.6 kg per person on the planet. Only 17.4% of this hazardously mixed-with-precious-materials electronic garbage was recognized as having been adequately collected, handled, and recycled. E-waste can be dangerous, is not biodegradable, and accumulates in the environment's air, water, and other living things. When open-air burning and acid baths are used to recover valuable elements from electronic components, toxic contaminants are discharged into the environment. Additionally, these processes could expose workers to high levels of pollutants causing cancer, miscarriages, neurological damage, and lower IQs, as well as arsenic, beryllium, lead, thallium, mercury and cadmium. Municipal waste management is difficult (Choudhury and Dutta 2017).

Out of 53.6 Mt of e-waste generated world-wide only 17 % has been documented as collected and recycled, as per the observation of the Global E-waste Monitor 2020. A worldwide view and estimation of e-waste generation (Figure 20.1) provides a clear view regarding the ever-increasing e-waste on a regular basis. According to UNEP 2015, e-waste valued USD 19 billion is annually dumped or sold illegally which is in between 60–90 % of the world's electronic waste.

20.3 E-WASTE GENERATION IN INDIA

E-waste is a problem for many developing and under developed countries especially India where the domestic production and illegal import is adding to the burden. When the materials are imported from waste streams, the emerging economies can access raw materials and components that can be used to refurbish or re-assemble affordable electronic devices. This business model provides access to low-cost equipment for local consumers as well as creates employment opportunities thereby stimulating economic growth in the recycling and refurbishing industry.

The anticipated pace of India's growth in the generation of electronic wastes has been enhancing steadily over the last decade (Figure 20.2). The rise and production of e-waste does not appear to be

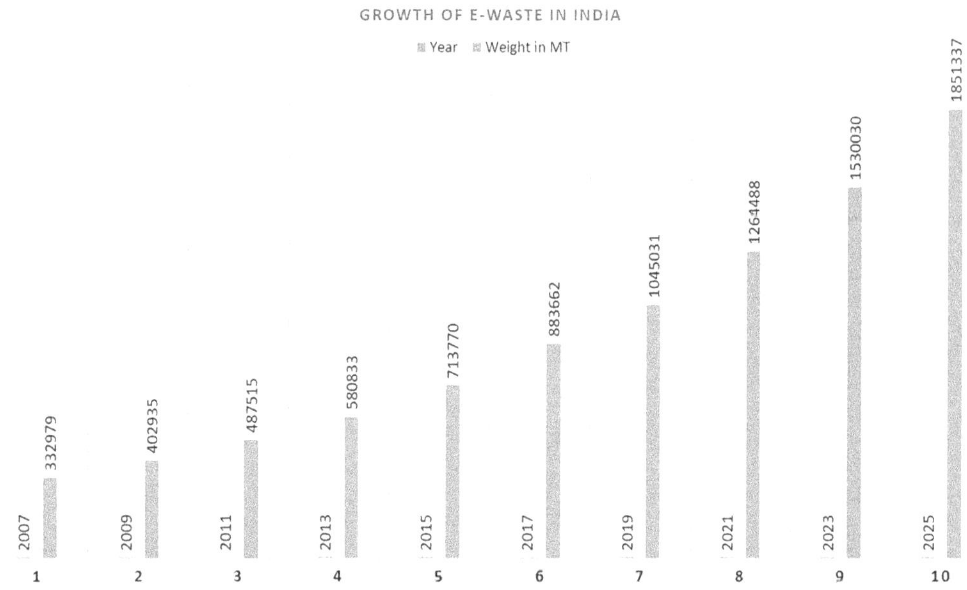

FIGURE 20.2 Growth of e-waste in India. (Ref: Mihai, (Ed.). (2016))

TABLE 20.1
E-waste Generation Amongst the Continents around the World

Continent	Total (Mt)	Per Capita (Kg)
North & South America	13.1	13.3
Europe	12.0	16.2
Asia	24.9	5.6
Africa	2.9	2.5
Oceania	0.7	16.1

Source: https://theroundup.org/global-e-waste-statistics/

decreasing in the ensuing decades, either, because of increasing economic development, a population explosion, and globalization.

The following table (Table 20.1) provides a quantitative value of the e-waste generation amongst continents around the world.

An emerging economic sector arises which focuses on the trade, repair, and recovery of materials from outdated electronic devices and presents opportunities for individuals living in urban and rural areas, to earn a livelihood. However, this sector also poses significant hazards to both the local population and the environment if recycling is done in primitive way. Many recyclers may lack awareness of the risks associated with handling e-waste, may not be familiar with safer techniques for recycling, or may face challenges accessing the necessary investment capital to implement better practices. As a result, unsafe practices in the handling and disposal of e-waste are leading to adverse consequences for human health and the environment.

Hence, focus should not only be on efficient material recovery but also on safe recycling of e-waste, implicating that our concern should aim at maximizing economic value along with minimizing the environmental and health risks associated with electronic waste (Rautela et al., 2021). E-waste contains valuable materials that can be retrieved and reused, as well as potentially harmful pollutants should be managed responsibly. To address these issues, it is crucial for all stakeholders, including the youth, to prioritize the proper management of e-waste. There is a pressing need to bridge the gap between the amount of electronic waste generated and the proportion that is effectively recycled.

20.4 OCCUPATIONAL HAZARDS DUE TO E-WASTE EXPOSURE

When e-waste is recycled, it can result in exposure to several dangerous compounds that are either present in EEE or created and discharged due to poor recycling processes. Direct exposure involves meeting dangerous substances through the skin, breathing in both fine and large particles, and eating contaminated dust. High amounts of direct, occupational exposure result from people recycling e-waste without adequate protection. The risk of exposure to hazardous compounds is frequently increased by unsafe recycling processes employed to recover valuable materials (Song et al., 2015). During such recycling processes, notably smelting, there is frequently a lack of adequate off-gas treatment.

On a daily, weekly, monthly, and annual basis, the amount of e-waste produced by various electrical and electronic devices—including laptops, refrigerators, microwaves, washing machines, stereos, 55printers, fax machines, various accessories, monitors, keyboards, computers, typewriters, mobile phones and chargers, remote controls, compact discs, headphones, batteries, LCD/Plasma TVs, i-pods, air conditioners, dryers—increases. Developing countries have been in the fore front towards growing demand of electric devices and generation of e-wastes.

The management of enormous and expanding amounts of electrical and electronic trash, which is emerging as one of the most significant environmental issues facing developing nations, particularly India, adds to this load. Due to the extremely serious environmental and health risks it poses, e-waste has grown to be a bigger issue than any other type of waste. If not properly disposed of in an environmentally responsible manner, the numerous hazardous components included in e-waste have the potential to seriously harm both the environment and both human and animal health. Numerous initiatives are made to address this expanding issue, but the time has come for a practical, scientific, lasting solution (Garlapati et al., 2016) that can be implemented and play a significant role in consumer education.

E-waste exposure and its negative health impacts require a lot of research. There is insufficient data to conclusively link occupational exposure (Zhang et al., 2019) to hazardous e-waste materials to negative health impacts. Further, in-depth investigation is necessary to determine the probable causal link between exposure and the adverse effects identified. Additionally, it is unknown how biological agents and secondary compounds from e-waste interact. Political and ethical concerns, a lack of funding, poor access to uncontrolled conditions, and other factors can all obstruct study on the dangers of e-waste. There is insufficient regulation and surveillance, especially of illegal e-waste recycling operations. Despite these research hindrances, more research is necessary. E-waste exposure risk evaluations are essential, but there is also a pressing need for research to support unauthorized local, regional, and global e-waste recycling strategies (Agarwal et al., 2023).

20.5 MINIMIZING E-WASTE FOLLOWING THE 3 R PRINCIPLE

Recycling electrical waste is crucial, but it needs to be done carefully and safely. E-waste should, whenever possible, be repaired and recycled as a whole piece as opposed to being disassembled. E-waste should be deconstructed at state-of-the-art recycling facility of electronic wastes worldwide (Perkins et al., 2014) by experienced, secured and reasonably compensated workers when refurbishment is not an option. Few fundamental principles should always be followed at the time of e-waste policy making.

It is not practicable to totally remove dangerous components from EEE, notwithstanding its effectiveness. Even while research is necessary, it is equally crucial to create and carry out educational and awareness campaigns regarding the potential risks associated with recycling e-waste. In developing countries, these programs are essential. Fighting to end child labor and enhance working conditions for all e-waste workers is essential. The regional culture, geography, and limitations of the most vulnerable communities should be taken into consideration while designing interventions. Globally, laws that promote safe, regulated, and paid e-waste recycling should be put into place (Okemo et al., 2019).

The exponential increase in the quantity of sophisticated end-of-life electronic equipment expediated the birth of an extremely significant issue termed as electronic waste or in short e-waste. Global production, consumption, and recycling levels produce substantial flows of both dangerous and useful elements.

20.6 CREATING AWARENESS

The Basel Convention-developed international laws, which primarily focus on a universal prohibition for transboundary movements of e-waste, appear to confront challenges in their successful implementation; nonetheless, a conclusive description of the status and trends is not yet available. In addition, the Mobile Phone Partnership Initiative (MPPI) was established in 2002 to handle mobile phones that are nearing the end of their useful lives in an environmentally sound manner (Basel Convention, n.d.). Five technical standards were created as part of the MPPI: increasing awareness about design issues, collecting used and end-of-life mobile phones, refurbishing used mobile phones, moving collected mobile phones across borders, and material recycling/ recovery of end-of-life mobile phones (Roy et al., 2022).

20.7 CONCLUSION

According to the present scenario, improper electronic waste management is leading to a massive loss of rare and significant raw materials, including valuable minerals used for various electronic gadgets. It is challenging to extract minerals from electronic wastes; sometimes, the recovery rate becomes as low as 20–30%. Recycling of metals found in e-waste (smartphones, laptops, etc.) not only helps meet their industrial demand but also offers environmental benefits. Recycled metals require significantly less energy to melt than metals produced from new ore, making recycling more environmentally friendly and energy-efficient. Moreover, it can help reduce greenhouse gas emissions, as mining used electronics for metals emits far less CO2 (80 %) per unit of gold than traditional mining processes. Electronic waste is not only corrosive, harmful, or impacts humanity, but it is also non-biodegradable and ultimately contaminates biosphere, lithosphere, hydrosphere, and sometimes atmospheric segments of the environment.

Various harmful substances are emitted into the open environment through burning in the open air, open incineration, and using methods like acid baths to extract precious minerals from the waste. The threat of technical waste in electronic waste exists in all countries (e-waste). E-waste impacts health and the environment by exposing individuals to dangerous substances, particularly polycyclic aromatic hydrocarbons and persistent organic pollutants, and by introducing toxic chemicals into the food chain. This study examines how e-waste impacts the environment and public health in India. At the end of the present study, producing affordable electronics with low levels of chemical toxicity will help showcase the e-waste problem, which has detrimental impacts on nature and humanity.

REFERENCES

Agarwal, S., Darbar, S., Saha, S., Choudhury, M., & Singh, R. P. (2023). E-waste management using different cost-effective, eco-friendly biological techniques: An overview *In Waste Management and Resource Recycling in the Developing World*, 205–235. https://doi.org/10.1016/B978-0-323-90463-6.00026-9

Basel Convention. (n.d.). *Basel Convention > Implementation > Technical Assistance > Partnerships > MPPI > Overview*. Copyright 2011 by Basel Convention. http://www.basel.int/Implementation/TechnicalAssistance/Partnerships/MPPI/Overview/tabid/3268/Default.aspx

Choudhury, M., & Dutta, J. (2017). A comparative study of municipal solid waste management status for three major towns of Upper Assam-India. *International Journal of Waste Resources*, 7(291), 2. https://doi.org/10.4172/2252-5211.1000291

Gangwar, C., Choudhari, R., Chauhan, A., Kumar, A., Singh, A., & Tripathi, A. (2019). Assessment of air pollution caused by illegal e-waste burning to evaluate the human health risk. *Environment international*, 125, 191–199.

Garlapati, V. K. (2016). E-waste in India and developed countries: Management, recycling, business and biotechnological initiatives. *Renewable and Sustainable Energy Reviews*, *54*, 874–881.

Mihai, F. C. (Ed.). (2016). *E-Waste in Transition: From Pollution to Resource*. BoD–Books on Demand.

Okeme, J. O., & Arrandale, V. H. (2019). Electronic waste recycling: occupational exposures and work-related health effects. *Current Environmental Health Reports*, 6, 256–268.

Perkins, D. N., Drisse, M. N. B., Nxele, T., & Sly, P. D. (2014). E-waste: A global hazard. *Annals of Global Health*, 80(4), 286–295.

Rautela, R., Arya, S., Vishwakarma, S., Lee, J., Kim, K. H., & Kumar, S. (2021). E-waste management and its effects on the environment and human health. *Science of the Total Environment*, 773, 145623.

Roy, J. J., Rarotra, S., Krikstolaityte, V., Zhuoran, K. W., Cindy, Y. D. I., Tan, X. Y., … & Srinivasan, M. (2022). Green recycling methods to treat lithium-ion batteries E-waste: a circular approach to sustainability. *Advanced Materials*, *34*(25), 2103346.

Song, Q., & Li, J. (2015). A review on human health consequences of metals exposure to e-waste in China. *Environmental Pollution*, *196*, 450–461.

Zhang, M., Shi, J., Meng, Y., Guo, W., Li, H., Liu, X., … & Hu, Q. (2019). Occupational exposure characteristics and health risk of PBDEs at different domestic e-waste recycling workshops in China. *Ecotoxicology and Environmental Safety*, *174*, 532–539.

Sengupta, D., Ilankoon, I. M. S. K., Kang, K. D., & Chong, M. N. (2022). Circular economy and household e-waste management in India: Integration of formal and informal sectors. *Minerals Engineering*, *184*, 107661.

21 Metal Extraction Process from Electronic Wastes

Arvind Laxmikant Chel and Abhay Vasantrao Wavare
MGM University, Aurangabad, India

Vijay Kumar Srivastava
The Maharaja Sayajirao University of Baroda, Vadodara, India

Geetanjali Kaushik
Hi Tech Institute of Technology, Aurangabad, India

21.1 CURRENT STATUS OF ELECTRONIC WASTES

E-wastes include electronic wastes which are generated from modern establishments. Usually these are described as discarded electrical or electronic devices that have completed their life period or can no longer serve their original purpose. Such devices or appliances include like computer monitors, keyboards, CPUs (central processing units), printers, batteries, mobile phones, headphones, mobile chargers, typewriters, LCD/Plasma TVs and their remotes, CDs (compact discs), refrigerators, ACs (air conditioners) and other useful appliances in the households (UNEP, 2007).

Generally, the unrecycled e-waste from developed countries reaches incinerators, landfills or developing countries (Rajpal et al., 2024). As per estimates, each year within the US market there is entry of about 20 million computers while over 12 million computers get disposed annually, and just 10% of these computers get recycled (Paul et al., 2019). Reports have claimed that the recycling industry within the USA exports about 80% of the received e-waste to Asian countries, and almost 90% of this waste was sent to China (Hicks et al. 2005). According to researchconducted by UNEP (United Nation Environmental Program), during 2019 the inherent value of the generated e-waste was valued at $57 billion while in 2014 it was about $ 52 billion (UNEP, 2019). Developed countries possess technologies for recycling e-waste, technical expertise and there exists a well-established boundary within the system. However, the situation with regard to the developing countries is not so encouraging (Maphosa & Maphosa, 2020). About 2% Asian populace depends on waste recycling for earning their livelihood. In 2019 on a global basis, formally only 17.4% of the total e-waste was recycled while the remaining 82.6% of the generated e-waste quantity was either not recycled or not tracked.

Composition of E-waste is generally categorized within two major categories: hazardous e-waste as well as non-hazardous e-waste. It mainly comprises of metals (ferrous and non-ferrous metals), wood, plywood, glass, plastics, PCBs, concrete, ceramics and rubber among others. 50% composition of e-waste is comprised mainly by iron and steel, almost 21% of content is comprised by plastics, and 13% of e-waste is comprised up of non-ferrous metals and others. Non-ferrous metals include aluminium and copper in addition to valuable metals like silver, gold, platinum, palladium, etc. Further the heavy metals presence such as those of mercury, cadmium, lead, arsenic, chromium, selenium as well as the flame retardants in quantities above the permissible make e-waste extremely hazardous to the environment. It has been reported that e-waste comprises greater than 1000 varied materials, most of these components are lethal in nature and may result in severe environmental pollution upon their disposal. E-wastes management hence, is crucial for the protection of human

DOI: 10.1201/9781003582311-21

health, environment and livelihood (Herat and Agamuthu, 2012). Recent estimates suggest that each year, there is generation of 50 million tons e-waste that is expected to twofold itself by the year 2050. In India during 2019–2020 almost 10 million tons of waste was generated while the formal recycling is as low as 2% of overall generated e-waste on an annual basis (CPCB, 2020). In the absence of a formal reverse e-waste sully chain and prevalence of crude methods for metal extraction, there is significant risk regards to the health and safety of recyclers as well as dealers in scrap within the informal sector.

WEEE (Waste Electronic and Electrical Equipment) Generation -has grown at an extraordinary rate. WEEE comprises of several types of heavy metals and hence, form a potential resource for metal recovery as well as for preventing environmental pollution resulting from inadequate recycling of WEEE. An approach that is sustainable for the recovery of metals from e- waste is urgently needed. PCBs present a significant challenge for recovery of heavy metals from the e-waste since it comprises of greater than 25 metals. In PCBs the presence of few prized metals (like silver, gold and palladium) and in addition to the presence of rare earth metals make it a useful solid waste. PCBs comprise of 40% metals, 30% of plastics and about 30% of ceramics (Lu & Xu, 2016). Amongst the metals, PCBs are composed of 8–38% iron, 10–27% copper, nickel ranging from 0.3–2%, 0.03–0.42% of zinc, 0.08–0.9% of tin, almost 2–19% content of aluminium. Precious metals such as gold (content ranging from 250–2050 ppm), silver (content varying from 110–4500 ppm) as well as palladium (40–4000 ppm) are also present within high grade PCB (Li et al., 2018; Korf et al., 2019). PCBs comprise of harmful flame retardants (Wang et.al, 2017). So, while waste PCBs are a high value resource in case of improper recycling and recovery there are hazardous drawbacks.

A major issue in e-waste recycling is on account of its highly diverse and complex composition, which make its processing difficult when it gets mixed with the common domestic waste. PCBS comprise of a useful component of e-waste owing to their valuable metal content however, they make up to only 3–6% of the entire e-waste. Further, similar to the issue with other types of e-waste PCBs are extremely diverse and their composition being dependent over the initial purpose (Hsu et al., 2019).

21.2 RECYCLING OF PRINTED CIRCUIT BOARDS WITHIN INFORMAL SECTOR – CASE STUDY FROM INDIA

There are several steps involved in the recycling of PCBs from computer monitors, CPUs, discs, printers, etc., as shown in Figure 21.1.

Procedure 1:	**Procedure 2:**
Mechanical Shredding + Physical Separation + Hydrometallurgical processing:	Smelting + Hydrometallurgical + Electrolysis
• **Pretreatment** • **Physical recycling of PCBs** • **Chemical recycling** • **Pyrometallurgical, Hydrometallurgical or Biotechnological process**	• **Smelting** • **Pyrometallurgy, Hydrometallurgy and Electrometallurgy**

FIGURE 21.1 Printed circuit boards recycling practiced globally.

As every motherboard core is made of flat plate of gold, firstly the gold-plated pins are manually removed. Laminated gold portions are slashed and sent to gold-smiths for recovery. Thereafter, heating is undertaken for the removal of components which can be resold such as ICs, bearings, condensers. Process of pre-heating includes leaving the motherboard on a stove at low heat. These boards which are pre-heated are further sent to the dealers for the recovery of solder. Conventional process is followed for recovery, it involves the use of a boiling gas kit of kerosene (enclosed within a water tub) for storing of the lead. Extracted Lead is released within the water tub and owing to its low density it floats on top. Post the process of solder removal, copper is retrieved through roasting or acid bath.

Open Burning – Once the components are separated, motherboards are left in open pit burning for the extraction of the copper layer laminated within the circuit board. Subsequent to charring, through the use of froth floatation method it is then distilled. Ash is washed and the copper containing few carbon impurity is sent to a different recycling unit. Even defective ICs, condensers, etc., which have no potential for resale are also set blaze within tiny enclosed spaces having hoods for extracting the metallic portions.

Acidic Bath – Collected motherboards are further dipped inside the acidic medium for long duration. Motherboard as well as the acid are heated up inside a large vessel for formulating copper sulphate crystal. Subsequently, the irons chips are added to acid. The sludge which contains copper is then extracted through ion exchange process for recovering copper from it.

It is important to highlight that both the processes, open burning as well as acid bath, involve occupational health risks to the workers involved in the extraction process as well as for the people residing within the surrounding areas.

21.3 ROLE IN TRADE CHAIN

Informal sector at this stage plays the role of manual dismantling-collected scrap which has potential for reuse is sold to second hand market, defective parts are repaired and then sold while the scrap which has no further use or cannot be repaired is dismantled (Borthakur & Sinha, 2013). The next level within the trade chain is a specialized task.

21.4 PROCESSING – RECYCLERS/PROCESSORS/MELTERS

At this level the E-waste is at final level of processing. Those components of e-waste which cannot be further repaired are processed at this stage. For extraction of sub parts shredding or size reduction is done so that it is suitable for recycling/smelting. It is clear from the above discussion regarding e-waste processing that the stakeholders which are engaged in this level of process gain experience in a particular activity or set of activities that finally become their main strength. Therefore, all the processors or recyclers within this level may not be competent enough in complete processing of the e-waste. Thus, from above studies we can conclude that in India most of the e-waste goes to informal sector. In unorganized sector there is no scientific way for proper processing or recycling of the e-waste properly.

21.5 E-WASTE MANAGEMENT: PRACTICES AND LEGISLATIONS

WEEE is becoming a priority stream within waste management on account of major challenges associated with it. Several European and Asian countries (like Japan and Taiwan) are formulating strict policies and legislations that focus on the manufacturer's responsibility for taking back the products once they complete their useful life span (Queiruga et al., 2012). The maximum focus is laid over brown goods (which include computer, mobiles, etc.), the white goods (which comprise of the refrigerators, air conditioners, etc.), the automobiles and the batteries that require distinct handling as well as disposal (Widmer et al., 2005; Langrova, 2002). In a similar manner legislations are in place in the USA (Gutowski et al., 2005). Even the European Union has formulated various policies like the WEEE (Directive 2002/96/EC) and the RoHS, (Directive 2002/96/EC) Directives.

21.6 RECYCLING TECHNOLOGIES FOR E-WASTE WITHIN DEVELOPED NATIONS

In case it is not possible to reuse or repair EEE it must be collected and recycled through three major steps: Collection of e-waste, pre-processing and processing (Figure 21.2). Since composition of e-waste is heterogonous and it consists of a wide range of materials, for achieving a circular economy it is crucial that all these materials are properly collected and then recycled (Van Yken et al., 2021).

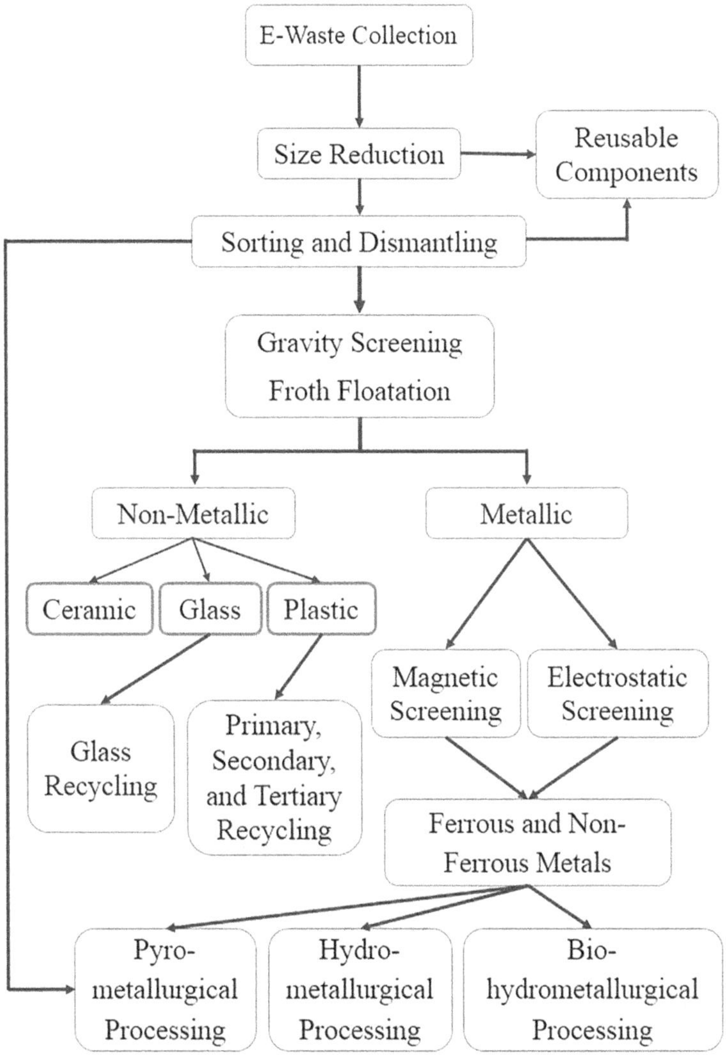

FIGURE 21.2 E-waste recycling and processing.

21.7 E-WASTE PROCESSING

21.7.1 E-Waste Collection and Pre-Treatment

Authorized collection of the e-waste is under the control of government legislations and policies. Almost, 78 countries worldwide have legislation regarding e-waste collection and recycling. Collection is either at facilities within public spaces or domestic businesses that collect e-waste. On e-waste collection, it is pre-processed. Prior to its processing e waste is sorted as few harmful components in e-waste need special preprocessing. Further stage in pre-treatment is known as dismantling which may be manual or mechanical dismantling. Improper dismantling poses threat to human health and the environment owing to the discharge of fumes and toxic substances (Mohite, 2013, Agarwal et al. 2023). Dismantling involves removing of fasteners (like screws, bolts) but in case the constituents have been fused by welding, coating, etc., the procedure gets complex thereby needing extra processing for materials separation. Components that are marked for recycling further are reduced in size with the use of shredders, mills, crushers, etc., reduction in the size involves the subjecting electrical components to heat greater than solder's melting point (240–250 °C). Further the substance is subjected to use of exterior forces like impact, shearing and vibration. With the use of this approach a disassembly ratio of 94% may be achieved. After the reduction in size, the material can be processed further for the separation of metallic as well as the nonmetallic parts. Processes may involve gravimetric separation, froth flotation, electrostatic or magnetic method of separation. Gravity separation can concentrate the particles on basis of specific gravity. Magnetic separation separates the magnetic (metallic) parts from those of non-magnetic parts. Electrostatic separation sorts materials of varied electrical conductivities. Froth flotation employs particle hydrophobicity for the separation of metallic as well as the non-metallic components. This process has high level of success in particle recovery from PCBs (95.6% recovery for base metals). Greater effectiveness for copper recovery it may lead to significant loss of gold. It is to be noted that anyone process, or a mix of procedures, may be useful for concentrating metals within e-waste for further recycling through hydrometallurgical and bio-hydrometallurgical processing.

21.8 METAL RECYCLING

Recycling metals from e-waste is of significant economic value (Wang et al., 2021). Commonly the Pyro-metallurgical as well as the hydrometallurgical approaches are utilized to extract the base and the precious metals (Jadhao et al., 2020). Even the use of microbes and the metabolites with the help of bio-hydrometallurgical processing, or bioleaching, have also been researched for extracting metals from E-waste.

21.9 PYRO-METALLURGY

The conventional and the most commonly utilized approach for metal recovery from E-waste is Pyro-metallurgy (Panda et al., 2021). It makes use of higher temperatures within either reductive or oxidative conditions for bringing the physical as well as the chemical transformations for recovering the concerned metals. The pyro-metallurgical treatment for E-waste is commonly utilized in smelting within furnaces, incineration and pyrolysis.

21.10 HYDROMETALLURGY

Hydrometallurgy involves the use of aqueous solution for the extraction of metals present with in a feedstock. Chemical leaching encompasses the use of alkali, acids or complexes termed as lixiviants. Several studies have highlighted on hydrometallurgical application for metal recovery from e-wastes (Panda et al., 2021).

21.11 BIO-HYDROMETALLURGY

Bio-hydrometallurgy is included within hydrometallurgical processing. It utilizes microbes like bacteria and fungi for extracting and recovering metals from the ores, metal concentrates as well as waste materials which are present in a water environment (Baez et al., 2022).

21.12 GREENER PHYSICAL TREATMENT

Cryo-milling is another novel, recent approach for recycle of the metals from waste PCBs. It involves the freezing of waste. This method while being ascendable and eco-friendly in approach owing to lower operating temperatures as well as minimum waste generation, it needs greater energy as compared to the hydrometallurgical methods.

21.13 CONCLUSION

The increasing volumes of electronic waste lead to several environmental and health issues if not appropriately addressed. Electronic waste recycling in India is still low; the nation is able to officially recycle only 5% of its total e-waste, which is generated, and the rest 95% is managed by the informal sector. Waste PCBs (printed circuit boards) contain various valuable metals, such as gold, paladallium, silver and copper, in addition to hazardous metals, such as lead. Traditional methods of PCB recycling are not sustainable Hence, it is of significance to use relevant and sustainable approaches for metal recovery from e-waste.

REFERENCES

Agarwal, S., Darbar, S., Saha, S., Choudhury, M., & Singh, R. P. (2023). E-waste management using different cost-effective, eco-friendly biological techniques: an overview. In *Waste Management and Resource Recycling in the Developing World*, 205–235. https://doi.org/10.1016/B978-0-323-90463-6.00026-9

Baez, A. G., Muñoz, L. P., Garelick, H., & Purchase, D. (2022). Characterization of industrially pre-treated waste printed circuit boards for the potential recovery of rare earth elements. *Environmental Technology & Innovation*, *27*, 102481.

Borthakur, A., & Sinha, K. (2013). Generation of electronic waste in India: Current scenario, dilemmas and stakeholders. *African Journal of Environmental Science and Technology*, *7*(9), 899–910.

Gutowski, T., Murphy, C., Allen, D., Bauer, D., Bras, B., Piwonka, T., … & Wolff, E. (2005). Environmentally benign manufacturing: observations from Japan, Europe and the United States. *Journal of Cleaner Production*, *13*(1), 1–17.

Herat, S., & Agamuthu, P. (2012). E-waste: a problem or an opportunity? Review of issues, challenges and solutions in Asian countries. *Waste Management & Research*, *30*(11), 1113–1129.

Hicks, C., Dietmar, R., & Eugster, M. (2005). The recycling and disposal of electrical and electronic waste in China—legislative and market responses. *Environmental Impact Assessment Review*, *25*(5), 459–471.

Hsu, E., Barmak, K., West, A. C., & Park, A. H. A. (2019). Advancements in the treatment and processing of electronic waste with sustainability: a review of metal extraction and recovery technologies. *Green Chemistry*, *21*(5), 919–936. 10.1039/C8GC03688H

http://www.indiaenvironmentportal.org.in/content/469512/cpcb-report-on-e-waste-management-in-india-18122020/. CPCB report on e-waste management in India (2020).

Jadhao, P. R., Ahmad, E., Pant, K. K., & Nigam, K. D. P. (2020). Environmentally friendly approach for the recovery of metallic fraction from waste printed circuit boards using pyrolysis and ultrasonication. *Waste Management*, *118*, 150–160.

Korf, N., Løvik, A. N., Figi, R., Schreiner, C., Kuntz, C., Mählitz, P. M., Rösslein M., Wäger P. & Rotter, V. S. (2019). Multi-element chemical analysis of printed circuit boards–challenges and pitfalls. *Waste Management*, *92*, 124–136. 10.1016/j.wasman.2019.04.061.

Langrova, V. (2002) *Comparative analysis of EPR programmes for small consumer batteries: Case study of the Netherlands, Switzerland and Sweden.* IIIEE Report 2002:9. Sweden, Lund: The International Institute for Industrial Environmental Economics, IIIEE, Lund University.

Li, H., Eksteen, J., & Oraby, E. (2018). Hydrometallurgical recovery of metals from waste printed circuit boards (WPCBs): Current status and perspectives–A review. *Resources, Conservation and Recycling, 139*, 122–139. 10.1016/j.resconrec.2018.08.007

Lu, Y., & Xu, Z. (2016). Precious metals recovery from waste printed circuit boards: A review for current status and perspective. *Resources, Conservation and Recycling, 113*, 28–39. 10.1016/j.resconrec.2016.05.007.

Maphosa, V., Maphosa, M., & Tan, A. W. K. (2020). E-waste management in Sub-Saharan Africa: a systematic literature review. *Cogent Business and Management 7*(1), 1814503.

Mohite, D. B. (2013). Issues and strategies in managing e-waste in India. *Indian Journal of Research in Management, Business and Social Sciences (IJRMBSS), 1*(1), ISSN, (2319–6998), 46–51.

Paul, S., Choudhury, M., Deb, U., Pegu, R., Das, S., & Bhattacharya, S. S. (2019). Assessing the ecological impacts of ageing on hazard potential of solid waste landfills: A green approach through vermitechnology. *Journal of cleaner production*, 236, 117643. https://doi.org/10.1016/j.jclepro.2019.117643

Panda, R., Pant, K. K., Bhaskar, T., & Naik, S. N. (2021). Dissolution of brominated epoxy resin for environment friendly recovery of copper as cupric oxide nanoparticles from waste printed circuit boards using ammonium chloride roasting. *Journal of Cleaner Production, 291*, 125928.

Queiruga, D., Benito, J. G., & Lannelongue, G. (2012). Evolution of the electronic waste management system in Spain. *Journal of cleaner production, 24*, 56–65.

Rajpal, A., Choudhury, M., Goswami, S., Chakravorty, A., & Raghavan, V. (Eds.). (2024). *Waste Management and Treatment: Advances and Innovations* (1st ed.) CRC Press. https://doi.org/10.1201/9781003258377

United Nations Environmental Programme 2007, E-waste volume I: Inventory assessment manual, https://wedocs.unep.org/bitstream/handle/20.500.11822/7857/EWasteManual_Vol1.pdf?sequence=3&isAllowed=y.

United Nations Environmental Programme 2019, Future E-waste scenarios, https://www.unep.org/ietc/resources/publication/future-e-waste-scenarios?_ga=2.144152695.311835634.1669204805-517553945.1669204805.

Van Yken, J., Boxall, N. J., Cheng, K. Y., Nikoloski, A. N., Moheimani, N. R., & Kaksonen, A. H. (2021). E-waste recycling and resource recovery: A review on technologies, barriers and enablers with a focus on oceania. *Metals, 11*(8), 1313.

Wang, H., Zhang, S., Li, B., Pan, D. A., Wu, Y., & Zuo, T. (2017). Recovery of waste printed circuit boards through pyrometallurgical processing: A review. *Resources, Conservation and Recycling, 126*, 209–218. 10.1016/j.resconrec.2017.08.001

Wang, R., Zhang, C., Zhao, Y., Zhou, Y., Ma, E., Bai, J., & Wang, J. (2021). Recycling gold from printed circuit boards gold-plated layer of waste mobile phones in "mild aqua regia" system. *Journal of Cleaner Production, 278*, 123597.

Widmer, R., Oswald-Krapf, H., Sinha-Khetriwal, D., Schnellmann, M., & Böni, H. (2005). Global perspectives on e-waste. *Environmental Impact Assessment Review, 25*(5), 436–458.

22 Electronic Wastes and Circular Economy

A Path towards Sustainable Management

Shubhansh Bajpai
S.M.S. Govt. Model Science College, Gwalior, India

Laxmi Kant Bhardwaj
Amity Institute of Environmental Toxicology, Safety & Management, Amity University Noida, India

Prangya Rath
Amity Institute of Environmental Sciences, Amity University, Noida, India

Farhana Rahman
University of Technology Sydney (UTS), Sydney, New South Wales, Australia

Aditya Kumar
National Institute of Mental Health, Prague-East, Czech Republic

Mukhtar Iderawumi Abdulraheem
Henan Agricultural University, Zhengzhou, China

22.1 INTRODUCTION

A wide variety of electrical and electronic devices that have been discarded are collectively known as electronic waste (e-waste). In general, an electrical device is considered e-waste if it is either broken, has passed its expiration date, or is no longer wanted by its owner (EPA, 2021; Bhardwaj et al., 2023; Bhardwaj et al., 2023). As technology improves quickly and technological item usage continues to climb, generating electrical debris has become a significant environmental problem. However, according to Ali and Shirazi (2023), the environment and human health are at risk when e-waste is not adequately removed and managed. A potential answer to the problem of e-waste management in the long run is the concept of the circular economy. Everything from old cell phones and computers to televisions (TVs) and other home appliances may be considered electronic garbage. Proper disposal of these products is sometimes challenging due to dangerous substances, including flame-retardant chemicals and heavy metals. A major global problem with broad implications is the increasing accumulation of e-waste. The growing buildup of e-waste is a significant worldwide issue

DOI: 10.1201/9781003582311-22

with far-reaching consequences. Toxic compounds are released into the environment due to incorrect electrical device disposal, harming ecosystems and human communities. Exposure to these electronic poisons poses significant health concerns, emphasizing the necessity of e-waste treatment in achieving environmental sustainability (WHO, 2019). The extent of this problem is broad, necessitating a thorough examination of its implications on human, animal, and environmental health. This chapter aims to comprehensively understand the complex interaction between e-waste management and circular economy ideas. The following section will explore the impact of e-waste on human health, wildlife, and the environment, thoroughly examining its consequences. The chapter will also address how the circular economy may be used as a successful paradigm for the long-term management of electronics. The discussion will encompass key principles, challenges, and opportunities associated with adopting circular practices in the electronics sector. Through case studies and real-world examples, the chapter will illustrate successful models and provide recommendations for policymakers, manufacturers, and consumers to contribute to a paradigm shift toward a more sustainable and resilient future.

22.2 GLOBAL PATTERNS IN E-WASTE MANAGEMENT

A complicated interaction between technical advancement, consumer preferences, and shifts in the disposal of e-waste affects the worldwide problem of electronic trash. This analysis outlines pivotal trends highlighting the contemporary situation of e-waste. A significant observation is the swift escalation in e-waste production. Ongoing technological advancements and increasing consumer appetite have led to a notable rise in e-waste worldwide. The decreasing lifespan of electronic devices, mainly due to rapid technological obsolescence, further accelerates the accumulation of e-waste (Ongondo et al., 2011).

Accompanying this surge is a significantly changing composition of e-waste. While traditional electronic products persist, newer categories such as smart devices, wearables, and Internet of Things (IoT) devices are becoming increasingly prevalent. This shift towards smaller, more intricate devices presents challenges in recycling and recovering valuable materials (Rocchetti et al., 2018). The Cross-Border Movement and Global Trade of e-waste present complex dynamics. Developed countries frequently export e-waste to developing nations for processing, introducing environmental and health risks in the receiving countries. The worldwide transportation of hazardous material, which includes certain types of e-waste, is supervised and managed by many international frameworks, most notably the Basel Convention (Grant et al., 2013). According to the findings of recent research (Bharathi et al., 2023), the yearly generation of electronic garbage throughout the globe is around fifty million tons. It is estimated that only around twenty percent of electronic trash is handled each year, despite the Basel Convention encompassing regions with legislation for electronic waste that accounts for sixty-six percent of the global population, which indicates that forty million tons of electronic garbage are either burned to recover resources or transferred illegally via markets. Illegal exporting and inadequate methods for disposing of electronic waste have an impact on a large number of nations. According to Zeng et al. (2013), the United States of America (USA) disposes of over 100 million electronic devices annually. In contrast, China may be responsible for disposing of over 160 million gadgets, making China the primary destination for e-waste disposal. According to the findings of a study conducted by Borthakur and Sinha in 2013, e-waste constitutes between 5 10 % of the total rubbish created throughout the globe, with India alone producing 146,000 tons of this waste each year.

Research has also shown that developed nations provide developing nations like Nigeria, Pakistan, Ghana, India, and China with significant electronic garbage (Mmereki et al., 2016). In 2003, the Extended Producer Responsibility (EPR) program collected about 70 % of South Korea's electronic trash. Twelve percent of it was put to use for other reasons; sixty-nine percent went into recycling; and the remaining nineteenth percent was either disposed of in a landfill or burned.

22.3 EFFECTS OF E-WASTE ON BIODIVERSITY, HUMAN HEALTH, AND THE ENVIRONMENT

In this part, we take a comprehensive look at some of the negative effects that the incorrect disposal of electronic trash may have on human health, animal preservation, and the environment. Improper disposing of electronic trash sets off a chain reaction of environmental consequences, perpetuating a cycle of degradation that has a detrimental effect on ecosystems and contributes to greater environmental issues. In addition to this, it poses significant public health hazards, so forming a complex relationship between the advancement of technology and the welfare of the general population. According to Mondal et al. (2023), practices such as burning or landfilling of electronic devices discharge harmful compounds into the air, water, and soil, which ultimately leads to the pollution of these environments. According to Awasthi et al. (2016) and Bourgeois et al. (2020), this pollution has a negative influence on the health of the communities of microbes, as well as the cycling of nutrients, the structure of the soil, and the health of plant and animal life. Environmental pollution of soil and water has far-reaching implications that extend beyond the local proximity of disposal sites for electronic trash. These actions are common in informal recycling activities, especially in poor countries with inadequate or nonexistent rules, and they are responsible for a significant amount of environmental damage (Awasthi et al., 2016).

Harmful substances from e-waste can permeate into water bodies and accumulate in aquatic creatures, which may lead to the phenomena of bioaccumulation and biomagnification (Horton et al., 2017; Bhardwaj and Sharma, 2021; Bhardwaj, 2023; Bhardwaj et al., 2024; Bhardwaj et al., 2025. This makes aquatic ecosystems very vulnerable to pollution. The introduction of contaminants from electronic trash into aquatic environments may result in changes in the diversity of species, population dynamics, and possibly the disintegration of the ecosystem. There is a correlation between improper disposal of electrical waste, which can sometimes be accomplished through methods such as cremation or landfilling and being exposed to brominated flame retardants (BFRs), cadmium (Cd), lead (Pb), mercury (Hg), polychlorinated biphenyls (PCBs), and persistent organic pollutants (POPs) that are found in electronic devices. These substances have been linked to adverse effects on the health of both humans and wildlife. According to Mairaj et al. (2022), Alam et al. (2023), Bhardwaj et al. (2023), and Bhardwaj et al. (2023), exposure to Pb may have a role in the formation of neurological diseases, particularly in children. Both renal and respiratory problems have been related to exposure to Hg. There is evidence that polycyclic aromatic hydrocarbons (PAHs) and POPs are harmful to reproduction and growth mechanisms (UNEP, 2019; Bhardwaj and Jindal, 2019; Bhardwaj and Jindal, 2020; Bhardwaj et al., 2021; Bhardwaj and Jindal, 2022). According to Dong et al.'s research from 2021, BFRs, which are often embedded in electronic devices, have been associated with endocrine disruption as well as neurodevelopmental problems. It is important to mention that the uncontrolled practice of informal recycling of electronic trash in developing nations offers extra-occupational health dangers for workers who are exposed to toxic compounds (Grant et al., 2013). When these hazardous compounds are released into the environment, they exacerbate existing health disparities and have a more significant effect on vulnerable populations that are located in proximity to sites where electronic trash is disposed of.

Several case studies highlight the detrimental effects that the pollution of electronic trash has on human populations as well as animal populations. It is important to note that the Agbogbloshie e-waste dump in Ghana is notorious for its negative impact on the locals' health. According to a study conducted in the Guiyu region of China, which is a center for the recycling of electronic trash, significant quantities of heavy metals have been discovered in sediments of local rivers and fish. This has brought about adverse effects on the health of fish and their ability to reproduce (Han et al., 2019). In addition, Leung (2019) found that the same area had a higher prevalence of pulmonary and dermatological conditions than other regions (Alabi et al., 2021). Research conducted in the USA has shown that there is a correlation between the pollution caused by electronic trash and reproductive abnormalities in birds and animals that are located in close proximity to dumping places. These people have been shown to have elevated amounts of Pb along with other harmful components in their blood (Mittal, 2011). These individuals live in close proximity to these places.

As environmental and health dangers connected with e-waste become more widely recognized, there has been a surge in the amount of awareness and legislative action that has been taken. The implementation of increasingly stricter restrictions is being carried out by governments and international organizations. EPR programs are now being adopted by a number of nations. These programs give producers the duty for the lifespan of their goods during their full existence.

22.4 E-WASTE RECYCLING AND SUSTAINABLE MANAGEMENT

While it is true that mishandling e-waste can have health impacts on humans, wildlife, and the environment, some argue that implementing efficient recycling and disposal methods could help mitigate these issues to a certain extent. Yet, it's crucial to recognize that effective e-waste management practices need to be consistently applied or enforced, especially in developing nations with more relaxed regulations. Unofficial recycling activities and insufficient infrastructure for appropriate disposal lead to ongoing environmental contamination with hazardous substances (Paul et al., 2019; Agarwal et al., 2023; Gupta et al., 2023). Additionally, even when e-waste is disposed of properly, the production, transportation, and disposal processes of electronic devices still result in a considerable carbon footprint. Consequently, while efficient recycling and disposal are vital, they alone are inadequate in fully addressing the extensive risks and impacts associated with e-waste.

The environmental impact of electronics extends throughout their lifecycle, going beyond disposal. Manufacturing processes, often involving resource-intensive extraction and processing of raw materials, result in greenhouse gas (GHG) emissions (Bhardwaj, 2023; Bhardwaj et al., 2023). Effective e-waste management, encompassing recycling and proper disposal, is crucial not only for mitigating immediate environmental harm but also for reducing the overall environmental impact of electronic technology (UNEP, 2017).

E-waste presents a mix of challenges and opportunities. The challenges encompass inefficient collection systems, inadequate facilities for safe recycling, and limited consumer awareness. Conversely, opportunities arise in resource recovery, employment generation, and fostering innovation. Implementing policy measures and regulations, such as extended producer responsibility schemes and principles of eco-design, is vital to incorporate e-waste management within the circular economy framework. Consumer awareness and active participation are indispensable in promoting the circular economy.

Tackling e-waste within the circular economy context necessitates a comprehensive approach encompassing policy measures, technological advancements, consumer involvement, and global collaboration (Ali and Shirazi, 2023). By re-evaluating product life cycles, enhancing resource efficiency, and encouraging stakeholder collaboration, the negative impacts of e-waste can be mitigated. Also, it may be used to its full capacity as a secondary raw material source in a circular economy. Sustainable electronic trash management activities are becoming more popular on a global scale (Awasthi et al., 2016).

Innovations in recycling technology are always playing a crucial role. The extraction of valuable elements from electronic gadgets is becoming more efficient and less harmful to the environment. Baldé et al. (2017) mentioned urban mining and improved separation techniques as two ways to improve resource recovery from e-waste. There is also a clear trend towards supporting circular economy initiatives. Products are designed with durability, repairability, and recycling in mind using this technique. Increasing the usefulness of electronics and encouraging their proper disposal at the end of their useful lifespans are two ways to reduce the generation of e-waste (MacArthur, 2013).

22.5 THE CIRCULAR ECONOMY FRAMEWORK AND E-WASTE: PROSPECTS AND IMPLEMENTATION

When it comes to attaining sustainable development and tackling the environmental difficulties linked with e-waste, the research on e-waste and the use of circular economy concepts in managing electronic trash are both significant. Because technological advancements are still being made and the need for electronic devices is growing, it is necessary to investigate possible future directions and suggestions in this area (Pan et al., 2022).

- **Extended Producer Responsibility (EPR)**: EPR is a policy that advocates for the safe disposal and recycling of devices while promoting the smooth recycling, repair, and upgrading of such products. EPR transfers responsibility for managing e-waste from the public and the government to producers.
- **Design for Environment (DfE)**: The integration of Design for Environment principles into product development can significantly reduce environmental impact, focusing on factors like material selection, energy efficiency, recyclability, and ease of disassembly.
- **Adoption of Circular Economy Models**: Shifting towards a circular economy in managing electronic products focuses on preserving resources by employing repair, refurbishment, remanufacturing, and recycling strategies. This approach minimizes the dependence on new, unprocessed materials and enhances overall resource efficiency.
- **Strengthening Recycling Infrastructure**: Investing in recycling infrastructure, improving technologies, and ensuring proper treatment of hazardous substances are crucial for managing e-waste and promoting a circular economy.
- **Education and Awareness**: Raising awareness about e-waste's environmental impact and circular economy practices is crucial for driving change. Educational initiatives can target consumers, businesses, policymakers, and industry professionals, promoting responsible consumption and sustainable e-waste management.
- **International Cooperation**: International cooperation is crucial for addressing global e-waste issues, fostering harmonized regulations, standards, and best practices, and promoting technology transfer to developing countries while sharing knowledge and expertise.

22.6 SUGGESTIONS/RECOMMENDATIONS

We suggest several recommendations for the mitigation of e-waste, and these are given below:

- Governments worldwide should strengthen and strictly enforce e-waste management regulations and policies, including implementing EPR strategies, limiting e-waste export to developing countries, and providing incentives for sustainable practices.
- Given the worldwide breadth of e-waste challenges, increased international collaboration is critical. Countries must band together to develop standard rules, share best practices, and encourage responsible e-waste management internationally.
- It is essential to raise awareness among the population regarding the dangers of improper e-waste disposal. Informative marketing can empower individuals to make well-informed choices. These choices may include supporting eco products and participating in e-waste recycling programs.
- The development of technologies for recycling e-waste and the exploration of potential new materials are vital for achieving sustainable solutions. Through projects, collaboration among governments, industry sectors, and academic institutions is necessary to support and promote these initiatives.
- Manufacturers and producers should embrace corporate responsibility by designing products with a circular economy mindset. This includes prioritizing recyclability, reducing the

use of hazardous materials, and promoting product longevity through modular design and upgradability.

- Engaging local communities in e-waste management initiatives is essential. Community-based programs for collection, recycling, and awareness can contribute to effective and sustainable e-waste management practices.
- Advocacy groups and non-governmental organizations play a crucial role in raising awareness and advocating for responsible e-waste management. Grassroots movements can bring about societal change and influence industry practices.

22.7 CONCLUSION

To emphasize the significance of putting into practice long-term sustainable management solutions, it is essential to carry out an extensive study on waste materials and learn the consequences of these materials on human health, wildlife, and maintaining ecological balance. As the globe grapples with rising levels of e-waste, numerous crucial findings and recommendations emerge. E-waste has many consequences for human health, animal habitats, and the environment. Incorrect electronic device disposal procedures provide significant health risks and the accompanying ecosystem pollution and hazardous material discharge cause environmental impact.

The issue of e-waste transcends national boundaries, including international commerce and mobility, and thus calls for international collaboration. Because of the disproportionate effect that improper disposal of e-waste has on developing nations, there is a pressing need for waste management practices that are both equitable and sustainable. In addition, the ever-changing composition of electronic trash reflects the always-advancing technical capabilities. The influx of newer electronic categories poses challenges in terms of recycling and highlights the need for adaptive waste management strategies. The growing recognition of the environmental and health risks linked to e-waste has prompted worldwide regulatory reactions. EPR frameworks and initiatives promoting sustainable practices are critical steps towards responsible e-waste management. Embracing circular economy principles is essential for mitigating the adverse effects of e-waste. Designing products for longevity, reparability, and recyclability can significantly reduce the environmental footprint of electronic devices. The circular economy aims to minimize waste and maximize resource value by promoting sustainable production, consumption, and recycling practices.

REFERENCES

Agarwal, S., Darbar, S., Saha, S., Choudhury, M., & Singh, R. P. (2023). E-waste management using different cost-effective, eco-friendly biological techniques: an overview. In *Waste Management and Resource Recycling in the Developing World*, 205–235. https://doi.org/10.1016/B978-0-323-90463-6.00026-9

Alabi, O. A., Adeoluwa, Y. M., Huo, X., Xu, X., & Bakare, A. A. (2021). Environmental contamination and public health effects of electronic waste: an overview. *Journal of Environmental Health Science and Engineering*, *19*, 1209–1227.

Alam, S., Bhardwaj, L. K., Mallick, R., & Rai, S. (2023). Estimation of heavy metals and fluoride ion in vegetables grown nearby the stretch of River Yamuna, Delhi (NCR), India. *Indian Journal of Environmental Protection*, *43*(1), 64–73.

Ali, S., & Shirazi, F. (2023). The Paradigm of Circular Economy and an Effective Electronic Waste Management. *Sustainability*, *15*(3), 1998.

Awasthi, A. K., Zeng, X., & Li, J. (2016). Environmental pollution of electronic waste recycling in India: A critical review. *Environmental Pollution*, *211*, 259–270.

Baldé, C. P., Forti, V., Gray, V., Kuehr, R., & Stegmann, P. (2017). The global e-waste monitor. *United Nations University (UNU), International Telecommunication Union (ITU) & International Solid Waste Association (ISWA), Bonn/Geneva/Vienna*, 1–109.

Bharathi, S. D., Dilshani, A., Rishivanthi, S., Khaitan, P., Vamsidhar, A., & Jacob, S. (2023). Resource recycling, recovery, and xenobiotic remediation from E-wastes through biofilm technology: a review. *Applied Biochemistry and Biotechnology*, *195*(9), 5669–5692.

Bhardwaj, L. K., & Jindal, T. (2019). Contamination of lakes in Broknes peninsula, East Antarctica through the pesticides and PAHs. *Asian-Journal of Chemistry, 31*(7), 1574–1580.
Bhardwaj, L. K., & Jindal, T. (2020). Persistent organic pollutants in lakes of Grovnes Peninsula at Larsemann Hill area, East Antarctica. *Earth Systems and Environment, 4*, 349–358.
Bhardwaj, L. K., Sharma, S., & Jindal, T. (2021). Occurrence of polycyclic aromatic hydrocarbons (PAHs) in the Lake water at Grovnes Peninsula Over East Antarctica. *Chemistry Africa, 4*, 965–980.
Bhardwaj, L.K., & Sharma, A. (2021). Microplastics (MPs) in Drinking Water: Uses, Sources & Transport. https://doi.org/10.20944/preprints202104.0498.v1
Bhardwaj, L., & Jindal, T. (2022). Polar ecotoxicology: Sources and toxic effects of pollutants. *New Frontiers in Environmental Toxicology*, 9–14.
Bhardwaj, L.K. (2023). Occurrence of Microplastics (MPs) in Antarctica and Its Impact on the Health of Organisms. *Maritime Technology and Research*, 6(2):265418. https://doi.org/10.33175/mtr.2024.265418
Bhardwaj, L.K., Rath, P., & Choudhury, M. (2023). Assessment of e-waste pollution level in soil and water and its impact on human health: A review. https://doi.org/10.20944/preprints202311.1714.v1
Bhardwaj, L.K., Rath, P., & Tokas, R. (2023). E-Waste Management in Developing Countries: Current Practices, *Challenges, Disposal, and Impact on Human Health & Environment*. https://doi.org/10.20944/preprints202312.0582.v1
Bhardwaj, L. K., Kumar, D., & Kumar, A. (2023). Phytoremediation potential of ocimum sanctum: A sustainable approach for remediation of heavy metals. In *Phytoremediation Potential of Medicinal and Aromatic Plants* (pp. 46–57). CRC Press.
Bhardwaj, L. K., Sharma, S., & Jindal, T. (2023). Estimation of physico-chemical and heavy metals in the Lakes of Grovnes & Broknes Peninsula, Larsemann Hill, East Antarctica. *Chemistry Africa*, 1–18, https://doi.org/10.1007/s42250-023-00668-6.
Bhardwaj, L.K., Rath, P., & Yadav, P. (2023). A comprehensive review on the impact of climate-induced livelihood, health and migration on women and girls. https://doi.org/10.20944/preprints202312.2271.v1
Bhardwaj, L. K., Rath, P., Yadav, P., & Gupta, U. (2024). Microplastic contamination, an emerging threat to the freshwater environment: a systematic review. *Environmental Systems Research, 13*(1), 8.
Bhardwaj, L. K., Rath, P., Rehman, F., Yadav, P., Massahi, T., Mittal, R., & Jain, H. (2025). Status of Micro/Nano-Plastic (MNP) Pollution in the Aquatic System. In *Societal and Environmental Ramifications of Plastic Pollution* (pp. 43–62). IGI Global.
Borthakur, A., & Sinha, K. (2013). Electronic waste management in India: a stakeholder's perspective. *Electronic Green Journal, 1*(36).
Bourgeois, D., Lacanau, V., Mastretta, R., Contino-Pépin, C., & Meyer, D. (2020). A simple process for the recovery of palladium from wastes of printed circuit boards. *Hydrometallurgy, 191*, 105241.
Dong, L., Wang, S., Qu, J., You, H., & Liu, D. (2021). New understanding of novel brominated flame retardants (NBFRs): Neuro (endocrine) toxicity. *Ecotoxicology and Environmental Safety, 208*, 111570.
Environmental Protection Agency (EPA) (2021). Electronic waste management. Retrieved from https://www.epa.gov/international-cooperation/cleaning-electronic-waste-e-waste
Grant, K., Goldizen, F. C., Sly, P. D., Brune, M. N., Neira, M., van den Berg, M., & Norman, R. E. (2013). Health consequences of exposure to e-waste: a systematic review. *The Lancet Global Health, 1*(6), e350–e361.
Gupta, P., Sharma, A., & Bhardwaj, L.K. (2023). Solid Waste Management (SWM) and Its Effect on Environment & Human Health. https://doi.org/10.20944/preprints202309.0384.v1
Han, Y., Tang, Z., Sun, J., Xing, X., Zhang, M., & Cheng, J. (2019). Heavy metals in soil contaminated through e-waste processing activities in a recycling area: Implications for risk management. *Process Safety and Environmental Protection, 125*, 189–196.
Horton, A. A., Walton, A., Spurgeon, D. J., Lahive, E., & Svendsen, C. (2017). Microplastics in freshwater and terrestrial environments: Evaluating the current understanding to identify the knowledge gaps and future research priorities. *Science of the Total Environment, 586*, 127–141.
Leung, A. O. W. (2019). Environmental contamination and health effects due to E-waste recycling. *Electronic Waste Management and Treatment Technology*, 335–362, https://doi.org/10.1016/B978-0-12-816190-6.00015-7
MacArthur, E. (2013). Towards the circular economy. *Journal of Industrial Ecology, 2*(1), 23–44.
Mittal, A. (2011). High blood lead levels in e-waste recyclers. *Epidemiology, 22*(1), S293.
Mairaj, S., Nagar, R. D., Bhardwaj, L. K., Rehman, F., Punnakal, A., Chauhan, A., & Jindal, T. (2022). Antioxidant responses of arsenite-induced oxidative stress in rice (Oryza sativa L.) and its modulation by eugenol (extracted from Ocimum sanctum). *Indian Journal of Agricultural Research*, 56(4), 396–400.

Mmereki, D., Li, B., Baldwin, A., & Hong, L. (2016). The generation, composition, collection, treatment and disposal system, and impact of E-waste. *E-waste in Transition-from Pollution to Resource*, *10*, 61332.
Mondal, T., Choudhury, M., Kundu, D., Dutta, D., & Samanta, P. (2023). Landfill: An eclectic review on structure, reactions and remediation approach. *Waste Management*, 164, 127–142. https://doi.org/10.1016/j.wasman.2023.03.034
Ongondo, F. O., Williams, I. D., & Cherrett, T. J. (2011). How are WEEE doing? A global review of the management of electrical and electronic wastes. *Waste Management*, *31*(4), 714–730.
Pan, X., Wong, C. W., & Li, C. (2022). Circular economy practices in the waste electrical and electronic equipment (WEEE) industry: A systematic review and future research agendas. *Journal of Cleaner Production*, *365*, 132671.
Paul, S., Choudhury, M., Deb, U., Pegu, R., Das, S., & Bhattacharya, S. S. (2019). Assessing the ecological impacts of ageing on hazard potential of solid waste landfills: A green approach through vermitechnology. *Journal of Cleaner Production*, 236, 117643. https://doi.org/10.1016/j.jclepro.2019.117643
Rocchetti, L., Amato, A., & Beolchini, F. (2018). Printed circuit board recycling: A patent review. *Journal of Cleaner Production*, *178*, 814–832.
UNEP (United Nations Environment Programme) (2017). Waste and climate change: Global trends and strategy framework. Retrieved from https://www.unep.org/resources/report/waste-and-climate-change-global-trends-and-strategy-framework
UNEP (United Nations Environment Programme) (2019). E-waste management: From waste to resource. Retrieved from https://www.unep.org/ietc/what-we-do/e-waste-management
World Health Organization (WHO) (2019). Children and the environment: Electronic waste. Retrieved from https://www.who.int/teams/environment-climate-change-and-health/settings-populations/children/e-waste
Zeng, X., Li, J., Stevels, A. L. N., & Liu, L. (2013). Perspective of electronic waste management in China based on a legislation comparison between China and the EU. *Journal of Cleaner Production*, *51*, 80–87.

23 Managing Electronic Waste in Developing Nations

An Overview

Kalyani S. Kondekar, Rishikesh S. Aher, R. F. Siddiqui, and Geetanjali Kaushik
Hi-Tech Institute of Technology, Aurangabad, India

Chaudhery Mustansar Hussain
New Jersey Institute of Technology, Newark, New Jersey, United States

Govind Dhage
Hi-Tech Institute of Technology, Aurangabad, India

23.1 INTRODUCTION

India has emerged as a leading producer of e-waste following the United States and China. As outlined in the ICEA report, in India administration of electronic waste is primarily handled through unorganized channels, using approximately 90% of collection of e-waste and 70% of recycling being conducted by a competitive informal sector. This informal sector demonstrates proficiency in salvaging older electronic devices to extract spare parts and conduct profitable repairs. Notably, industrial centers such as Moradabad specialize in the processing of printed circuit boards (PCBs) to recover valuable materials like gold and silver (Mor et al., 2021). E-waste is a broad expression encompassing all forms of outdated, conclusion phase, or discarded electrical and electronic equipment, including domestic appliances and workplace communications and information devices. E-waste contains a plethora of harmful material, notably metals like lead, cadmium, mercury, and nickel. Currently, India holds the third position globally in e-waste generation, trailing only the United States and China (Liu et al., 2023). In India the quantity of electronic waste has notably intensified to 1.6 million tons in the 2021–2022 periods. Among 65 cities within India, over 60 percent of the electronic waste is produced, and 70 percent of the total electronic waste is produced through 10 states. Despite a fourfold increase in e-waste collection and processing within four years, a staggering 95% e-waste is managed via illegal channels within the informal sector in India. Informal waste pickers, often referred to as kabadiwalas, operate outside established environmental standards. They resort to burning materials that are not recyclable or direct them to landfills, posing significant risks of environmental harm and health hazards. Moreover, informal sector recyclers employ basic and often unsafe recycling methods that can release toxic pollutants immersing it in the nearby surroundings (Mondal et al., 2023). E-waste comprises numerous dangerous elements that can present significant threats to both human health and the ecosystem. In the fiscal year 2020–2021, India managed to process 3.4 lakh tons of e-waste. As indicated by the Central Pollution Control Board (CPCB, 2022), the yearly increase in production of plastic waste is around 3%, but the rise in e-waste generation is even steeper. E-waste production reached 7.1 lakh tons in 2018–2019

DOI: 10.1201/9781003582311-23

and escalated to 10.14 lakh tons in 2019–2020, reflecting a substantial 31% annual increase. Presently, there are 468 authorized recyclers and 2,808 collection points across 22 states. However, the combined capacity of these 468 recyclers is 13 lakh tons, which falls short of effectively managing India's growing e-waste production.

As per the UN's Global E-waste Monitor 2020, a historic million metric tons (Mt) of e-waste were generated worldwide in 2019, recording a 21 percent intensified in only 5 years. The document also predicted the worldwide electronic waste products with a plug or battery will soar to 74 Mt up to 2030, nearly doubling within a span of 16 years. This alarming increase positions electronic waste, as the rapidly expanding global household waste branch concern, primarily driven by heightened utilization of electronic and electric devices, their brief lifespan, and limited repair options. Regrettably, in 2019 only 17.4 percent of the electronic waste produced was gathered and reproduced (Forti et al., 2020). As a result, precious resources such as gold, copper, platinum and other recoverable valuable components, cautiously valued at a staggering US $57 billion an amount exceeding the Gross Domestic Product of various nations, valuable materials were predominantly disposed of or incinerated instead of being collected for proper processing and reuse. As outlined in the report, Asia was the largest contributing to e-waste generation in 2019, considering a significant volume of 24.9 Mt. This was subsequent to the Americas with 13.1 Mt and Europe with 12 Mt. In comparison, E-waste produced in Africa was 2.9 Mt and in Oceania it was 0.7 Mt during the same year. To put this into perspective, the e-waste produced in 2019 outweighed the collective weight of all adults in Europe and equaled the weight of about 350 cruise ships equivalent in size to the Queen Mary 2, forming a queue spanning 125 km. electronic waste poses substantial environmental and health risks due to its harmful additives and dangerous substances like mercury, can cause harmful to the human brain and coordination mechanism. According to the United Nations, the worldwide e-waste production per person will reach 8 kg in 2023, resulting in a staggering 61.3 million tons of electronic waste discarded in a single year. This amount exceeds the weight of the Great Wall of China. Shockingly, only 17.4% of this e-waste, containing both harmful substances and high value materials, will be roughly gathered, treated, and reproduced on a global scale (Park et al., 2017). The remaining 50.6 million tons will end up in landfills, be incinerated, traded illegally, treated inadequately, or stored within households (Baldé et al., 2017; Paul et al., 2019). As per the Global E-waste Monitor 2020, roughly 53.6 million MT of Waste Electrical and Electronic Equipment (WEEE) were generated in 2019. This marked a significant 21% rise over the span of five years since 2014. Additionally, the report indicates that e-waste is anticipated to further escalate; projecting a trajectory toward 74 Mt by the year 2030. In the year 2021, the cumulative global weight of WEEE is projected to reach approximately 57.4 million tons.

China, the United States, and India are foremost producers of electronic waste. The following Figure 23.1 displays the top 10 nations globally that generates e-waste, based on the total volume produced (expressed in KiloTons).

Each country in the top ten also showcases its respective recycling rate based on the most recent available data (Updated October 2023).

23.2 WHAT IS ELECTRONIC WASTE?

E-waste, akin to hazardous waste, presents pressing immediate and long-term concerns due to uncontrolled accumulation and inadequate recycling, posing significant environmental risks and jeopardizing human health. The advent of information technology has profoundly transformed our lifestyle, work dynamics, and communication methods, delivering a plethora of advantages and prosperity to its users. The advancements in technology and global integration of the economy have made a wide array of products accessible and cost-effective, altering individual's way of life considerably (Adhana, 2020).

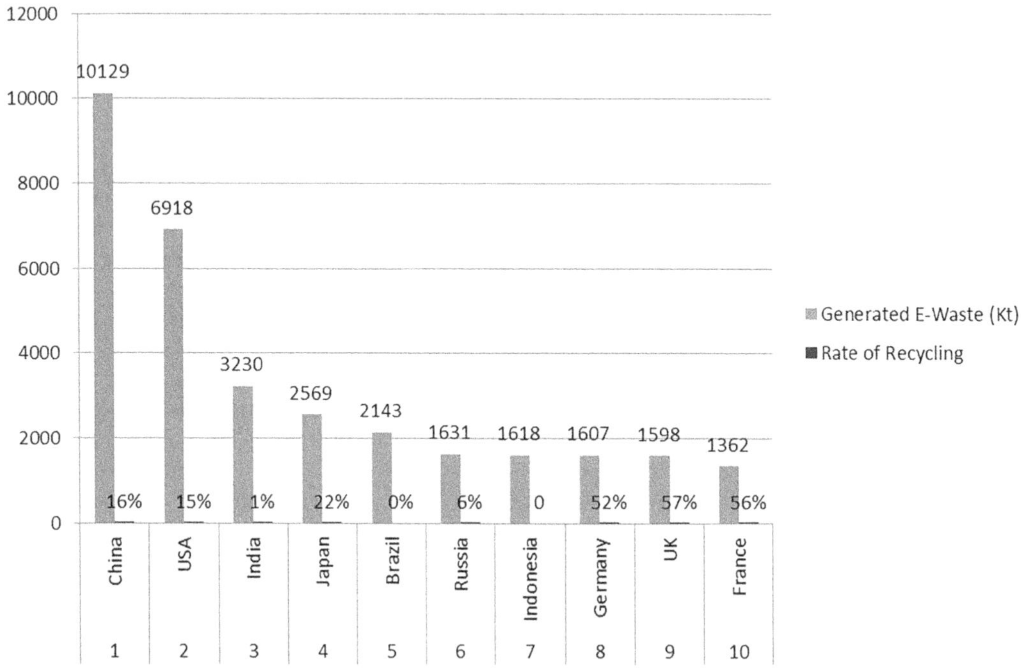

FIGURE 23.1 Top 10 E-waste generation in countries.

Modern electronic devices have seamlessly integrated into our daily routines, offering enhanced comfort and safety, and expedited obtaining and sharing of information. However, this surge in technological consumption has also resulted in unbridled resource usage and a concerning upswing in waste generation. This issue is prevalent in addition to developed nations; this is also applicable in developing countries like India (Richard & Jones, 2014). The swift pace of technological growth, continuous technical enhancements, and elevated rate of product obsolescence within the electronics sector has resulted in the emergence of one of the world's rapidly expanding waste streams — conclusion phase electronic and electrical equipment.

This waste category encompasses a wide spectrum of electronic and electrical items, ranging from washing machines, refrigerators, printers and computers to mobile devices, televisions and iPods, many of which contain harmful substances. Numerous patterns in usage and production processes prove not sustainable and present a substantial menace to the human and environment well-being. Addressing these challenges requires optimizing natural resource use, waste minimization, designing cleaner products, and implementing environmentally sustainable regenerating and waste discarding practices, all while fostering economic growth and enhancing life quality.

Nations within the European Union (EU) and other developed regions have taken proactive measures to tackle the e-waste issue. These initiatives involve policy interventions and the adoption of scientific approaches for regenerating and proper discarding of such type of waste. The EU describes this burgeoning stream of waste as WEEE. Their guidelines key aspects of WEEE, including the definition of 'EEE' (Electrical and Electronic Equipment), its categorization into ten distinct groups, and voltage-based classifications. Conversely, in India, environmental regulations lack a specific definition for WEEE, and thus, it is known as 'e-waste.'

In essence, e-waste encompasses loosely disposed, excess, outdated, malfunctioning, or damaged electronic or electrical devices. Table 23.1 presents the major sources of e-waste.

TABLE 23.1
Sources of E-Waste

Origins of e-wastes	Component	Health Impacts
Solder in PCBs, glass panels, and monitor seals.	Lead (PB)	Impact on the peripheral nervous and central systems, blood systems and kidneys can harm the Children's brain growth.
Semiconductors and Resistor chips.	Cadmium (CD)	Toxic harm to health, kidney and liver buildup, neural damage.
Switches, Relays and PCBs.	Mercury (Hg)	Long-term brain damage, respiratory problems, and skin issues from fish bioaccumulation toxins.
Protecting untreated and galvanized steel plates from corrosion, acting as a Coating or strengthener for steel enclosures.	Hexavalent chromium (Cr) VI	DNA harm, Asthma symptoms with bronchitis.
Electronic devices and circuit boards often feature Plastic casing	Brominated flame retardants (BFR)	Impairs the functions of the endocrine system
The frontal interface of CRTs	Barium (Ba)	Brief exposure results in muscle Weakness; harm to the liver,heart and spleen.
Main circuit board	Beryllium (Be)	Lung cancer risk due to inhaling fumes and dust. Leads to chronic conditions like beryllium disease or beryllicosis as well as skin disorders.

23.3 STATISTICS OF E-WASTE IN INDIA

In the contemporary world, our dependence on a wide array of electrical and electronic equipment (EEE) has turned into an ingrained habit. This incessant demand for EEE is coupled with a concerning trend of increasingly shorter lifespans, often attributed to the perceived difficulty and expense of repairing older devices. Leading companies and renowned brands allocate enormous financial resources to continually churn out novel devices incorporating the latest technological advancements. Consequently, many individuals opt to purchase new gadgets, swiftly discarding their old ones (Yoheeswaran, 2013).

The inevitable retirement of these older devices necessitates responsible disposal methods in order to safeguard human well-being and environment. The disposal of EEE culminates in what is commonly known as e-waste for short. E-waste comprises a hazardous amalgamation of diverse toxic materials, including mercury, lead, cadmium, iron, steel, glass, and various metals. The presence of these substances poses a significant threat to the well-being of both living organisms and the environment.

As of 2021, a staggering 57.4 million MT of electronic waste were created worldwide. Within this context, it is noteworthy that India ranks as the third-largest contributor to this growing issue, following the United States and China. Through this article, we aim to provide the latest statistics and factual insights to shed light on the ramifications of the escalating electronic waste problem.

23.4 COLLECTION STRATEGIES IN INFORMAL SECTOR

The main challenge in electronic waste administration has been the presence of informal sector in the current situation. Despite playing significant role in managing of the waste, this sector has not been considered for its role. The organizational structure and extensive network established nationwide grant the informal sector distinctive advantage ion accessing e-waste from business and households. Wide network and hierarchical business operations from gathering, conveyance, waste consolidation, and decomposition and extraction of material provide business opportunities in the informal sector at every level (Reena & Sangita, 2011). The presence of waste and the existence of poverty offer distinctive opportunities for innovative business models that demand thorough examination to enhance societal benefits and sustainability within this sector (Agarwal et al. 2023).

Collectors of informal waste in India are generally known by the name 'Kabadiwalas'. To collect waste from house to house is an old rule of the Indian society. A considerable number of individuals still opt to discard their waste through kabadiwalas with the capacity to operate on a nationwide scale. Nevertheless, the number of individuals discarding of their e-waste to kabadiwalas is unknown.

Most Indians store their e-waste at home because they are concerned about data safety issues. There are some general e-waste discarding practices in India including discarding e-waste with other domestic waste into bins, selling it to kabadiwalas, dismantling of replacement parts and selling the same parts to informal e-waste recyclers or shops (Sharma et al., 2015).

There are various stakeholders in administration of electronic waste in informal sector in which role of agent are vital. Collection of household Electronic waste and mixed waste is done by collectors who commonly known as 'kabadiwala' in India. These kabadiwalas are appointed by agent on the basis of daily wages or quantity of waste collected. Three wheeler vehicles or four wheeler vehicles (Hatgadi) are provided by the agent to kabadiwala to collect the waste. Kabadiwala travel house to house in assigned area by the agent to collect household waste. Kabadiwala gives the money to disposers and collect the waste. Generally the timing of waste collection is on day time i.e. from morning to evening and they working long hours in weekends, as most of the house cleaning activities are on those days. Once the collection vehicle gets full to its limit, kabadiwala dump that household mixed waste on agent's storage site and cycle is go on for whole day. The wages and exchanged money are given by agent to the kabadiwala after completing his work either on daily basis or weekly. This dumped mixed waste is then segregated, dismantled and stored on storage site. As per the requirement this segregated and dismantled waste is then delivered to the respective dealers who further sale this raw material to the manufacturers, refurbishers or value added recycling units. Figure 23.2 presents the informal e-waste management (India).

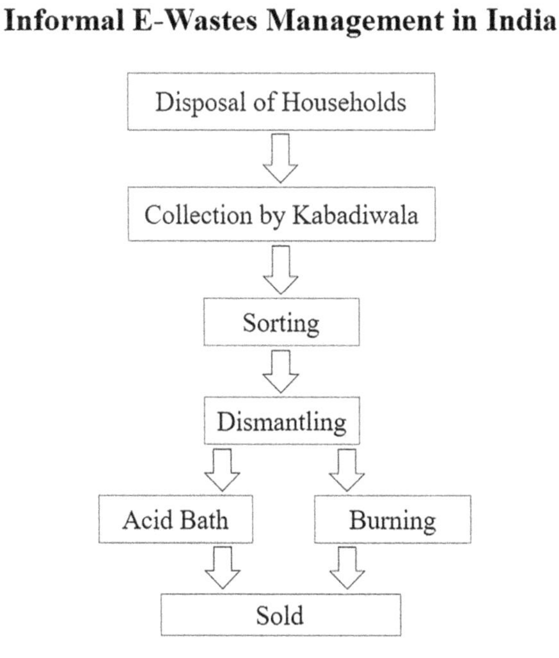

FIGURE 23.2 Process of informal E-waste management industry (India).

TABLE 23.2
Metal Inventory Located in Electronic Waste and Its Effect on Human Well-being and Environment

Sr. No.	Metal	Impact to Human Health	Impact to Environment
1	Lead	High BP, kidney problems, anemia, reproductive issues, weakness in extremities, miscarriage, behavioral disturbances, and concentration issues in children.	Severe and persistent harm to ecosystems, aquatic toxicity, and lasting environmental damage and pollution.
2	Mercury	Cancer risk, brain and fetal damage, shyness, tumors, vision and memory problems, reproductive issues, Parkinson's, hearing loss, headaches, and depression.	Terrestrial and aquatic ecological networks and living beings are negatively impacted.
3	Cadmium	Carcinogenic effects, significant lung damage, kidney disorders, brittle bones, gene mutations, chromosomal abnormalities, DNA damage.	Air, water and soil contamination, detrimental impacts on plants, posing threats to wildlife species, harmful to microorganisms, and adverse effects on ecosystems.
4	Chromium	Asthma, cough, difficulty breathing, skin ulcers, intense redness, swelling of the skin, liver damage, kidney damage, nerve tissue damage, cancer in the lungs and nasal cavity, stomach cancer, and harm to the male reproductive systems.	Soil, water, and air contamination and pollution, along with significant damage to crops.
5	Barium	Vomiting, cramps, diarrhea, breathing issues, blood pressure changes, paralysis, death risk.	Pollution of air, soil and water.
6	Arsenic	Harmful to living things, carcinogenic, skin lesions, nausea, vomiting, irregular heart rhythm, blood vessel damage, DNA synthesis inhibition, potentially fatal.	Contamination and pollution of the soil, water, and air environments, acting as a catalyst for the modification of genetic material in fish and in all animals that consume contaminated fish.
7	Nickel	Significant effects on reproductive systems, developmental toxicity, heart problems, lung issues, asthma, chronic bronchitis, allergies, hyperglycemia.	Toxic disrupts enzyme activities, hinders the growth of plants.

Most activities related to e-waste, such as sorting, collection, disassembly, regeneration and discarding, rely on manual labor. Due to the lack of suitable technologies and equipment, many of the strategies employed for e-waste recycling and processing are rudimentary and pose inherent risks. Electronic waste includes various metallic contaminants that are highly dangerous to both human well-being and the environment (Mohana et al., 2019). Table 23.2 contains the metal and its details.

23.5 SCIENTIFIC APPROACHES TO MANAGE E-WASTE IN LEGAL MANNER

The present study emphasizes that a crucial objective in establishing effective management of e-waste is to cultivate a community which comprehends the significance of environmental recycling and complies with particular regulations. Various techniques and strategies are currently employed worldwide to address e-waste (Yang et al., 2023). These approaches include extended producer responsibility (EPR), material flow analysis (MFA), life cycle assessment (LCA), and multi-criteria analysis (MCA). The problem of electronic waste originated in developed countries. The primary reasons behind the presence of electronic waste issues in certain nations is the substantial quantities of e-waste being exported to these developing regions.

23.6 LIFE CYCLE ASSESSMENT

LCA serves as a valuable instrument for the creation of environmentally conscious electronic devices and the reduction of e-waste issues. It aids in the recognition of potential environmental consequences and facilitates the development of eco-friendly products, processes, or services. LCA further delineates environmental impact categories, including carcinogens, climate change, ozone depletion, toxicity, land utilization, and acidic transformation. These various usages of LCA are harnessed to enhance the environmental sustainability of products and promote the production of eco-friendly goods (Awasthi & Li, 2017).

23.7 MULTI CRITERIA ANALYSIS

MCA is a decision-support instrument designed to address tactical decisions and tackle intricate, temporal-intensive challenges involving both quantitative and qualitative factors. It is instrumental in determining the optimal approach and method for production of a specific product. MCA examines multiple conditions and incorporates a range of criteria to pinpoint the most suitable and lucrative approach for producing and distributing a product.

Solving issues related to the strategy and choosing of the technological framework of a waste management process is a multi-step process that encompasses recognizing disparities and shared elements among potential solutions, taking the most advantageous options, and evaluating the outcomes. This tool has found extensive application in the manufacturing of different items in various cities, including Poland, Szczecin.

23.8 MATERIAL FLOW ANALYSIS

MFA serves as another valuable decision-making tool in the realm of management of e-waste. MFA encompasses the evaluation of E-waste flows, considering their environmental, financial and social implications. This process involves the utilization of software-based simulation modeling.

23.9 EXTENDED PRODUCER RESPONSIBILITY

EPR is an environmental approach that broadens a manufacturer's accountability for a product to encompass its post-consumer stage, encompassing its ultimate disposal. The objectives for product designers under this approach encompass various aspects, such as reducing toxicity, minimizing energy consumption, optimizing materials and product weight, and recognizing opportunities for enhanced reusability, among other goals. Manufacturers are tasked with enhancing product designs by replacing hazardous materials like lead, mercury, cadmium, etc. with safer and more effective alternatives. They are also obligated to take steps to recognize and facilitate the reuse of elements and substances, particularly plastics, and to actively encourage plastic regenerating. Moreover, producers are encouraged to offer incentives to customers who trade in their old products for new ones. To ensure effective E-waste collection, collection systems must be established originating, facilitating the efficient routing of electronic waste to recycling facilities.

23.10 LEGAL PERSPECTIVE

In the contemporary world, waste management, particularly the management of e-waste, presents a significant and formidable challenge for every country. The proliferation of technological advancements in the electronics industry has led to a staggering rise in the volume of discarded electronic goods, making e-waste the largest and rapidly expanding waste stream globally. The worldwide-waste generation has given rise to serious environmental and health concerns. Therefore, it has become of utmost importance for policymakers to found robust systems for management of e-waste, alongside the formulation of relevant regulations. Different nations and unions have adopted distinct strategies to e-waste management, varying not only in the scope of their legislative measures but also in the effectiveness of the instruments they employ.

23.11 ELECTRONIC WASTE REGULATIONS IN EUROPEAN COUNTRIES

The European Union of Countries (EU) has been actively addressing electronic waste problems since the 1990s, implementing comprehensive and forward-thinking e-waste regulations. Waste Shipment Regulation (WSR) was initially enacted in 1993 and subsequently revised in 2007, marking the EU's first committed electronic waste regulation. The WSR underscores the prohibition of EU member states exporting e-waste categorized as harmful to non-OECD (Organization for Economic Cooperation & Development) nations. Nevertheless, certain e-waste components that did not meet the WSR's criteria for hazardous substances continued to be shipped internationally to non-OECD countries under alternative supplies.

In 2003, the European Union introduced the e-waste Directive, a legislative framework aimed at transforming items designs and elevating rates of recycling for discarded WEEE. Additionally, the Restriction of the Use of Certain Hazardous Substances (RoHS) instructions was enacted. The RoHS guideline concentrates on the early stages of the EEE life span by striving to eradicate harmful materials like lead, mercury and using fire retardants at home manufactured or brought in electronic and electrical products (Forslind, 2005; Morris & Metternicht, 2016).

In response to the shortcomings in the guideline's efficiency and effectiveness, the European Commission initiated a revision in 2008. This revision brought about various changes, which proved instrumental in curtailing illicit e-waste exports to non-OECD nations. Key modifications included the introduction of more stringent compulsory collection targets for e-waste, the establishment of minimum surveillance prerequisites for WEEE shipments, and the implementation of legally conclusive provisions to distinguish between new, used, or waste products. These measures were put in place to combat the mislabeling of WEEE as used EEE.

23.12 ELECTRONIC WASTE REGULATIONS IN THE UNITED STATES

In the United States, the development and implementation of directives, legislation, and regulations related to electronic waste management have followed a segmented approach. This is due to the absence of a comprehensive federal legislation specifically designed to regulate the national administration and shipped internationally of WEEE. Presently, there are only 2 national regulations that address electronic waste and it's shipped internationally: Act of 1976 RCRA (The Resource Conservation and Recovery Act) and CRT Rule of the Environmental Protection Agency (EPA).

RCRA is limited in its effectiveness to regulating the discard of municipal electronic waste within the United States and it's shipped internationally. Additionally, the current CRT rule, due to certain vulnerabilities, has demonstrated reduced efficiency. This rule governs the shipped internationally of uncategorized CRT glass and CRTs intended for regeneration, but it does not impose restrictions on the export of non-used, intact CRTs designed for reutilization of recycling. As the national government typically does not classify most of the e-waste as harmful, American reproducers are permitted to export electronics overseas devoid of significant restrictions (Althaf et al., 2019). Furthermore, individual U.S. states have initiated their own measures to address e-waste concerns, enacting regulations and implementing effective administrative systems. Many states enacted laws to restrict the discard of specific types of electronic waste.

23.13 ELECTRONIC WASTE REGULATIONS IN CHINA

In the last decade, the Chinese government has introduced a diverse set of environmental rules and regulations concerning the administration of electronic waste (Yang et al., 2023). These regulations are presented in Table 23.3.

TABLE 23.3
E-Waste Regulations in China

Sr. No.	Name of the Law/Regulation	Year of Introduction	Highlighted Regulations
1	Waste import management Catalogue	2000 (amendment in 2009)	• Prohibition on import of e-waste
2	Technical guidelines on Pollution Prevention & Control of WEEE	2006	• Stipulates Ecological design Establish principles for '3R' & 'Polluter pays principle' • Set provision for environmentally sound collection, reproducing, & discard of WEEE
3	Guidelines for Managing and Preventing Pollution from Information and Electronic Products.	2007	• Specifications for eco design of product • Limitations on the use of dangerous substances
4	Managerial Measures on Pollution Prevention of WEEE	2008	• Appeal to Prevent Pollution from Disassembly, Reproduction, and Disposal of E-Waste. • License guidelines for firms involved in the regeneration of e-waste.
5	Guidelines on administration of the reproduction and discard of waste electronic and electrical equipment	2011	• Makes e-waste reuse compulsory • Enact Producer responsibility extended. • Institutes a dedicated financial resource to support electronic waste recycling subsidize

23.14 E-WASTE REGULATIONS IN INDIA

Electronic waste consists of various elements, few elements are harmful in nature and some are valuable substances. Because of the evaluation of technology, advancement amount of e-waste is rising daily and causes damages to environment and essentials.

Mostly electronic waste handling is done by the unorganized industry in India (Dutta & Goel, 2021). Moreover, other reason for this problem is that inadequate execution of the electronic waste administrative rules (Manish & Chakraborty, 2019). There are various regulations in India which deals with the e-waste management directly and indirectly given in Table 23.4.

TABLE 23.4
Regulations in India Which Deal with the E-waste Management

Sr. No.	Name of the Law/ Regulation	Year of Introduction	Highlighted Regulations
1	The Environmental Protection Act	1986	• Ensure the safeguarding and enhancement of environment • Facilitates for the formation of Pollution Control Boards • Institute authorities with sufficient powers to regulate and safeguard the environment
2	The Rule of Hazardous Waste	1989	• Ensure the safe and responsible management of waste from generation to disposal
3	The Rule of Batteries	2001	• Endure the secure getting rid of used lead-acid batteries
4	The Rule of Hazardous Waste	2008	• Describes the registration procedure for dangerous waste recycler CPCB, India
5	Guidelines for Environmentally sound management of E-waste	2008	• Offer guidance on identifying of diverse sources of waste

(Continued)

TABLE 23.4 (Continued)
Regulations in India Which Deal with the E-waste Management

Sr. No.	Name of the Law/ Regulation	Year of Introduction	Highlighted Regulations
6	The Rule of Plastics	2009 (amendment in 2022)	• Specify the duties of local bodies and various stakeholders in overseeing plastic waste management while enforcing a ban on single-use plastic
7	The Rule of E-Waste	2011	• EPR is the main feature • Decided the producers' liability for regenerating and minimizing e-waste in the nation
8	The Rule of E-waste	2016	• Producers to follow EPR

23.15 CONCLUSION

The growing population generates vast quantities of waste. Over the past few years, E-waste has become a significant concern for urban local bodies, waste managers, and policymakers. This issue is critical to address for residents staying near the waste management sites owing to adverse health impacts from this waste. The approaches to address e-waste include EPR, MFA, LCA, and MCA. China and India have taken the lead in this domain and have a set of comprehensive legislations that deal with the management of e-waste. Other developing countries can take guidance from them in framing their policies. However, in addition to policies and legislations more implementation is needed at the ground level so that we are able to tackle the problems in a sustainable manner related to the management of e-waste.

REFERENCES

Adhana, D.K. (2020). E-waste management in India: A study of current scenario. *International Journal of Management, Technology and Engineering*, IX, 2791–2803.

Agarwal, S., Darbar, S., Saha, S., Choudhury, M., & Singh, R. P. (2023). E-waste management using different cost-effective, eco-friendly biological techniques: An overview. In *Waste Management and Resource Recycling in the Developing World*, pp. 205–235. https://doi.org/10.1016/B978-0-323-90463-6.00026-9

Althaf, S., Babbitt, C. W., & Chen, R. (2019). Forecasting electronic waste flows for effective circular economy planning. *Resources, Conservation and Recycling*, 151, 104362.

Awasthi, A. K., & Li, J. (2017). Management of electrical and electronic waste: A comparative evaluation of China and India. *Renewable and Sustainable Energy Reviews*, 76, 434–447.

Baldé, C. P., Forti, V., Gray, V., Kuehr, R., & Stegmann, P. (2017). The global e-waste monitor. United Nations University (UNU), International Telecommunication Union (ITU) & International Solid Waste Association (ISWA), Bonn/Geneva/Vienna, 1–109.

Dutta, D. Goel, S. (2021). Understanding the gap between formal and informal e-waste recycling facilities in India. *Waste Management*, 125, 163–171, 10.1016/j.wasman.2021.02.045

Forslind, K. H. (2005). Implementing extended producer responsibility: the case of Sweden's car scrapping scheme. *Journal of Cleaner Production*, 13(6), 619–629.

Forti V, Balde CP, Kuehr R, Bel G, 2020. The global E-waste monitor 2020: Quantities, flows and the circular economy potential. *United Nations University (UNU)/United Nations Institute for Training and Research (UNITAR) – co-hosted SCYCLE Programme, International Telecommunication Union (ITU) & International Solid Waste Association (ISWA), Bonn/Geneva/Rotterdam*

EPR portal for e-waste management, 2022. https://eprewastecpcb.in/pdf/e-waste_rules_2022.pdf.E-wastemanagementsystem

Liu, K., Tan, Q., Yu, J., & Wang, M. (2023). A global perspective on e-waste recycling. *Circular Economy*, 2 (1), 100028.

Manish, A. & Chakraborty, P., 2019. E-Waste Management in India: Challenges and opportunities. https://www.teriin.org/article/e-waste-management-india-challenges-and-opportunities

Mohana, R., Turaga, R., & Bhaskar, K., (2019). E-Waste Management in India: Issues and Strategies. *The Journal for Decision Makers*, 44, 3, 127–162.

Mondal, T., Choudhury, M., Kundu, D., Dutta, D., & Samanta, P. (2023). Landfill: An eclectic review on structure, reactions and remediation approach. *Waste Management*, 164, 127–142. https://doi.org/10.1016/j.wasman.2023.03.034

Mor, Rahul S., Sangwan, Kuldip S., Singh, S., Singh, A. & Kharub, M. (2021). E-waste Management for Environmental Sustainability: An exploratory study. *28th CIRP Conference on Life Cycle Engineering*, pp. 193–198.

Morris, A., & Metternicht, G. (2016). Assessing effectiveness of WEEE management policy in Australia. *Journal of Environmental Management*, 181, 218–230.

Park, J.K., Hoerning, L., Watry, S., Burgett, T. & Matthias, S. (2017) Effects of electronic waste on developing countries. *Advances in Recycling Waste Management*, 2, 128. https://doi.org/10.4172/2475-7675.1000128.).

Paul, S., Choudhury, M., Deb, U., Pegu, R., Das, S., & Bhattacharya, S. S. (2019). Assessing the ecological impacts of ageing on hazard potential of solid waste landfills: A green approach through vermitechnology. *Journal of Cleaner Production*, 236, 117643.

Reena, G. & Sangita, V.K., (2011). Electronic Waste: A Case Study. *Research Journal of Chemical Sciences*, 19, 49–56.

Richard & Jones, C. (2014). Understanding e-Waste Management in Developing Countries: Strategies, Determinants, and Policy Implications in the Indian ICT Sector.

Sharma, B.K., Bhardwaj, A., & Bhardwaj, R. (2015). E-Waste Management in India: Opportunities and Predicaments under Swachh Bharat Abhiyan, *2nd international conference on science, technology and management*, pp. 2470–2481.

Yang, M., Chen, L., Wang, J., Msigwa, G., Osman, A. I., Fawzy, S., Rooney, D. W., & Yap, P.-S. (2023). Circular economy strategies for combating climate change and other environmental issues. *Environmental Chemistry Letters*, 21(1), 55–80. https://doi.org/10.1007/s10311-022-01499-6

Yoheeswaran, E. (2013). E-Waste Management in India. *Global Research Analysis*, 2, 4, 54–55.

24 Significance of Mass Media in Addressing the Electronic Waste Pollution

Muhammad Umair Chaudhary
Islamia University, Bahawalpur, Pakistan

Sobia Usman
SZABIST University Karachi, Pakistan

24.1 INTRODUCTION

The widespread use of electronic gadgets and the quick progress in information technology have significantly altered several aspects of our daily lives, professional endeavors, and modes of communication. The advent of these technical advancements has yielded many advantages, but it has also engendered a growing ecological apprehension: electronic garbage, sometimes referred to as e-waste. The disposal of electronic trash is a significant issue, and mass media and information technology may play a crucial role in providing viable solutions. Scholars are increasingly concerned about environmental contamination and health concerns due to electronic trash. Citizens' insensitivity to these wastes' life-threatening effects compounds the problem. According to the research, media coverage of e-waste concerns could have been more dangerously adequate. The broadcast media and the internet were important sources of information, but citizens seldom saw e-waste concerns in the news. Even though e-waste was not considered a severe threat, inhabitants were unaware of its health and environmental impacts. E-waste caused a lackluster approach to tackling the issue. In addition to conventional and internet media, interpersonal communication channels should be used to educate and encourage individuals to embrace global e-waste management standards (Ben-Enukora et al., 2017). Human innovations advance humanity daily. Though innovations improve human existence, they may also harm the environment and humans. Thus, protecting nature from contemporary technology's degradation is a major task. Computer and internet use are rising daily. The need for computers and electronics is rising daily. E-waste is a serious issue nowadays. People need to be protected against e-waste. Since internet and blog usage is rising quickly, blogs may raise awareness. According to the research, blogging, an essential component of the internet, handles our e-waste issues and how to manage them (Kumar, 2014). Technology, gazettes, and consumer gadgets change rapidly. Information technology is the world's biggest manufacturing sector. Since upgrading and obsolescence are part of this sector, experts are worried about e-waste. E-waste has grown faster than any other industry, much like manufacturing. Electronic items, including microwaves, air conditioners, refrigerators, TVs, mobile phones, music systems, computers, and life-saving medical packages like ECG and BP monitors, are all considered e-waste. Electronic product categories are large and increasing fast. Electronic items include harmful materials like lead and mercury, which harm the environment and humans (PR & Narula, 2018). In recent years, there has been growing concern regarding the detrimental impacts of electronic trash (e-waste) on human health. China is among the nations grappling with significant pollution and human exposure to electronic trash (e-waste). A considerable portion of the population is exposed to potentially

 DOI: 10.1201/9781003582311-24

harmful compounds resulting from informal practices of recycling e-waste. The toxicities investigated include those affecting the endocrine, respiratory, reproductive, developmental, neurological, and genetic systems. The study also demonstrated the adverse effects of toxic heavy metals (such as lead, cadmium, chromium, mercury, and nickel) and organic pollutants (including polybrominated diphenyl ethers, polychlorinated biphenyl, polycyclic aromatic hydrocarbons, polyhalogenated aromatic hydrocarbons) on human health (Xu et al., 2015). This chapter analyzes the intricate relationship between mass media and the problem of e-waste pollution in Pakistan. The chapter examines how these factors contribute to the generation of electronic waste, raise public awareness of the issue, and promote the formulation of sustainable solutions.

24.2 E-WASTE GENERATION

The topic of electronic trash, usually known as e-waste, is gaining prominence and poses both challenges and possible economic opportunities. The reason for this phenomenon may be attributed to the significant volumes of electronic trash being generated, along with the existence of both dangerous and valuable components inside these materials. The composition of electronic waste (e-waste) is mostly comprised of metals, including iron, copper, aluminum, gold, and several other elements. These metals constitute more than 60% of the entire percentage of e-waste. Conversely, pollutants make up around 2.70% of the overall composition. Due to the significant levels of toxicity exhibited by these pollutants, particularly when subjected to combustion or unregulated recycling processes, the Basel Convention has designated electronic trash (e-waste) as a hazardous category.

As a result, the convention has developed a thorough framework for controlling and overseeing the transboundary flow of these waste products. The Basel Ban is an addition to the Basel Convention that still needs to be implemented. By restricting the flow of electronic garbage (e-waste) from developed to developing nations, it seeks to tighten rules (Widmer et al., 2005). There is an exponential rise in the production and consumption of electronic equipment, such as laptops, cell phones, and consumer electronics, and a clear link between this growth and the generation of electronic waste. Electronic waste, or "e-waste," is a rapidly expanding worldwide commodity with substantial economic potential. Governments everywhere are gradually realizing the importance of the problem of electronic garbage, or "e-waste," and enacting laws and regulations to combat it as the world community grows more aware of it. It is crucial to remember that sizable amounts of e-waste must be made aware of accurate tracking and reporting. A sizable amount of electronic waste worldwide is dumped in open landfills and disposal sites. Notable environmental and public health problems have arisen as a result of this circumstance (Paul et al., 2019; Andeobu et al., 2023).

24.3 CONNECTION TO MASS MEDIA

Mass media has been influential on consumer behavior, specifically focusing on the role of advertising and the promotion of consumer culture in shaping the need for new technological gadgets. Mass media advertisements and technology inflicted on customer behavior. Advertising is a mode of communication that aims to persuade an audience, comprising viewers, readers, or listeners, to engage in the acquisition or utilization of products, information, or services, among other possibilities. The study reveals that the emotional reaction of consumer purchase behavior exhibits a particularly high link with consumer buying behavior. It is indeed accurate that individuals tend to acquire brands to which they have developed an emotional connection (Saleem & Abideen, 2011).

24.4 MASS MEDIA'S ROLE IN RAISING AWARENESS

24.4.1 Advocacy and Documentary Films

Mass media could be utilized by advocacy organizations and documentary filmmakers to raise awareness about the e-waste dilemma, emphasizing the ecological and societal ramifications resulting from inadequate e-waste disposal practices.

24.4.2 E-Waste Awareness Campaigns

The potential of social media and online platforms to facilitate the initiation of e-waste awareness campaigns could be effectively utilized for mass awareness campaigns. The primary objective of such campaigns should be to educate the general public on appropriate disposal and recycling techniques for electronic trash.

24.4.3 Environmental Journalism

Mass media plays a significant role in disseminating information and increasing public consciousness of the environmental ramifications associated with e-waste pollution. Further, environmental journalists also play a key role in the investigation and dissemination of information pertaining to electronic waste.

24.5 ROLE OF MASS MEDIA IN THE ISSUE OF E-WASTE POLLUTION IN THE CONTEXT OF PAKISTAN

The world is currently grappling with a substantial and increasing issue related to the rapid proliferation of electronic and electrical garbage, known as e-waste. Large quantities of this material are being shipped from developed countries to impoverished countries, including Pakistan, to reduce the expenses related to the proper disposal and recycling of electronic trash (e-waste). Adopting an appropriate management strategy to address the growing e-waste problem in Pakistan is crucial. A recent study has shown that over 50 kilotons (kt) of electronic waste (e-waste) is being imported as scrap, in addition to the local production of around 38 kt annually. Electronic garbage (e-waste) within the country is being handled in a primitive way without the deployment of adequate safety protocols and protective gear. Prompt and effective execution of surveillance and regulation strategies to tackle the urgent problem of informal electronic waste management in Pakistan (Sajid et al., 2019).

To address the growing e-waste issue in Pakistan, it is necessary to enact potential legislative initiatives and governmental actions. Research indicates a widespread requirement for further reliable data, inventories, and research studies in Pakistan explicitly focusing on e-waste problems and environmental and human health. Consequently, it is imperative to broaden the existing knowledge base by incorporating research from other nations that have encountered similar difficulties. Additional investigation into these matters in Pakistan is crucial (Iqbal et al., 2015). From a Pakistani perspective, the role of mass media in addressing e-waste pollution is highly significant since it involves both its impact on the problem and its capacity to raise awareness about the issue. Pakistan faces unique challenges related to electronic waste (e-waste) due to its growing consumer electronics sector and inadequate infrastructure for e-waste management. This discussion will analyze the importance of mass media in the context of e-waste pollution in Pakistan.

24.5.1 Promotion of Consumer Electronics

The use of mass media, including television, radio, and internet platforms, catalyzes and facilitates the widespread acceptance and integration of novel technological technologies. Advertisements and technology reviews often promote the acquisition of cutting-edge electronic devices, fostering the accelerated obsolescence of such products.

24.5.2 Inadequate Regulation and Enforcement

Mass media can influence the enforcement of regulations related to e-waste. In Pakistan, regulatory frameworks often need to be stronger and more effectively enforced. Media coverage can either draw attention to these regulatory gaps or, in some cases, indirectly contribute to their perpetuation.

24.5.3 Raising Awareness

The use of mass media may serve as a potent mechanism for increasing public consciousness about the ramifications associated with the pollution resulting from electronic waste. Media outlets can enlighten the general public on the environmental and health risks linked to inadequate disposal of electronic trash (e-waste) via investigative journalism, documentaries, and news reporting.

24.5.4 Environmental Journalism

Environmental journalists can conduct investigations and provide comprehensive coverage of matters pertaining to electronic waste (e-waste). This includes bringing attention to the unlawful disposal of e-waste, the prevalence of informal recycling methods, and the inadequate availability of appropriate disposal facilities. The act of reporting on these matters has the potential to elevate these concerns to a prominent position in public discussions and may result in demands for regulatory restructuring.

24.5.5 Advocacy and Campaigns

Advocacy groups and environmental organizations use mass media platforms to execute campaigns and raise awareness on the appropriate disposal of electronic trash. These initiatives have the potential to incentivize individuals to engage in the practice of recycling their electronic equipment while also raising awareness about the significance of effectively managing electronic waste.

24.5.6 Government Initiatives

Mass media may also function as a mechanism for governmental entities to distribute information pertaining to e-waste rules, recycling initiatives, and conscientious consumer conduct. Public service announcements and government-sponsored awareness initiatives have shown efficacy in effectively communicating these messages.

24.5.7 Challenges and Opportunities

The mass media can shed attention on the distinct issues encountered by Pakistan in effectively managing electronic trash (e-waste), including the existence of informal recyclers and the absence of adequate recycling infrastructure. Additionally, it has the potential to highlight potential avenues for the nation to cultivate sustainable methods in the handling of electronic trash.

In a nutshell, it is possible to summarize the multifaceted role that the media has played in the e-waste pollution situation in Pakistan. Technology may make the problem worse by encouraging people to buy things. However, it can also help change things by raising awareness, encouraging responsible behavior, and putting pressure on the government to deal with the e-waste problem. In the context of Pakistan, the efficient management of e-waste pollution necessitates a joint endeavor, including the media, government, and civil society.

24.6 CONCLUSION

The mass media and information technology are inextricably linked to the crisis of e-waste pollution. While they contribute to the issue through consumption promotion and technological advancements, they also have the potential to play a significant role in raising awareness and driving solutions to mitigate e-waste pollution. This chapter elucidates the intricate interplay between these components and underscores the need for a synchronized endeavor to tackle the e-waste predicament. This paper summarizes the complex relationship between mass media, information technology, and the dilemma of e-waste pollution. It highlights their influence's dual nature, increasing the problem and offering potential remedies. This paper examines probable future developments, explicitly focusing on the role of new technologies in mitigating the issue of electronic trash (e-waste).

Additionally, it emphasizes the need for sustained mass media coverage and public awareness to address this problem. The importance of mass media in tackling the e-waste pollution situation in Pakistan is of great magnitude. The media plays a vital role in the fight against e-waste pollution due to its ability to effectively distribute information, increase public consciousness, and shape policy-making processes. This abstract highlights the need to foster collaborative endeavors among governmental entities, industries, and media to address the imperative of sustainable e-waste management. By doing so, the adverse environmental and health risks connected with electronic trash in Pakistan may be effectively alleviated.

REFERENCES

Andeobu, L., Wibowo, S., & Grandhi, S. (2023). Informal E-waste recycling practices and environmental pollution in Africa: What is the way forward? *International Journal of Hygiene and Environmental Health, 252*, 114192.

Ben-Enukora, C., Okorie, N., Oresanya, T., & Ekanem, T. (2017). Awareness and perception of media campaign on e-waste effects among residents of Ado Odo-Ota, Nigeria. *Covenant Journal of Communication, 4*(2), 18–32.

Iqbal, M., Breivik, K., Syed, J. H., Malik, R. N., Li, J., Zhang, G., & Jones, K. C. (2015). Emerging issue of e-waste in Pakistan: a review of status, research needs and data gaps. *Environmental Pollution, 207*, 308–318.

Kumar, C. S. (2014). Internet and E-waste Awareness. *Universal Journal of Environmental Research & Technology, 4*(5), 227–234.

Paul, S., Choudhury, M., Deb, U., Pegu, R., Das, S., & Bhattacharya, S. S. (2019). Assessing the ecological impacts of ageing on hazard potential of solid waste landfills: A green approach through vermitechnology. *Journal of Cleaner Production, 236*, 117643. https://doi.org/10.1016/j.jclepro.2019.117643

PR, S. K., & Narula, D. S. (2018). Educational Strategies for all Eviating the Environmental Impact of E-Waste for Sustainable Development. *Cosmos An International Journal of Management, 7*(2), 1–9.

Saleem, S., & Abideen, Z. (2011). Effective advertising and its influence on consumer buying behavior. *European Journal of Business and Management, 3*(3), 55–67.

Sajid, M., Syed, J. H., Iqbal, M., Abbas, Z., Hussain, I., & Baig, M. A. (2019). Assessing the generation, recycling and disposal practices of electronic/electrical-waste (E-Waste) from major cities in Pakistan. *Waste management, 84*, 394–401.

Widmer, R., Oswald-Krapf, H., Sinha-Khetriwal, D., Schnellmann, M., & Böni, H. (2005). Global perspectives on e-waste. *Environmental Impact Assessment Review, 25*(5), 436–458.

Xu, X., Zeng, X., Boezen, H. M., & Huo, X. (2015). E-waste environmental contamination and harm to public health in China. *Frontiers, 9*(2), 220–228.

25 Innovative Electronic Waste Recycling and Systematic Management Strategies – Concerns and Challenges

Banani Ray Chowdhury
Bengal Institute of Technology, Kolkata, India
Baranagar Baghajatin Social Welfare Organisation, Baranagar, India

Saptaswa Chakraborty, Induja Samanta, and Soutrik Bhattacharya
Bengal Institute of Technology, Kolkata, India

Pallabi Roy Chowdhury
Baranagar Baghajatin Social Welfare Organisation, Baranagar, India

Tirtha Pratim Mandal
College of Staten Island, City University of New York, New York, New York, United States

Sayari Bhunia
March Friechel's Lab, Pharmakologische Institut, Medizinischen Fakultät Heidelberg, Germany

25.1 INTRODUCTION

The industrial revolution of the 18th century, brought about by scientific and technological advancements, signaled the beginning of a new era of human civilization. The 20th century's revolution in information and communication has completely transformed the foundation of our institutions, economies, and daily routines. Undeniably, these incredible advancements have greatly enhanced our standard of living. However, with progress comes consequences, and one of the major concerns is the overwhelming amount of toxic waste and pollutants generated by electronics. The public health is seriously threatened by these and other trash. For this reason, waste management is essential to maintaining the environment, human health, and standard of living (Rajpal et al., 2024). It poses a significant issue to contemporary societies, and in order to achieve sustainable development, it must be addressed with concerted efforts (Dwivedy et al., 2010). The electronic industry reigns as the largest and most rapidly expanding sector in the world of manufacturing. Today, India's information technology (IT) is widely recognized throughout the world, partly because of the software industry. Leading multinational corporations (MNCs) have been flooding India with headquarters,

DOI: 10.1201/9781003582311-25

R&D centers, and software development facilities as a result of regulatory reforms that have occurred more recently. From just 13 IT companies in 1991, Bangalore presently has over 3000 IT companies offering top-notch infrastructure. Bangalore's phenomenon is being reproduced in a number of Indian cities, including Gurgaon, Hyderabad, Chennai, Mumbai, and Pune. The IT industry's exponential growth has resulted in a corresponding increase in waste disposal issues (Wang et al., 2019).

25.1.1 Economic Growth and the Digital Revolution

The digital revolution began in 1980 and has continued ever since. The advent of the digital revolution brought with it an array of affordable and user-friendly goods that quickly took over our homes. Replacing them is now more convenient and easier than having them repaired. In developed nations, a computer's average lifespan is two years. The amount of e-waste increases as new electronic devices replaces outdated PCs, laptops, and other electronic equipment. At present, consumer electronics make up the most fleetly expanding order of external solid waste in both developed and developing countries. Today, e-waste is a global issue that, if left unchecked, would negatively affect the environment in a big way (Kumar et al., 2007).

Tables 25.1 and 25.2 display various components found in e-waste such as refrigerators, freezers, personal computers, and TVs.

TABLE 25.1
Explore the Average Weight and Composition of Our Carefully Chosen Appliances for Ultimate Understanding and Applications (Mishra et al., 2010)

Appliances	Refrigerators and Freezers	Personal Computer	TV Sets
Average Weight	48	29.6	36.2
Iron, %	64.4	53.3	5.3
Non-Ferrous metal, %	6	8.4	5.4
Glass, %	1.4	15	62
Plastic, %	13	23.3	22.9
Electronic components, %	–	17.3	0.9
Others, %	15.1	0.7	3.5

TABLE 25.2
A PC Usually Contains a Range of Metal Values (Nagpal et al., 1993)

Elements	Percent of Total Weight	Content (g)	Elements	Total Weight	Content
Plastics	23	6250	Zinc	2	55
Lead	6	1710	Beryllium	0.0157	4.26
Aluminium	-	3850	Gold	0.0016	0.434
Germanium	0.0016	0.4345	Cobalt	0.0157	4.2
Gallium	0.0013	0.3531	Palladium	0.0003	0.081
Iron	20	5570	Manganese	0.0315	8.55
Tin	1	271	Silver	0.0189	5.13
Copper	7	1880	Mercury	0.0022	0.597
Barium	0.0315	8.55	Arsenic	0.0013	0.35
Nickel	0.8503	231	Silica	24.8803	6770

Electronic garbage is the fastest-growing trash concern on a global scale. According to a recent report from IRG System, South Asia, India alone contributes a staggering 1,46,180 metric tons of e-waste every year through the disposal of outdated or faulty electronic equipment. This problem is exacerbated by the practice of wealthy countries offloading their e-waste, particularly computer trash, onto India. As a result, India has become the primary dumping ground for obsolete computers and other devices from industrialized nations. In fact, it is estimated that India receives an influx of almost 50,000 tons of discarded electronic equipment each month.

The continuous pursuit for innovative processing techniques to effectively handle diverse, out of use electronic scrap, while retaining valuable metals in an eco-friendly manner, will persist. Pyrometallurgy, hydrometallurgy, and electrochemistry have been proposed as potential methods for metal recovery. However, the utmost importance is placed on the mechanical processing and physical separation stages in determining the most efficient processing route. Typically, the first step in e-waste processing is mechanical, aimed at reducing bulk and segregating materials. This involves various unit operations such as hand disassembly, hammer milling, air classification, magnetic and eddy current separation, and high-tension separation. Recyclers place great emphasis on this stage.

Personal computers are commonly divided into three main sections: the body, keyboard, and monitor. These parts are made up of various materials, such as plastics, iron, aluminum, and printed circuit boards. Through a process of crushing, cutting, and melting, the plastic, iron, and aluminum components are sold as recyclable scrap. The success of recycling PCs and mobile phones largely relies on the use of cost-effective and efficient methods for extracting valuable metals from the printed circuit boards (Baba et al., 2010).

25.2 E-WASTE RECYCLING

E-waste recycling is the process of separating useful materials from the waste after it has been broken up into small enough pieces to be put back into a new electronic device. But several present obstacles are keeping the electronic recycling sector from growing.

Recycling e-waste has several advantages beyond safeguarding the environment and public health. Since non-renewable minerals make up the majority of the materials used to create computers and smartphones, recycling these materials can keep the supply of consumer goods, which are an inescapable part of life, from being stopped until viable alternatives are found. Recycling common but non-renewable minerals nevertheless provides economic benefits, even though in some circumstances the non-renewable resources are not necessarily uncommon. Lithium is a non-renewable material that is rather ubiquitous and nearly universally available, but its price has been skyrocketing. Although lithium is widely utilized in many other industries, its significance in the creation of rechargeable batteries for electric cars is what makes it most well-known. The demand for lithium skyrocketed due to the growing public interest in electric vehicles as a means of reducing carbon emissions in transportation. However, because of the market's inability to meet the abrupt increase in demand, there is a shortage of lithium—not because of a shortage, but rather because of the sluggish rate of extraction and refinement. Recycled lithium-ion batteries will add to the market's supply of lithium, enabling companies to make batteries and electric cars that are more affordable while still meeting consumer and environmental standards (Arain et al., 2020).

A assortment of perilous components can be found in e-waste, counting arsenic, polychlorinated biphenyls (PCBs), chlorofluorocarbons (CFCs), hydrochlorofluorocarbons (HCFCs), nickel, and asbestos. These chemicals are known to as often as possible discharge harmful leachate and vapors, as seen within the vanishing of metals and dimethylene mercury. Moreover, uncontrolled fires in landfills can result within the emanation of amazingly destructive dioxins and furans, both complex compounds (Mondal et al., 2023). The nearness of polybrominated fire retardants in plastic includes to the dangers related with reusing, as these substances can break down into harmful polybrominated dibenzo furans (PBDF) and polybrominated dibenzo dioxins (PBDD) amid the transfer prepare. As a result, reusing of plastic presents unsafe and sensitive challenges. As a result, the generation of these chemicals was ended by the German chemical industry in 1986. Moreover, specialists in reusing shops have been found to have tall levels of PBDEs in their blood. A later consider in Sweden found that amid the reusing prepare of electronic gear such as computers and fax machines, tidy containing harmful fire retardants is discharged into the encompassing discuss. Specialists at these offices were found to have levels of one sort of fire retardant that were 70 times higher than those found in clinic cleaners. Furthermore, due to the predominance of these chemicals within the discuss, full-time representatives at computer defenses lines too had follows of fire retardants in their blood, in spite of the fact that somewhat lower than those found in cleaners. People may possibly retain PBDEs specifically, as they are transmitted from electronic circuit boards and plastic computer and tv cabinets. Shockingly, a critical sum of electronic squander closes up in unmonitored ways that lead to adjacent recyclers. Whereas the casual scrap industry gives work for people with changing levels of ability, it too presents risks and flimsiness. Without legitimate controls and securities input, these specialists are not ensured social security benefits or lawful security. In addition, the strategies utilized for extraction and reusing of profitable materials can posture genuine long-term wellbeing dangers for both specialists and the environment (Bandura et al., 1986).

25.3 MANAGEMENT OF E-WASTE

There isn't a single, perfect approach for managing e-waste in developing nations because every one of them has distinct environmental, social, technological, economic, and cultural circumstances. Reduction, reuse, and recycling—the three Rs—are recognized in environmentally effective WEEE management. Reusing electronic equipment until it is still functional, recycling damaged components, and smart production and maintenance are all ways to lessen the amount of electronic-waste produced.

An intelligent e-waste management system for developing nations must first evaluate the current state of e-Waste, acknowledge that e-waste is a complex mixture of both hazardous and non-hazardous materials and substances, and define an integrated e-waste management system that takes into account factors like the penetration of the EEE market, the life cycle of ICT equipment, financing mechanisms, etc. (Williams et al., 2005).

The primary rules and regulations to be followed when creating ICT waste management guidelines for developing countries are:

- The approach and controls for consequence and trade of electronic and electrical gear (EEE) and squander electronic and electrical hardware (WEEE) point to adjust with the laws of each nation and universal enactment.
- This involves outlining the roles and responsibilities of key stakeholders, including government, supply chain partners, consumers of ICT equipment, and waste disposal entities.
- In addition, the concept of extended producer responsibility (EPR) requires manufacturers to take ownership throughout the life cycle of their ICT equipment, including bearing the cost of managing it at the end of its use.
- To facilitate effective management, a responsible information system is needed with accurate data on ICT equipment in the market, the handling of unused EEE, and WEEE disposal, enabling better monitoring and future planning.
- Additionally, the policy aims to boost employment opportunities and provide training for informal sector workers involved in recycling activities (Bandyopadhyay et al., 2008).

25.4 TYPES OF E-WASTES

There exist multiple recognized categories of e-waste worldwide, which are based on the source and use of electronic products. Numerous factors, including as socioeconomic conditions, consumer behavior, population size, and the reliance of families and companies on electrical and electronic

TABLE 25.3
The Diverse Types of E-Waste and Their Corresponding Quantities Generated (Wong et al., 2007)

Number	Category	Example	Quantity (Million tonnes)
1	Huge Electronics	Dishwashing machines, washing machines, and huge printing machines	11.8
2	Equipment used for exchanging temperatures	Refrigerators, Freezers, heat pumps, and air conditioners	7.0
3	Tiny IT devices for communications	Printers, laptops, phones, and pocket calculators	3.0
4	Tiny Electronics	Vacuum cleaners, radios, toasters, toys, cameras, medical equipment and electrical tools	12.8
5	Lamps	Discharge lamps with high intensity, LED lamps, and fluorescent lamps	1.0
6	Mobiles and screens	Laptops, monitors, and televisions	6.3

equipment (EEE). The average percentage of each category of e-waste can be impacted. However, big household products account for the majority of e-waste (42.1%), with IT and telecommunications coming in third (33.9%), consumer electronics in second place (13.7%), and tiny domestic equipment in fourth place (4.7%). To enhance readability, there has been a decrease in contributions to other categories, such as medical, lighting, electrical, and electronic instruments (less than 2%), as well as automatic dispensers, toys, sports, monitoring, and control equipment (less than 1%). However, it is worth noting that in impoverished nations, the majority of e-waste is still comprised of mobile phones, TVs, and PCs (Table 25.3).

25.5 MATERIAL RECOVERY FACILITY (MRF) OF E-WASTES

25.5.1 Types of Material Recovery Facility (MRF)

A Material Recovery Facility (MRF) is a plant that sorts and processes recyclable materials from waste. They play a vital role in diverting waste from landfills and turning it into usable raw materials for new products.

There are two main types of MRFs based on how the incoming recyclables are sorted:

Single-Stream MRF: This is the most common type, where all the recyclable materials are mixed together in one bin. At the MRF, the materials are sorted using a combination of conveyor belts, screens, magnets, and eddy currents.

Multi-Stream MRF: Residents sort their recyclables into separate bins for different materials (e.g., paper, plastic, glass, metal). This allows for cleaner and more efficient sorting at the MRF.

Here's a general overview of the process at a Material Recovery Facility:

- **Delivery**: Recyclables are delivered to the MRF in trucks.
- **Sorting**:
  1. **Single-Stream MRF**: A combination of manual sorting and automated machinery separates the materials into different categories.
  2. **Multi-Stream MRF**: The pre-sorted recyclables are further refined using screens and manual sorting to remove any contaminants.

- **Processing**: The sorted materials are then processed into bales or chopped into smaller pieces, depending on the material.
- **Marketing**: The processed recyclables are then sold to manufacturers who will use them to create new products (Seif et al., 2024).

25.5.2 E-Recycling Methods

Electronic waste (e-waste) contains valuable metals that can be recovered through various techniques. Here are some common methods used for e-waste metal recovery:

Mechanical separation: This involves processes like shredding, crushing, and sieving to separate different materials based on their size and density. Magnets may be used to separate ferrous metals (iron and steel) from non-ferrous metals.

Hydrometallurgical processes: These methods involve using aqueous solutions (typically acids or bases) to dissolve metals from electronic waste. Commonly recovered metals include copper, gold, silver, and palladium. The process usually includes leaching, precipitation, and refining steps.

Pyrometallurgical processes: In this method, e-waste is heated to high temperatures in a controlled environment. Metals with high melting points, such as copper, lead, tin, and precious metals like gold and silver, can be recovered through smelting and refining.

Electrochemical processes: These involve using electrochemical cells to recover metals. For instance, copper can be extracted by electrolysis, where an electric current is passed through a solution containing copper ions.

Biotechnological methods: Recently, biotechnological approaches involving microorganisms or plants have been explored for their potential to recover metals from e-waste. This area of research is still developing but shows promise due to its environmentally friendly nature.

Mechanical and thermal processes for plastic and other materials: After metal recovery, techniques such as mechanical separation (for plastics) and thermal processes (for hazardous substances) are employed to further process the remaining e-waste components.

Each of these techniques has its advantages and limitations, and the choice of method often depends on the specific composition of the e-waste and the desired purity of the recovered metals. Environmental considerations, such as the generation of hazardous wastes and energy consumption, also play a significant role in determining the most appropriate recovery technique (Wilson et al., 2017).

E-waste metal recovery includes two main categories of techniques:

i. **Pyrometallurgy**: This method uses high temperatures (often exceeding 1,500°C) to separate metals from other materials in e-waste. Here are some common pyrometallurgical processes:
   a. **Smelting**: Shredded e-waste is melted in a furnace, where metals with lower melting points like copper and tin liquefy and separate from other materials like slag (impurities).
   b. **Incineration**: E-waste is burned at high temperatures, releasing metal oxides into the flue gas. These oxides can then be captured and further processed to recover the metals.

ii. **Hydrometallurgy**: This method uses aqueous solutions (liquids) to dissolve metals from e-waste. It's generally considered more environmentally friendly than pyrometallurgy. Here's a breakdown of some hydrometallurgical processes:
   a. **Leaching**: E-waste is crushed and exposed to acidic or caustic solutions that dissolve specific metals. The desired metals are then separated from the solution through various purification techniques.

b. **Bioleaching**: This emerging technique uses microbes to extract metals from e-waste. It's still under development but offers a potentially more sustainable option.

Apart from this main two categories, additional techniques include Electrostatic Separation, Electrochemical Separation, Supercritical Fluid Extraction (SCF), Vaccum Metallurgy, Remelting/Refining.

Factor controlling the selection of suitable metal recovery technique:

The selection of the most suitable metal recovery technique depends on several factors, including:

- **Type of metals**: Different techniques are better suited for extracting specific metals.
- **Environmental impact**: Hydrometallurgy generally has a lower environmental footprint compared to pyrometallurgy.
- **Economic considerations**: Factors like cost-effectiveness and energy consumption influence the choice of technique.

By employing these techniques, valuable metals from e-waste can be recovered and reintroduced into the manufacturing cycle, reducing reliance on virgin materials and minimizing environmental impact (Wang et al., 2019).

25.6 DIFFERENT E-RECYCLING PROCESSES

The process of e-recycling involves several crucial steps, such as collecting, extracting dust, shredding, sorting, separating with magnets and water, purifying waste streams, and preparing materials for sale (Bhutta et al., 2011).

<u>Steps of E-Recycling process</u>

25.6.1 Collection

E-waste should not be disposed of in regular waste, primarily. Electronic waste must be transported to designated facilities for proper disposal. These facilities are usually approved collection sites, major electronics retailers, or dedicated recycling containers.

The first step in the recycling process of electronic trash involves collecting electronic garbage through recycling bins, take-back programs, collection locations, or on-demand collection services. Afterwards, the mixed electronic waste is directed to certain electronic recyclers.

Due to the implementation of best practices, separating electronic trash by kind is necessary. As a result, numerous collection locations will continue to have many boxes and containers for different goods. It is crucial to handle e-waste with specific attention, especially concerning components like batteries, as they can be severely harmed if mixed with ordinary garbage.

25.6.2 Sorting and Shredded

After the e-waste has been gathered and transported, the following step is to sort and shred it. For a future separation to be successful, shredding is required. And for that reason, effectiveness is vital right now.

E-waste is shred, or broken into smaller bits, so that it can be properly sorted. These tiny parts are manually disassembled after being sorted with the help of hands. Because waste materials are now separated in order to recover different parts, this usually takes a lot of work.

After this point, materials are typically separated into components and core materials. After then, these things are divided up into other groups. Items that can be immediately reused and those that require further recycling procedures are frequently included in these categories.

E-waste is generally sorted by hand, although certain materials—like fluorescent lights, batteries, UPS batteries, and toner cartridges—shouldn't be physically crushed or shredded (Balabanic et al., 2011).

25.6.3 Dust Removal

The little waste particles are equally disseminated on the conveyor belt by means of a shaking operation. Then, more breakdown is performed on these finely divided e-waste fragments. At this point, all dust is cleaned and disposed of ecologically appropriate to prevent environmental degradation.

25.6.4 Magnetic Separation

After that, you can separate iron and steel from other wastes with the use of a powerful overhead magnet. In this manner, the steel has been effectively recycled out of the waste stream.

25.6.5 Water Separation

Then, in order to separate the glass from the plastic, the water separation technology is used. Leads containing glass can then be shipped to smelters to be used in the production of new batteries, X-ray tubes, and CRTs.

25.6.6 Cleaning Up the Waste Stream

Finding and removing any remaining metal from plastics is the next step in further purifying the waste stream.

25.6.7 Preparing Recycled Materials for Sale

Preparing recycled materials for sale is the last step. Here, the materials that were separated during the SSS are ready to be sold as components needed to make new devices (Lee et al., 2009) (Figure 25.1).

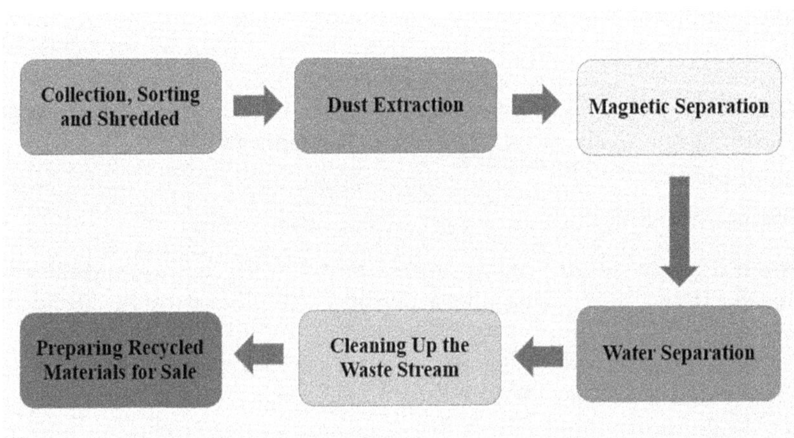

FIGURE 25.1 Flowchart of e-recycling process.

Application of E-waste Recycling:

1. **Use of Electronic Waste Concrete for Road Construction**: Natural aggregates like sand and gravel are commonly used in concrete. E-waste can partially replace these, reducing demand for virgin materials and conserving natural resources. E-waste is a growing global problem. Using it in concrete offers a way to divert e-waste from landfills and improper disposal, promoting a more sustainable solution. Studies suggest that incorporating processed e-waste in concrete, particularly at specific ratios; can improve its compressive and flexural strength compared to traditional concrete. This could lead to longer-lasting roads. Investigation on the use of E-waste and Waste plastic has been conducted as an environment friendly means of Road Construction. (Ingole et al., 2020), (Dimitrakakis et al., 2009).
2. **E-waste recycled materials as efficient catalysts for renewable energy technologies and better environmental sustainability**: Electronic devices are packed with valuable metals like rare earth elements, platinum, and palladium. Traditional mining of these materials can be environmentally destructive and resource-intensive. E-waste recycling offers a chance to recover these valuable resources and give them a second life. Recovered metals from e-waste can be used to create highly efficient catalysts for renewable energy technologies (Reed et al., 2016).
3. **Use of Electronic Waste for Toy Industry**: Toy e-recycling involves repurposing or responsibly disposing of electronic toys to reduce environmental impact. Components like batteries, plastics, and metals are extracted and recycled, minimizing landfill waste and conserving resources. This process not only helps in managing electronic waste but also promotes sustainability by reusing valuable materials. Additionally, it raises awareness about the importance of recycling and responsible consumption among children and adults alike, fostering eco-friendly habits and reducing the overall carbon footprint (Pérez-Belis et al., 2013).

25.7 ECONOMIC AND SOCIAL CONSEQUENCES OF E-WASTE RECYCLING

While the government has taken steps to combat the growing issue of e-waste, such as the implementation of the E-Waste Management (EWM) Rules in 2016, there is still ample room for improvement. These rules aim to prevent e-waste from reaching unapproved recyclers and ensure that hazardous waste is disposed of safely through the Extended Producer Responsibility (EPR) model. The government has the authority to enact laws like:

I. Growing recycling facilities and supporting start-ups and enterprises involved in recycling.
II. Establishing publicly accessible authorized collecting centers.
III. Forcing IT companies to abandon their deliberate obsolescence scheme.
IV. Enacting regulations that prohibit the use of dangerous materials in electronics while highlighting safe alternatives.
V. Throughout the lifecycle of the electronic item, consumer support needs to be made required.
VI. To prevent overproduction, put in place a policy of proportionate production.
VII. As quickly as feasible, pass the “Right to Repair” legislation in parliament.
VIII. Educate consumers on the issues that arise from e-waste.

Therefore, if we are successful in putting the ideas into practice, we will be able to tackle the expanding problem of electronic waste management, safeguard the environment, and steer ourselves toward a brighter future (Wang et al., 2019) (Figure 25.2).

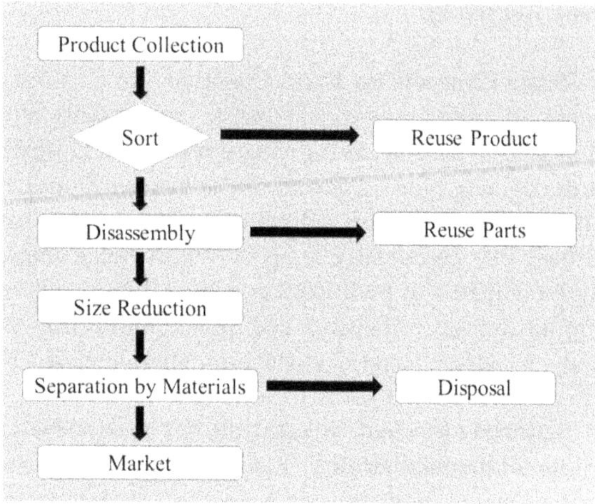

FIGURE 25.2 Structure for managing e-waste from manufacturing to disposal and related tasks (Hossain et al., 2021).

Prevailing economic systems are built on the premise that natural resources will perpetually be accessible, following the "take-make-use-dispose" linear paradigm (Centre 2015). In contrast, circular business models focus on reevaluating product-service systems to enhance firm capabilities and competitiveness in the market. Adopting circular economy models can help firms decrease their environmental footprint while uncovering untapped financial and economic opportunities. Did you know that a single tonne of discarded mobile phones can yield over 100 times the amount of gold compared to traditional gold mining methods? This astonishing fact, as cited by Ryder and Houlin in 2019, highlights the immense potential for corporations to recover valuable resources from electronic waste.

- **Electronics shared economy model**: The participated frugality model in electronics focuses more around furnishing services rather than dealing products. This innovative strategy offers are-evaluation of traditional value chains and transforms the usual patron and consumer connections. By exercising technology similar as block-chain and digital passports, companies can effectively tap into secondary requests for recyclable accoutrements and prognosticate force and demand dynamics, paving the way for a successful future (CEP 2022).
- **The electronics industry's Products-as-a-Service (PAAS) model:**
  The conception of Products- as-a-Service (PAAS) has revolutionized the electronics assiduity, furnishing a successful indirect business strategy. In traditional models consumers purchase and own electronic goods but the PAAS conception offers a different approach. Then, druggies pay a figure to use the products, without the burden of power. For case, an IT company can work the PAAS model for its computer tackle products, where druggies pay a yearly class figure. This encourages the company to prioritize form and re-manufacturing at the point of return. In addition, manufacturers, as they're financially motivated by the PAAS's figure- grounded system, are driven to produce modular products with proper end- of- life running and product life extension(PLE) designs (Ford et al., 2016).
- **Use of design for recyclability (DFR) and PLE in the product ownership model**
  In order to grant customers complete control over the product lifetime, a corporation may, at last, fall back on the conventional product ownership model. In this model, a producer is not as motivated to produce goods that support PLE and DFR principles, which are fundamental to the circular economy, as they are in the shared economy and PAAS models.

The cooperation and support of the general population are vital in order for government authorities and legislation to effectively address this important issue. This means that it is our responsibility to take concrete actions such as minimizing waste, reusing materials, recycling, and making informed choices. It is time for us to join forces and tackle this problem together because our actions as informed consumers can make a significant difference. Your contribution is valuable and can truly impact the world for the better.

25.8 ENVIRONMENTAL EFFECTS OF E-WASTE RECYCLING

Recent studies have demonstrated that the inclusion of excessive levels of lead and mercury in the production of these devices could potentially contaminate the land surrounding disposal sites. These components pose a significant threat to the environment as they are not properly contained and allowed to enter the soil, particularly during warmer months. The impact of electronic waste on our surroundings is further exacerbated by the release of harmful chemicals into the air when exposed to heat. This can ultimately lead to the infiltration of dangerous substances into groundwater, posing a risk to both aquatic and land-based organisms.

Excitingly, recycling gadgets is a straightforward process with many promising rewards. By increasing our recycling efforts, not only will we promote flourishing green industries, but we can also make a positive impact on our environment. Not only does e-waste recycling help conserve energy, it also allows for the reuse of natural resources like aluminum, silver, and copper. Many of the metals found in our electronic devices are rare earth metals that are in short supply. By reusing them, we can avoid the need for mining new materials. This not only benefits the environment, but it also helps to protect these scarce resources.

This aids in preventing contamination of the air and water. We all win from this.

The top five benefits of recycling e-waste include the following:

1. **Reduces Landfill Sites:** Our generation of waste decreases with increased recycling; this is true for all types of waste. Other land may be used for other worthwhile endeavors like home construction and agriculture when there are fewer waste sites (Paul et al., 2019).
2. **Conserving Our Natural Resources:** As we've already discussed, recycling e-waste allows us to reclaim valuable materials and repurpose them to make new products. This not only conserves energy, but also reduces our carbon footprint and preserves our planet's precious resources (Figure 25.3).
3. **Stops Toxic Chemicals from Degrading the Environment:** Our electronic bias contains an assembly of dangerous substances, hanging both our weal and the Earth's. Nickel, cadmium, lithium, mercury, and lead are some of the dangerous accoutrements. When these poisons end up in tips, they sluggishly transude into our water sources, soil, and wildlife, wreaking annihilation on creatures, fish, and crops. And as these adulterants ultimately make their way into our food force, the long- continuing impact on our health and the terrain becomes apparent.
4. **Creates New Business & Employment Opportunities:** The challenge of managing e-waste may seem daunting, but it also presents a world of untapped potential for businesses and employment. Creative minds and clever innovations can turn this issue into a lucrative opportunity. Professor Veena Sahaj Walla of UNSW, made history by creating the first ever e-waste micro factory, a compact modular system that can fit into a space as small as 50 square meters. This smart factory is equipped to crush electronic waste and utilize a robot to extract usable components, which can then be melted down in a mini furnace for future use. This brilliant solution is not only cost effective, but also easily transportable, allowing it to be placed at various disposal locations, making e-waste recycling more accessible and affordable for all.

FIGURE 25.3 Benefits of e-waste recycling (Abou et al., 2021).

5. **Encourages Conscientious Buying:** By regularly disposing of e-waste, we can attain a deeper understanding of the consequences our choices and actions have on the terrain. Let's move down from the extravagant and reckless throwaway intelligence and grasp careful consumerism by precisely considering our options before making purchases, and repairing or recovering particulars rather of simply tossing them out. With their expansive track record of environmentally-friendly electrical recycling, the Great Lakes Electronics is involved with disposing of bias similar as computers, circuit boards, phone outfit, and communication systems.

These things will be broken down into its component pieces by the Great Lakes Electronics team, and the parts that are still usable can be sold. Everything is recycled, and additional components are utilized to recover metals (Bandura et al., 1986).

25.9 CONCLUSION

Instead of simply being inherently harmful, the main environmental impacts of e-waste stem from inadequate handling. However, economically and environmentally, the most beneficial solution is to process e-waste to extract valuable metals properly. To reach the goal, the approach must change to the following: The importance of recycling WEEE cannot be overstated, as it plays a crucial role in waste management and reclaiming valuable materials. However, there are a few challenges; mechanical or physical processing offers an alternate method of recovering valuable materials. The industry's primary challenge is separating the many materials that make up WEEE. The issue at hand presents a variety of opportunities for enhancing processes. One particularly impactful method is the creation of separation techniques that consider the specific physical and chemical properties of materials. These techniques play a crucial role in making the recycling of WEEE components a financially sound endeavor. E-waste recycling is vital for environmental sustainability. It helps reduce electronic waste in landfills, prevents hazardous materials from polluting the environment, and conserves valuable resources. To combat the ever-increasing problem of electronic waste, it is imperative that we implement efficient e-waste recycling programs, raise awareness through campaigns, and enforce strict regulations. Public participation, industry responsibility, and government initiatives are essential to ensure the success of e-waste recycling efforts worldwide.

REFERENCES

Abou, B.N., Szabo-Müller, P., Handmann, U. (2021). Transfer learning-based method for automated e-waste recycling in smart cities. *EAI Endorsed Transactions on Smart Cities*, 5(16), 1–9.

Arain, A.L., Pummill, R., Adu-Brimpong, J., Becker, S., Green, M., Ilardi, M., Van Dam, E. and Neitzel, R.L. (2020). Analysis of e-waste recycling behavior based on survey at a Midwestern US University. *Waste Management*, 105, 119–127. https://doi.org/10.1016/j.wasman.2020.02.002.

Baba, A., Adekola, F., Ayodele, D. (2010). Study of metals dissolution from a brand of mobile phone waste. *Metalurgija – Journal of Metallurgy* 16(4), 269–277.

Balabanic, D., Rupnik, M., Klemencic, A.K. (2011). Negative impact of endocrine disrupting compounds on human reproductive health. *Reproduction, Fertility and Development*, 23 (3), 403–416.

Bandura, A., (1986). Social foundations of thought and action: A social cognitive theory. *Prentice-Hall series in social learning theory*. Englewood Cliffs, N.J: Prentice-Hall.

Bandyopadhyay, A. (2008). A regulatory approach for e-waste management: A cross-national review of current practice and policy with an assessment and policy recommendation for the Indian perspective. *International Journal of Environment and Waste Management* 2 (1–2):139–186.

Bhutta, K.S., Adnam, O., Xia, O.Y. (2011). Electronic waste: A growing concern in today's environment. *Economics Research International*; Article ID 474230, https://doi.org/10.1151/2011/474230.

Dimitrakakis, E., Janz, A., Bilitewski, B. and Gidarakos, E. (2009). Determination of heavy metals and halogens in plastics from electric and electronic waste. *Waste Management [online]* 29 (10), 2700–2706. https://doi.org/10.1016/j.wasman.2009.05.020.

Dwivedy, M., Mittal, R.K. (2010). Future trends in computer waste generation in India, *Waste Management*, 30, 2265–2277.

Ford, P., Santos, E., Ferrão, P., Margarido, F., Van Vliet, K.J., Olivetti, E. (2016). Economics of End-of-Life Materials Recovery: A Study of Small Appliances and Computer Devices in Portugal. *Environmental Science & Technology*, 50 (9), 4854–4862. https://doi.org/10.1021/acs.est.6b00237

Hossain, R., Mahmood, A.A., Sahajwalla. V. (2021). Facile Solution for Recycling Hazardous Flexible Plastic-Laminated Metal Packaging Waste to Produce Value-Added Metal Alloys. *ACS Sustainable Chemistry & Engineering*, 9 (50), 16881–16895. https://doi.org/10.1021/acssuschemeng.1c04368

Ingole, A.R., Gulhane, S.K., Shirbhate, S.S., Chavhan, M.P. (2020). Use of Electronic Waste Concrete for Road Construction. *International Journal of Engineering and Advanced Technology (IJEAT)*, 9 (5), 974–980.

Kumar, R., Tripathi, S. (2007). "Electronics Hi_Tech-Highly Toxic" India, Green Peace.

Lee, J.C., Yoo, J.M., Jeomg, J.K., Jha, M.K. (2009). *Novel pre-treatment process for liberation of metals from waste printed circuit boards using organic solution*. Washington, DC, USA; Patent Publication No.: WO/2009/064063.

Mishra, D., Rhee, Y.H. (2010). Current research trends of microbiological leaching for metal recovery from industrial wastes. *Current Research Topics in Applied Microbiology and Microbial Biotechnology* 2, 1289–1292.

Mondal, T., Choudhury, M., Kundu, D., Dutta, D., & Samanta, P. (2023). Landfill: An eclectic review on structure, reactions and remediation approach. *Waste Management*, 164, 127–142. https://doi.org/10.1016/j.wasman.2023.03.034

Nagpal, S., Dahlstrom, D., Oolman, T. (1993). Effect of carbon dioxide concentration on the bioleaching of a pyrite-arsenopyrite ore concentrate. *Biotechnology and Bioengineering* 41, 459–464. https://doi.org/10.1002/bit.260410409.

Paul, S., Choudhury, M., Deb, U., Pegu, R., Das, S., & Bhattacharya, S. S. (2019). Assessing the ecological impacts of ageing on hazard potential of solid waste landfills: A green approach through vermitechnology. *Journal of Cleaner Production*, 236, 117643. https://doi.org/10.1016/j.jclepro.2019.117643

Pérez-Belis, V., Bovea, M.D. and Gómez, A. (2013). Waste electric and electronic toys: Management practices and characterisation. *Resources, Conservation and Recycling* 77, 1–12. https://doi.org/10.1016/j.resconrec.2013.05.002.

Rajpal, A., Choudhury, M., Goswami, S., Chakravorty, A., & Raghavan, V. (Eds.). (2024). Waste Management and Treatment: Advances and Innovations (1st ed.) CRC Press. https://doi.org/10.1201/9781003258377

Reed, D. W., Fujita, D.L.Y., Daubaras, Y., Jiao, and Thompson, V. S., (2016). Bioleaching of rare earth elements from waste phosphors and cracking catalysts. *Hydrometallurgy*, 166, 34–40. https://doi.org/10.1016/j.hydromet.2016.08.006.

Seif, R., Salem, F.Z. & Allam, N.K. (2024). E-waste recycled materials as efficient catalysts for renewable energy technologies and better environmental sustainability. *Environment, Development and Sustainability* 26, 5473–5508. https://doi.org/10.1007/s10668-023-02925-7

Wang, N., Lei, H., Zhang, Z., Jianfeng Li, W., & Zhang, R. C. (2019). Electrocatalytic hydrogen evolution with gallium hydride and ligand-centered reduction. *Chemical Science*, 10, 2308–2314. https://doi.org/10.1039/c8sc05247f

Williams, E. (2005). In: Proc. Third Workshop on Materials Cycles and Waste Management in Asia. Tsukuba, Japan: National Institute of Environmental Sciences; International activities on E-waste and guidelines for future work.

Wilson, G. T., Smalley, G., Suckling, J. R., et al. (2017). The hibernating mobile phone: Dead storage as a barrier to efficient electronic waste recovery. *Waste Management*, 60, 521–533. https://doi.org/10.1016/j.wasman.2016.12.023

Wong, M.H., Wu, S.C., Deng, W.J., Yu, X.Z., Luo, Q., Leung, A.O. (2007). Export of toxic chemicals-A review of the case of uncontrolled electronic-waste recycling. *Environmental Pollution* 149, 131–140.

26 Sustainable Ways to Manage Electronic Waste Contaminated Soil Using Phytoremediation Techniques

Lekshmi S, Ardra Lekshmi A, and Smitha Chandran S
School of Physical Sciences, Amrita Vishwa Vidyapeetham, Amritapuri Campus, Kollam, India

Deepthi Achuthavarier
Modelling Program Division, Office of Science and Technology Integration, National Weather Service, NOAA, Silver Spring, United States

26.1 INTRODUCTION

India is facing a substantial problem with the accumulation of electronic waste, known as "e-waste." The proper handling of e-waste, which is the rapidly expanding category of official waste generated by municipalities globally, is increasingly becoming a crucial issue for both the atmosphere and the health of the public. E-waste, or Waste Electrical and Electronic Equipment (WEEE), refers to surplus, malfunctioning, outdated, and improperly discarded electrical and electronic devices. E-waste is composed of more than 60% gold, copper, iron, aluminum, and other metals, with contaminants accounting for 2.70% of its makeup. E-waste has been classified as hazardous by the Basel Convention because of the significant toxicity of these pollutants, particularly when they are burned or recycled in unregulated environments. The Basel Convention has additionally established a framework for imposing restrictions on the global transit of this substance. The Basel Ban, a version of the Basel Convention that has not yet come into effect, would prohibit the transfer of electronic waste from developed to industrializing nations. Phytoremediation, an acronym for cleaning up environmental pollution using naturally occurring or genetically engineered plants, has been researched since the 1990s. The growing interest in research on phytoremediation has made it a widely considered option for e-waste treatment. Bioremediation technology employs microorganisms and their enzymes to reinstate the polluted environment to its initial condition or enhance environmental well-being. The advantages of this include its affordability, ecological safety, and environmental compatibility. Bioremediation method is very suitable for the treatment of soil contaminated with organic pollutants. Nevertheless, the presence of high concentrations of toxic metals in the environment contaminated with both organic and metal pollutants can impede the ability of microorganisms to break down organic pollutants through biodegradation (Luo et al., 2018).

Most nations have three methods for managing the e-waste produced at the home level: (i) throwing it in the trash, (ii) having it collected formally by organizations approved by the government, and (iii) having it collected outside of the formal system by private garbage dealers and businesses (Chi et al., 2014). For the safe and efficient extraction of valuable elements, e-waste is frequently recycled in facilities equipped with state-of-the-art equipment, modern infrastructure, and modern

DOI: 10.1201/9781003582311-26

technology (formal recycling). E-waste is collected via approved commercial, municipal, and collection sites and services. Conversely, e-waste gathered outside the official system by private garbage firms or dealers is typically recycled and processed using antiquated methods and conditions, often with no safeguards to lessen the release of dangerous chemicals into the environment. A multifaceted strategy is required to address e-waste comprehensively and simultaneously promote resource conservation and the global economy. This is required to balance the negative impacts of recycling e-waste on the environment, human health, and resource reimbursement (financial). The local environment is frequently exposed to toxic substances produced during recycling activities, such as polychlorinated dibenzo-p-dioxins, polycyclic aromatic hydrocarbons, as well as acids used to extract metals, such as HCl, H_2SO_4, HNO_3, $HClO_4$, and NaClO. These substances can also result from informal recycling, including heavy metals, brominated flame retardants, and polychlorinated biphenyls. The ecosystem's concentration of hazardous contaminants may rise due to these releases.

Excessive costs are associated with other methods of repairing e-waste accumulated sites using conventional procedures, including physical treatment, excavation of following disposal and various chemical additive treatments. Thus, developing novel, inexpensive, and environmentally benign techniques to remove toxins from locations contaminated by e-waste is an increasingly important global problem. Biological treatment, particularly phyto/bioremediation, is vital in this process. One exciting technique that may be used successfully for in-site remediation of organic and trace metal pollutants is phytoremediation, which uses plants to absorb, store, and break down toxins from the soil. Thus, the application of phytoremediation to clean up locations contaminated by e-waste will be covered in detail in this chapter.

26.2 TYPES OF ELECTRONIC WASTES

The origin and use of electronic products can be used to categorize e-waste. As per several studies, there are eight distinct categories of e-waste internationally recognized (Widmer et al., 2005) (Table 26.1) (Widmer et al., 2005). Various factors can influence the average percentage of each kind

TABLE 26.1
Effect of E-Waste on Human Health (Rautela et al., 2021)

Toxicants	Sources	Health Effects
Mercury	Relays and switches, printed circuit boards	• Chronic damage to the brain • Respiratory and skin disorders in fishes due to bioaccumulation
Lithium	Lithium-ion battery	• Lithium can pass into breast milk and may harm a nursing baby • Inhalation of the substance may cause lung edema
Nickel	Nickel-cadmium rechargeable batteries	• Allergy of skin to nickel results in dermatitis while allergy of the lung to nickel results in asthma
Beryllium	Motherboard	• Carcinogenic (lung cancer) • Inhalation of fumes and dust causes chronic beryllium disease or berylliosis
Cadmium	Chip resistors and semi-conductors	• Irreversible damage to human health • Accumulates in the kidney and liver • Causes damage to nerves • Teratogenicity
Barium, phosporous	Front panel of CRTs	• Cause muscle weakness and damage due to heart, liver, spleen.
Lead	Solder in printed circuit boards, glass panels and gaskets in computer monitors	• Damage to central and peripheral nervous systems, brain and kidney • Affects brain development in children

of e-waste, such as population growth, consumer behavior, socioeconomic status, and the reliance of households and businesses on electrical and electronic equipment (EEE) (Balde et al. 2017). As per European 2012, the primary source of e-waste is large household items, accounting for 42.1% of the total. This is followed by IT and communications, consumer devices, small household equipment, and consumer devices, in that order. Medical, lighting, electrical, and electronic instruments have contributed less than 2%, while automatic dispensers, toys, sports, monitoring, and control devices have contributed less than 1% (. This is because the categories are easier to understand. E-waste in developing nations is primarily generated by computers, TVs, and smartphones (Needhidasan et al., 2014; Kalia et al., 2022; Nnorom & Osibanjo, 2008). Valuable metals are included in e-waste, along with a variety of hazardous contaminants. E-waste contains many hazardous elements (Cd, Hg, Pb, and Cr) that could raise the toxicity level in the environment (Qu et al., 2019). Hazardous substance exposure may rise due to e-waste's long-term persistence in the environment. These dangerous substances may be linked to significant groundwater contamination and human health. The soil-crop-food pipeline is a significant mechanism via which hazardous particles in the electronic waste reaches the human body (Gogoi et al., 2024). Toxicity, adverse environmental effects, and monetary compensation from e-waste prompt the requirement for efficient recovery of metals from e-waste.

26.3 HEALTH IMPACTS

Improper disposal of "e-waste," can result in serious health hazards. Hazardous compounds, including lead, mercury, cadmium, and brominated flame retardants, can leak into the surroundings due to incorrect recycling and disposal of electronic gadgets. These compounds can contaminate water, soil, and the air, endangering human health and the environment. Various health effects, such as respiratory troubles, skin conditions, neurological damage, and developmental problems, particularly in children, can result from exposure to these toxic compounds. In addition, workers in informal recycling operations—which are sometimes common in some areas—may come into direct touch with hazardous materials without the proper protection, which can increase health hazards. The dangerous levels of organic pollutants and HMs discovered in Guiyu, China's dust, soil, river sediment, surface water, and groundwater samples have been the subject of numerous scientific studies. Skin damage, headaches, nausea, vertigo, persistent gastritis, and duodenal and stomach ulcers were also prevalent among the same populations. It was also discovered that there was no noticeable distinction in blood lead levels between boys and girls and that children's levels were more significant than the national average in China. Appropriate procedures for e-waste management, such as recycling and secure disposal techniques, are essential to reducing these health hazards. Lowering the detrimental impact of electronic waste on health can also be achieved by increasing consumer responsibility and educating people about the dangers and aftereffects of e-waste (Figure 26.1).

26.4 GLOBAL E-WASTE STATISTICS

Modern civilization is experiencing a growing demand for Electrical and Electronic Equipment (EEE) due to rising disposable income, accelerated urbanization and industrialization, and population expansion. Most EEE is short-lived and frequently thought to be costly or challenging to fix. To make older technology obsolete and unwelcome even when still functional, brands spend billions inducing consumer desire for the newest models. Electronic garbage, or e-waste, is what remains of EEE after it is used and disposed of. In addition to hazardous compounds like lead, mercury, and cadmium that can endanger human health and the environment, e-waste also contains valuable materials that can be recycled but are typically not. FY 2021–2022 saw 1.6 million electronic garbage generated and 0.5 million electronic wastes collected and processed. In 2021, e-waste amounted to 57.4 Mt (million metric tons). An average of 2 Mt is added to the total annually. In 2023 the world's unrecycled e-waste capacity will exceed 347 million tons. The top producers of e-waste are China,

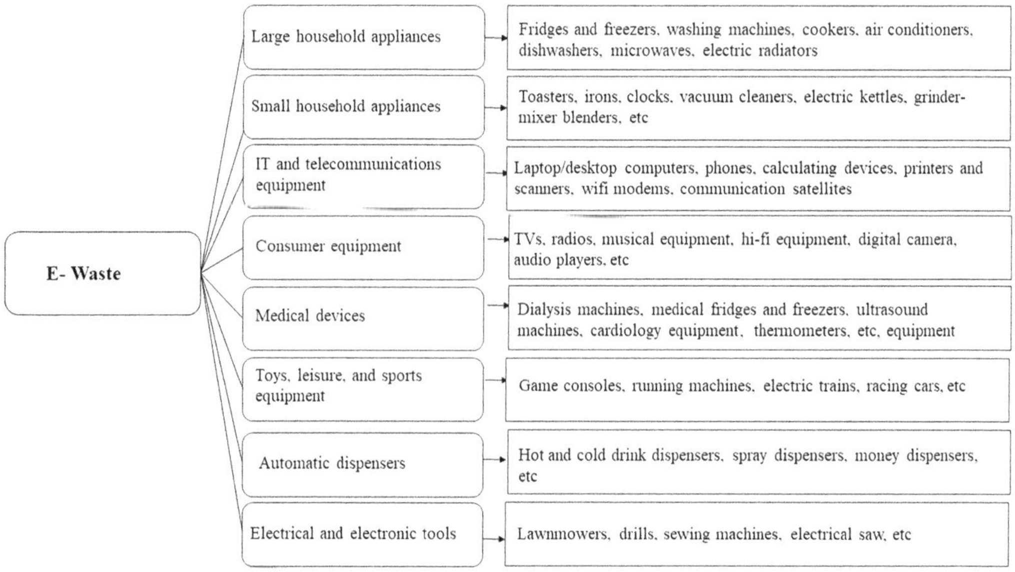

FIGURE 26.1 Source of E-waste generation (Shahabuddin et al., 2023).

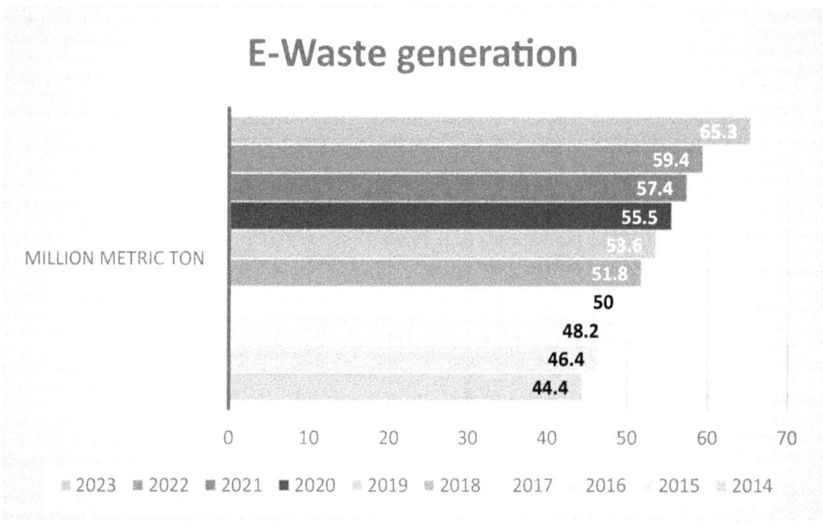

FIGURE 26.2 The e-waste generation all over the world from the year 2014–2023 (As per the statistics published by WHO).

the US, and India. Just 17.4% of e-waste is reportedly gathered and recycled correctly. There are only 78 nations that have e-waste-related laws in place. The countries that recycle e-waste the most include Estonia, Norway, and Iceland. The market for recycling e-waste was estimated to be worth $49,880 million in 2020 (Figure 26.2).

26.5 E-WASTE MANAGEMENT

Many management techniques are used to limit the amount of e-waste produced each day. E-waste recycling, thermal treatment procedures, the use of acid baths, and garbage disposal or dumping into dumpsites are the main ways frequently employed for managing e-waste.

26.5.1 Recycling

Recycling removes particles from antiquated electronic equipment and reassembles them afterward. According to Asante et al. (2012), recycling comprises a few subprocesses, like separating ferrous and nonferrous metals and removing them from hazardous materials, such as printed circuit boards and cathode ray tubes (Asante et al., 2012). Certain valuable metals, including lead, copper, and gold, can be recovered by using an effective e-waste recycler. Reusing a variety of electrical devices, including computers, phones, keyboards, CPUs, displays, cables, and connecting wires, can be facilitated by recycling, which can be done mechanically or manually (Asante et al., 2019). Employees classify the waste based on its types and models after the electronic garbage is gathered and brought to the recycling centers. The still usable components will be removed from all the electronic gadgets once inspected and they will be sold separately or successively joined to form a new computer or phone. The recycling process was carried out for non-functional e-waste.

Breaking down a product into its parts is known as de-mantling, and it is the step that comes before e-waste is thrown into a massive machine and shredded into tiny pieces. This process eliminates any potentially dangerous elements from electronic gadgets that, if dumped in a landfill, could damage the equipment or poison the surrounding area. Metals, the precious components that make recycling e-waste lucrative, will be removed from the debris after it has been shredded. In contrast to the previous sessions, this procedure does not call for manual sorting. All ferromagnetic materials, such as steel and iron, with high susceptibilities to magnetization will first be drawn to a massive magnet. The leftover metals and alloys are then separated by further mechanical method that makes use of the eddy current physical principle. Not magnetized materials, such as plastic, will continue, but paramagnetic materials, or those that are weakly attracted to magnets, will be bounced away. After that, water is utilized to further separate the trash. At this stage, almost all of the remaining materials are non-magnetic. They will go through another machine that is filled with water, through which low-density materials (mainly plastic) will flow while heavier materials (glass) will sink. Finding out if any precious materials are stick to the plastic is the final stage before recycling materials are sold.

26.5.2 Landfills

Sanitary landfills are considered the most popular method for disposing of electronic trash to minimize or reduce environmental and public health threats. Ordinarily, landfills are located where the dominant land features can act as organic barriers between the surrounding environment and the dump. Excavated soil is used to create trenches, and before burying E-waste, impermeable liners are created to keep the dangerous materials from escaping. Furthermore, as a substitute for sanitary landfills, controlled dumps are employed. They have several characteristics in common with sanitary landfills, where waste is mixed and disposed of (Mondal et al., 2023). Despite their well-organized capacity, these dumps are unrelated to cell planning. Due to the combined nature of the pollutants and the lack of any gas management or appropriate covering, the pollution occurrences may become complex and diverse (Paul et al., 2019). However, landfills and dumps are not ecologically healthy practices because harmful compounds may leak into the soil and groundwater. Groundwater pollution is mainly caused by receiving dumps and landfills for e-waste (Kasassi et al., 2008).

Along with dissolved and suspended organic and inorganic materials, the leachates that percolate from E-waste sites contain noticeably more significant amounts of trace metals (Spalvins et al., 2008). The pollutants have the potential to travel through food chains and accumulate within living organisms, ultimately impacting human health. Sanitary landfills offer relatively low threats to human health and the environment, but they also come with considerably greater starting costs.

26.5.3 Incineration

One technique for handling e-waste is incineration, which entails burning the material at a high temperature. There are concerns because of possible adverse effects on the environment and human health, even if it can be utilized to produce energy and reduce trash. Here's a condensed summary:

a. **Collection:** Gather materials of e-waste for burning, taking care to identify and manage any dangerous components.
b. **Preparation:** Take out any non-combustible items and divide the garbage into distinct categories according to the materials.
c. **Incineration:** The e-waste is burned at high temperatures in a regulated incinerator. Combustible materials and biological materials are broken down by heat.
d. **Recovery of Energy:** By producing steam from the heat produced during incineration, turbines may be powered to provide electricity.
e. **Remedy for Gas:** There are contaminants in the gases that are created during combustion. Before being released into the atmosphere, they must be treated to eliminate dangerous materials. Filters and scrubbers are standard techniques.
f. **Ash Residue:** Following burning, metals, and other leftovers might be found in the left-over ash. This ash may need additional treatment and handling attention to control any contaminants.

Although incineration can recover energy and minimize the amount of e-waste produced, it is not without problems. Pollutant-containing air emissions, the release of hazardous materials, and incomplete combustion that produces dioxins and furans are among the issues that need to be addressed. As a result, instead of burning electronic garbage, several nations prioritize other strategies like recycling and appropriate disposal.

26.6 PHYTOREMEDIATION OF E-WASTE

E-waste is a diverse range of materials including metals, glass, plastics, and ceramics. Together with non-hazardous elements like steel and iron, it also includes valuable metals like platinum (Pt), silver (Ag), gold (Au), palladium (Pd), and REMs (neodymium, praseodymium, tantalum, and indium). In addition, there are dangerous elements such as arsenic, cobalt, mercury, lead, mercury, copper, nickel, and cadmium. Thus, to remove heavy metals, electronic trash, and persistent organic pollutants from soil and water, an effective green technology known as phytoremediation is being developed (Dutta et al., 2023; Agarwal et al. 2022). roots, stems, leaves, buds, and flowers accumulate these contaminants. The soil framework, fertility, and activity of enzymes will not be impacted by this. The phytoremediation method uses solar energy to remove the pollutants. Using green plants, phytoremediation extracts and removes traces of pollutants from soil and water. By using a range of phytoremediation strategies, plants eliminate contaminants. Some of the cleanup methods include the following: phytostimulation, phytostabilization, phytoextraction, phytodegradation, rhizofilteration, and phytovolatalization. Every strategy has a method for eliminating the pollutants in the soil. The plant uses a variety of methods, either separately or in conjunction, to clean up soil surface and underground water, based on the kinds, forms, and media of the contaminants.

26.6.1 Acid Bath

Using the acid bath approach, some essential components are eliminated from electronic circuit boards by submerging them for a set amount of time in solutions containing sulfuric, hydrochloric, or nitric acid. (Sivaramanan, 2013). Copper, silver, and gold are some extractable metals that precipitate after being dissolved in the concentrated acid solution during the soaking time. Waste acid is released into the environment, but precipitated metals are collected and used to make other goods.

As they wind up in groundwater, these dangerous chemicals and acid wastes find their way to water supplies. To clean up contaminated areas caused by e-waste, creative, affordable, and environmentally friendly remediation techniques need to be looked into. Physicochemical and biological methods are commonly employed to clean up pollutants from electronic waste. Green methods that reduce environmental danger are now practical options.

The technique known as "bioremediation" uses natural biological processes to eliminate or degrade dangerous environmental pollutants. This strategy depends on low-tech, affordable techniques that are widely accepted to provide contaminated sites with on-site cleanup. Using their unique metabolic pathways, a vast array of bacteria, fungi, and plant species that have evolved to eliminate or degrade a particular range of toxins are used in bioremediation.

The potential of bioremediation to remove potential contaminants from E-waste has been the subject of numerous investigations (Jiang et al., 2019; Kang et al., 2016). The phrase "phytoremediation" is specifically used to describe instances in which plant species are employed as bioremediation agents.

26.6.2 Rhizofiltration

Plants employ their roots to remove dangerous contaminants during filtration. Rhizofiltration known as phytofiltration, can be applicable for get rid of dangerous pollutants from groundwater, surface water, and wastewater. Rhizofiltration involves both absorption and adsorption. Rhizofiltration decreases the mobility of the pollution and obstructs groundwater movement, hence decreasing the contaminant's bioavailability for assimilation into food chains. To sum up, this approach makes use of plants that may take up toxins from the contaminated surroundings (Dushenkov 1995). Plant species offer an excellent opportunity to help clean up the environment. By using the mechanical and biological properties of plant roots, rhizofiltration purifies wastewater. The effectiveness of rhizofiltration is dependent on roots' capacity. Metals precipitate onto the surfaces of roots as a result of rhizosphere pH variations and exudates from the roots. Hydroponically grown plants absorb and accumulate metals in their root systems and branches after being planted in water tainted with metals. Plant cells possess specialized genes that enable them to be efficient metal accumulators. The metabolism of glutathione and organic acids greatly influences a plant's ability to withstand metals (Abhilash et al., 2009). Temperature, pH, and light are other environmental factors that have a significant influence on the consumption of heavy metals.

26.6.3 Phytoextraction

Toxic metals are taken up by plant roots and carried up shoots where they accumulate at non-metabolizable regions of the plant tissues such as cell walls, vacuoles, and membranes. This process is known as phytoextraction or phytoaccumulation. Redox reactions that alter the mobility of metals in soil and their propensity to be absorbed by roots are accelerated by bacteria (Suman et al., 2018). Moreover, root mycorrhizal interaction affects the metal absorbing rate. For a plant species to be considered suitable for phytoextraction, it needs to possess the following qualities: active heavy metal accumulators in easily collected components, high biomass production, and metal resistance to hazardous metals. The absorption of metal cations is a major step leading to the build-up of dangerous heavy metals. The plant cell then creates complexes known as metal-ligand or metal-phytochelatin (M-PC). After forming the M-PC, these complex molecules are moved to the plant's vacuole where they are stored. The heavy metals (HMs) in biomass and above-ground plant tissues are the main factor influencing a species' capability for extraction (Marchiol et al., 2004).

26.6.4 Phytodegradation

This approach can be employed for treating water as well as soil. It can be used effectively to reduce chemical compounds to their more basic forms. Phytodegradation is also known as phytotransformation.

Within the body of the plant, organic contaminants are broken down with the help of enzymes and/or microbes. Phytodegradation, the process by which organic contaminants are broken down by plants, can render them non-toxic. These contaminants are mostly electronic waste and industrial chemicals. The term "green liver" refers to the phytodegradation process, in which the plant performs liver-like tasks. After being broken down into smaller pollutants, toxins eventually become metabolites (Arthur et al., 2005). Therefore, it is essential to identify and analyze the metabolites created in the biota due to their unknown bioavailabilities and toxicities. Soil-found bacteria, fungi, and enzymes contribute to phytodegradation. Several factors impact the phytodegradation processes, including temperature, soil moisture, organic matter concentration, pH, and aeration.(Al-Baldawi et al., 2015). Additionally, a variety of soil characteristics and atmospheric factors have an impact on the contaminants found in the rhizosphere.

26.6.5 Phytostabilization

Metals-tolerant plant species are used in phytostabilization to reduce heavy metal bioavailability and render them inert below ground. By doing this, metals are prevented from moving into the natural environment and the chance of them entering the food chain is reduced(Kumari et al., 2021). Therefore, it is best to select a plant that can increase soil microbial activity. The three mechanisms that contribute to phytostabilization are absorption, the process of precipitation, and the complexation process. Where native biodiversity is unable to flourish because of high metal concentration contamination, this technique can aid in the restoration of vegetation. By employing metal-tolerant plant species as photo stabilizers, pollutants caused by rainfall, winds, and groundwater contamination can be controlled as they migrate (Regvar et al., 2006).

26.6.6 Phytostimulation

In phytoremediation, the phytostimulation procedure is essential for improving plants' capacity to clean up areas that are polluted. Utilizing plants to absorb, gather, and even change contaminants, the method purifies the soil, water, or air. The microbe life in the rhizosphere can improve the mechanism of phytostimulation activity. Through their roots, plants expel organic pollutants that improve nutrition availability and promote plant development. The activation of e-waste pollutants might also be attributed to the mycorrhizal connections. It is also known as bioremediation with plant support. Herbicides can help remove the contamination from the affected area. Therefore, it is best to select a plant that can increase soil microbial activity. These methods are used to remove the majority of dangerous electronic waste. Only lower-level contamination is possible with this approach (Hussain & Hasnain, 2011).

26.6.7 Phytovolatilization

For volatile and semi-volatile organic compounds (VOCs), phytovolatilization can be an important loss mechanism. There are a lot of research results about the phytovolatilization of inorganic as well as organic pollutants. Some of the inorganic compounds that can evaporate from plants are selenium (Se), arsenic (As), and mercury (Hg). Both indirect and direct phytovolatilization are two possible situations. Direct phytovolatilization is a rational and well-researched process where plants absorb and transport contaminants, causing the compounds to volatilize from the stem/trunk and leaves. The similarity between the channel and the water's transpiration vascular system has resulted in the use of the terms "phytovolatilization" or "transpired" pollution. Many of the chemicals that undergo phytovolatilization are somewhat hydrophobic and can pass through hydrophobic barriers such as cutin in the epidermis. However, the specific process of phytovolatilization often differs from transpiration. Since direct phytovolatilization necessitates that the component be absorbed, translocated, and volatilized by the plant, it does not apply to substances that the plant creates or changes.

Apart from moving enormous volumes of water, plants actively scout large areas of soil. These activities have a significant impact on the fate and transit of chemicals below the surface.

26.7 ADVANTAGES OF PHYTOREMEDIATION FOR E-WASTE REMOVAL

When comparing the environmental effects of traditional disposal techniques with those of phytoremediation—a sustainable and environmentally friendly process that uses botanicals to absorb and collect toxins and heavy metals from electronic waste. In locations where alternative cleanup techniques are prohibitively expensive or impossible, it may be more economical. When compared to high-tech techniques, plants are comparatively inexpensive (Gunarathne et al., 2020). Adding vegetation to e-waste cleanup can enhance green areas and landscaping, improving the process' aesthetic appeal and offering further ecological advantages. By incorporating local communities in this process, it is possible to engage the community in sustainable practices and increase awareness of the environmental problems related to e-waste. By establishing homes for a variety of creatures, some plants used in phytoremediation might improve biodiversity. This strategy supports the objectives of restoration of ecosystems and sustainability. Over time, phytoremediation can reduce the bioavailability of pollutants and improve soil quality, providing a long-term, permanent solution (Luo et al., 2017). Phytoremediation is an energy-efficient choice since the energy needs are typically lower than those of standard remediation procedures, which require a large amount of energy input. Because phytoremediation is adaptable to different environmental conditions, it can be used in a variety of settings with a range of e-waste issues (Lee, 2013).

26.8 NANOPHYTOREMEDIATION OF E-WASTE

Globally, the accumulation of electronic waste is leading to heavy metal poisoning of the environment, which is a serious ecotoxicological problem. Heavy metals are extremely dangerous for human health and can contaminate land and water since they are permanent and non-biodegradable. In place of traditional physicochemical cleanup techniques, phytoremediation has shown to be a less expensive and more environmentally responsible method of treating e-waste contamination. These methods disturb the soil microbiota, change the characteristics of the soil, and need a significant labor commitment. Nanoparticles can improve the process of phytoremediation in soil that is contaminated with lead, zinc, cadmium, chromium, and nickel (Dodamani et al., 2021). The regulation of heavy metal tolerance in plant cells involves processes such as chelation, where peptides that bind to metals are stimulated, the production of metal complexes by activating ion transport into the vacuole, and the binding of heavy metals from electronic waste to the cell wall. E-waste can lead to the production of phytochelatins and metallothioneins in plant tissues due to exposure to heavy metals. The peptides form a bond with the HMs. A stable compound is produced, which mitigates the deleterious impact of metals. Metals can be immobilized in the root system through their interaction with pectins in cell walls and the negatively charged interfaces between the cytoplasm and the cell membrane. Phytochelatin is exhibited by plants that absorb harmful metals. Plants produce organic acids that have the capacity to remove and reduce the presence of metals in the rhizosphere. Glutathione is used to create these peptides, which are subsequently transformed into peptides (Mijovilovich et al., 2009). Some plants possess the ability to synthesize redox enzymes and also have the capacity to mitigate the harmful effects of heavy metals by transforming them into less toxic forms. Various types of nanoparticles have the ability to directly extract toxins from the soil in a phytoremediation system, thereby alleviating the burden on plants to do this task. Nanoparticles may immediately remove contaminants through two major mechanisms: adsorption and redox reactions. nZVI is the most studied nanomaterial for using redox reactions to remove pollutants. For several procedures, including the reductive reduction of heavy metals, ZVI can supply electrons. Reports state that nZVI's reduction property is combined with its ability to absorb inorganic cations and anions (Pillai & Kottekottil, 2016).

Carbon nanoparticles have been extensively studied as an adsorption-based pollution removal method for e-waste. Nitrogen-doped carbon nanotubes are remarkably effective at adsorbing a variety of pollutants. An investigation found that heavy metals can be retained by carbon nanotubes by electrostatic interaction, physical adsorption the process of complexation and precipitation on their surface. Several interactions during adsorption can maintain a rather constant mixture of pollutants and carbon nanotubes (Ma & Wang, 2010). Carbon nanoparticles have been extensively studied as an adsorption based pollution removal method for e-waste. Nitrogen-doped carbon nanotubes are remarkably effective at adsorbing a variety of pollutants (Tripathi et al., 2015). The plant's resistance to chromium (VI) stress was reinforced by the introduction of nanoparticles, which also increased antioxidant activity, lowered the formation of ROS, and improved the efficiency of photosynthesis. In phytoremediation, zinc oxide nanoparticles act as a physiologically controller on the plants which lowers the toxic effects of cadmium. Tolerance is increased by zinc oxide nanoparticles because they control the expression of enzyme genes. Ryegrass may be able to efficiently remove heavy e-waste elements from the environment while lowering phytotoxicity by using nanoparticles like nano-hydroxyapatite. As well as the nanoparticles TiO_2 nanoparticles have been shown to accelerate photosynthetic activity in a soybean-based Cd phytoextraction process, hence aiding in the promotion of plant development (Singh & Lee 2016; Upadhyay & Chakma, 2022)). Regarding e-waste, this could improve plants' capacity to take up and retain heavy metals from electronic trash. The efficiency of nanophytoremediation and its effects on the environment in e-waste scenarios are being actively researched and are still under investigation.

26.9 CONCLUSION

Nanomaterials have the potential to enhance plants' ability to absorb and accumulate heavy metals, hence augmenting their efficacy in removing contaminants from electronic waste. Nanomaterials may improve phytoremediation methods' effectiveness, enabling the quicker and more thorough removal of pollutants By utilizing natural processes that cause little ecological disruption, nanophytoremediation might lessen the negative environmental effects of conventional remediation techniques. By engineering nanomaterials for particular pollutant uptake, remediation strategies can be more focused and selective. It is possible to customize different plant species and nanomaterials to target different types of pollutants, offering flexibility in managing a variety of electronic waste pollutants. In comparison to traditional techniques, nanophytoremediation might provide a more affordable option, particularly for large-scale applications. Nano-phytoremediation, which combines nanomaterials with plants' inherent capabilities, is a more environmentally responsible and sustainable method of cleaning up e-waste

REFERENCES

Abhilash, P. C., Pandey, V. C., Srivastava, P., Rakesh, P. S., Chandran, S., Singh, N., & Thomas, A. P. (2009). Phytofiltration of cadmium from water by Limnocharis flava (L.) Buchenau grown in free-floating culture system. *Journal of Hazardous Materials*, *170*(2–3), 791–797. https://doi.org/10.1016/j.jhazmat.2009.05.035

Agarwal, S., Albeshr, M. F., Mahboobb, S., Atique, U., Pramanick, P., & Mitra, A. (2022). Bioaccumulation Factor (BAF) of heavy metals in green seaweed to assess the phytoremediation potential. *Journal of King Saud University-Science*, *34*(5), 102078.

Al-Baldawi, I. A., Abdullah, S. R. S., Anuar, N., Suja, F., & Mushrifah, I. (2015). Phytodegradation of total petroleum hydrocarbon (TPH) in diesel-contaminated water using Scirpus grossus. *Ecological Engineering*, *74*, 463–473. https://doi.org/10.1016/j.ecoleng.2014.11.007

Arthur, E. L., Rice, P. J., Rice, P. J., Anderson, T. A., Baladi, S. M., Henderson, K. L. D., & Coats, J. R. (2005). Phytoremediation—An Overview. *Critical Reviews in Plant Sciences*, *24*(2), 109–122.

Asante, K. A., Agusa, T., Biney, C. A., Agyekum, W. A., Bello, M., Otsuka, M., Itai, K., Takahashi, S., & Tanabe, S. (2012). Multi-trace element levels and arsenic speciation in urine of e-waste recycling workers from Agbogbloshie, Accra in Ghana. *Science of the Total Environment*, *424*, 63–73.

Asante, K. A., Amoyaw-Osei, Y., & Agusa, T. (2019). E-waste recycling in Africa: risks and opportunities. *Current Opinion in Green and Sustainable Chemistry, 18*, 109–117. https://doi.org/10.1016/j.cogsc.2019.04.001

Balde, C.P., et al. (2017). The Global E-Waste Monitor 2017: Quantities, Flows and Resources. United Nations University, International Telecommunication Union, and International Solid Waste Association, 2017

Chi, X., Wang, M. Y. L., & Reuter, M. A. (2014). E-waste collection channels and household recycling behaviors in Taizhou of China. *Journal of Cleaner Production, 80*, 87–95. https://doi.org/10.1016/j.jclepro.2014.05.056

Dodamani, S., Kurangi, B., Teradal, N., & Kurjogi, M. (2021). Nanobiotechnology for E-waste management. In *Nanobiotechnology* (pp. 271–281). Elsevier. https://doi.org/10.1016/B978-0-12-822878-4.00017-1

Dutta, D., Rautela, R., Gujjala, L. K. S., Kundu, D., Sharma, P., Tembhare, M., & Kumar, S. (2023). A review on recovery processes of metals from E-waste: A green perspective. *Science of the Total Environment, 859*, 160391. https://doi.org/10.1016/j.scitotenv.2022.160391

Dushenkov, Viatcheslav., Kumar, P. B. A. Nanda., Motto, Harry., & Raskin, Ilya. (1995). Rhizofiltration: The Use of Plants to Remove Heavy Metals from Aqueous Streams. *Environmental Science & Technology, 29*(5), 1239–1245. https://doi.org/10.1021/es00005a015

Gunarathne, V., Gunatilake, S. R., Wanasinghe, S. T., Atugoda, T., Wijekoon, P., Biswas, J. K., & Vithanage, M. (2020). Phytoremediation for E-waste contaminated sites. In *Handbook of Electronic Waste Management* (pp. 141–170). Elsevier. https://doi.org/10.1016/B978-0-12-817030-4.00005-X

Gogoi, N., Sarma, A., Choudhury, M., Samanta, P., Faizan, M., & Sadhak, S. (2024). Soil heavy metal pollution and ecological risk assessment in disturbed and undisturbed soil of Morigaon, Assam. *Discover Environment*, 2(1), 13. https://doi.org/10.1007/s44274-024-00039-z

Hussain, A., & Hasnain, S. (2011). Phytostimulation and biofertilization in wheat by cyanobacteria. *Journal of Industrial Microbiology & Biotechnology, 38*(1), 85–92.

Jiang, B., Adebayo, A., Jia, J., Xing, Y., Deng, S., Guo, L., Liang, Y., & Zhang, D. (2019). Impacts of heavy metals and soil properties at a Nigerian e-waste site on soil microbial community. *Journal of Hazardous Materials, 362*, 187–195. https://doi.org/10.1016/j.jhazmat.2018.08.060

Kalia, P., Zia, A., & Mladenović, D. (2022). Examining country development indicators and e-waste under the moderating effect of country development levels and e-waste policy. *International Journal of Quality & Reliability Management, 39*(7), 1601–1616. https://doi.org/10.1108/IJQRM-09-2021-0335

Kang, J. J., Lee, J. S., Yang, W. S., Park, S. W., Alam, M. T., Back, S. K., Choi, H. S., Seo, Y. C., Yun, Y. S., Gu, J. H., Saravanakumar, A., & Kumar, K. V. (2016). A study on environmental assessment of residue from gasification of polyurethane waste in e-waste recycling process. *Procedia Environmental Sciences, 35*, 639–642. https://doi.org/10.1016/j.proenv.2016.07.056

Kasassi, A., Rakimbei, P., Karagiannidis, A., Zabaniotou, A., Tsiouvaras, K., Nastis, A., & Tzafeiropoulou, K. (2008). Soil contamination by heavy metals: Measurements from a closed unlined landfill. *Bioresource Technology, 99*(18), 8578–8584. https://doi.org/10.1016/j.biortech.2008.04.010

Kumari, P., Singh, A., & Kharwar, R. N. (2021). Phytostimulation and ISR responses of fungi. In *Fungi Bio-Prospects in Sustainable Agriculture, Environment and Nano-Technology* (pp. 459–473). Elsevier. https://doi.org/10.1016/B978-0-12-821394-0.00018-4

Lee, J. H. (2013). An overview of phytoremediation as a potentially promising technology for environmental pollution control. *Biotechnology and Bioprocess Engineering, 18*(3), 431–439. https://doi.org/10.1007/s12257-013-0193-8

Luo, J., Cai, L., Qi, S., Wu, J., & Sophie Gu, X. (2017). A multi-technique phytoremediation approach to purify metals contaminated soil from e-waste recycling site. *Journal of Environmental Management, 204*, 17–22. https://doi.org/10.1016/j.jenvman.2017.08.029

Luo, J., Wu, J., Huo, S., Qi, S., & Gu, X. S. (2018). A real scale phytoremediation of multi-metal contaminated e-waste recycling site with Eucalyptus globulus assisted by electrical fields. *Chemosphere, 201*, 262–268. https://doi.org/10.1016/j.chemosphere.2018.03.018

Ma, X., & Wang, C. (2010). Fullerene nanoparticles affect the fate and uptake of trichloroethylene in phytoremediation systems. *Environmental Engineering Science, 27*(11), 989–992. https://doi.org/10.1089/ees.2010.0141

Marchiol, L., Assolari, S., Sacco, P., & Zerbi, G. (2004). Phytoextraction of heavy metals by canola (Brassica napus) and radish (Raphanus sativus) grown on multicontaminated soil. *Environmental Pollution, 132*(1), 21–27. https://doi.org/10.1016/j.envpol.2004.04.001

Mijovilovich, A., Leitenmaier, B., Meyer-Klaucke, W., Kroneck, P. M. H., Götz, B., & Küpper, H. (2009). Complexation and Toxicity of Copper in Higher Plants. II. Different Mechanisms for Copper versus Cadmium Detoxification in the Copper-Sensitive Cadmium/Zinc Hyperaccumulator *Thlaspi caerulescens* (Ganges Ecotype). *Plant Physiology, 151*(2), 715–731. https://doi.org/10.1104/pp.109.144675

Mondal, T., Choudhury, M., Kundu, D., Dutta, D., & Samanta, P. (2023). Landfill: An eclectic review on structure, reactions and remediation approach. Waste Management, 164, 127–142. https://doi.org/10.1016/j.wasman.2023.03.034

Needhidasan, S., Samuel, M., & Chidambaram, R. (2014). Electronic waste – an emerging threat to the environment of urban India. *Journal of Environmental Health Science and Engineering*, *12*(1), 36. https://doi.org/10.1186/2052-336X-12-36

Nnorom, I. C., & Osibanjo, O. (2008). Overview of electronic waste (e-waste) management practices and legislations, and their poor applications in the developing countries. *Resources, Conservation and Recycling*, *52*(6), 843–858. https://doi.org/10.1016/j.resconrec.2008.01.004

Pillai, H. P. S., & Kottekottil, J. (2016). Nano-phytotechnological remediation of endosulfan using zero valent iron nanoparticles. *Journal of Environmental Protection*, *07*(05), 734–744. https://doi.org/10.4236/jep.2016.75066

Paul, S., Choudhury, M., Deb, U., Pegu, R., Das, S., & Bhattacharya, S. S. (2019). Assessing the ecological impacts of ageing on hazard potential of solid waste landfills: A green approach through vermitechnology. Journal of Cleaner Production, 236, 117643. https://doi.org/10.1016/j.jclepro.2019.117643

Pillai, H. P., & Kottekottil, J. (2016). Nano-phytotechnological remediation of endosulfan using zero valent iron nanoparticles. *Journal of Environmental Protection*, *7*(5), 734.

Qu, Y., Wang, W., Liu, Y., & Zhu, Q. (2019). Understanding residents' preferences for e-waste collection in China - A case study of waste mobile phones. *Journal of Cleaner Production*, *228*, 52–62. https://doi.org/10.1016/j.jclepro.2019.04.216.

Rautela, R., Arya, S., Vishwakarma, S., Lee, J., Kim, K.-H., & Kumar, S. (2021). E-waste management and its effects on the environment and human health. *Science of the Total Environment*, *773*, 145623. https://doi.org/10.1016/j.scitotenv.2021.145623

Regvar, M., Vogel-Mikuš, K., Kugonič, N., Turk, B., & Batič, F. (2006). Vegetational and mycorrhizal successions at a metal polluted site: Indications for the direction of phytostabilisation? *Environmental Pollution*, *144*(3), 976–984. https://doi.org/10.1016/j.envpol.2006.01.036

Shahabuddin, M., Uddin, M. N., Chowdhury, J. I., Ahmed, S. F., Uddin, M. N., Mofijur, M., & Uddin, M. A. (2023). A review of the recent development, challenges, and opportunities of electronic waste (e-waste). *International Journal of Environmental Science and Technology*, *20*(4), 4513–4520. https://doi.org/10.1007/s13762-022-04274-w

Spalvins, E., Dubey, B., Townsend, T., (2008). Impact of electronic waste disposal on lead concentrations in landfill leachate. Environmental Science & Technology 42, 7452–7458.

Sivaramanan, S. (2013) E-Waste Management, Disposal and Its Impact on the Environment. Universal Journal of Environmental Research and Technology, 3, 531–537.

Suman, J., Uhlik, O., Viktorova, J., & Macek, T. (2018). Phytoextraction of Heavy Metals: A Promising Tool for Clean-Up of Polluted Environment? *Frontiers in Plant Science*, https://doi.org/10.3389/fpls.2018.01476

Singh, J., & Lee, B.-K. (2016). Influence of nano-TiO_2 particles on the bioaccumulation of Cd in soybean plants (Glycine max): A possible mechanism for the removal of Cd from the contaminated soil. *Journal of Environmental Management*, *170*, 88–96. https://doi.org/10.1016/j.jenvman.2016.01.015

Tripathi, D. K., Singh, V. P., Prasad, S. M., Chauhan, D. K., & Dubey, N. K. (2015). Silicon nanoparticles (SiNp) alleviate chromium (VI) phytotoxicity in Pisum sativum (L.) seedlings. *Plant Physiology and Biochemistry*, *96*, 189–198. https://doi.org/10.1016/j.plaphy.2015.07.026

Upadhyay, P., & Chakma, S. (2022). Remedial technologies for future waste management. In *Hazardous Waste Management* (pp. 305–322). Elsevier. https://doi.org/10.1016/B978-0-12-824344-2.00003-3

Widmer, R., Oswald-Krapf, H., Sinha-Khetriwal, D., Schnellmann, M., & Böni, H. (2005). Global perspectives on e-waste. *Environmental Impact Assessment Review*, *25*(5), 436–458. https://doi.org/10.1016/j.eiar.2005.04.001.

27 Case Study I

A Short Review on the Impact of Electronic Wastes on Environment, Animal, and Human Health

Ayan Chatterjee
School of Health Science and Technology, Medhavi Skills University, Sikkim, India

Parna Chatterjee and Ashi Ramavat
Rishihood University, Sonipat, India

27.1 INTRODUCTION

The emergence of e-waste is a result of increasing number of populations of a country and purchasing power, which is aided by technological advancement. Anything that has been used only once or is damaged, useless or otherwise undesirable is considered waste. It has been also reporte in earlier literature that, the byproduct of these e waste having an adverse impact to human ad animal health. Therefore, the landfill of these E–waste also increasing day by day, India is no exceptional to it. But there should be a strict guidelines and proper strategies to minimize the chance of adverse impact. This chapter provides a short review on overall impact of WEEE. It also points out the possible strategies for WEEE disposal and management.

27.2 RESEARCH METHODOLOGY

An electronic search has been conducted for the study using the Google, Google Scholar, and PubMed databases. This search was limited to 1996–2021 as a time period. Industrial waste, origins, consumption, GDP, global waste, pollutants, greenhouse gases, pollution, disorders, recycling, and reuse were among the keywords used.

27.3 FINDINGS

Recycling process of these E–waste is not only expensive but it also a severe threat to public health as well as environmental sustainability. E-waste is a threat because the toxic chemicals from electronic and electrical equipment (EEE) end up leaching into groundwater and contaminating food and other products. Another recent study reported that, E-waste byproduct possess a serious health throughout the world, especially in developing countries in Africa, Europe and Asia. Pollution and other environmental impacts account for approximately 23% of deaths in these countries (Lakshmi et al., 2017).

DOI: 10.1201/9781003582311-27

27.4 ELECTRONIC WASTES: THE EMERGING PROBLEM

Both the quickly advancing technology and the rising spending power of consumers are factors contributing to the growing use of digital items. Increased consumption has an impact on the environment in terms of trash produced when products are used and the extraction of raw resources. Regardless of size or purpose, any electrically powered equipment that the user no longer requires might be included in this category. The term "no longer desired" describes how e-waste can arise due to a number of factors, such as consumer behavior or antiquated technology that reduces the useful life of electrical and electronic equipment (EEEs).

27.5 ARE USERS THE ONLY SOURCE OF E-WASTE?

Manufacturing Industries: Manufacturers aren't the only important sources of e-waste. Various e wastes are including motherboards, CRTs, IC chips, and other production-related consumables. It also includes defective PCs that you have purchased as replacements under warranty. Studies show that the majority of PCs sold nationwide are rebuilt with used parts from the secondary market, accounting for about 50% of the total volume. The remaining market share is occupied by Indian brands (20%) and multinationals (30%).

Consumers or the Generators: From Indian households, over 22 percent of garbage computers are produced. Exchanging from shops or giving the outdated computers to friends or family are common methods of getting rid of them. In India almost 25.8% installed PCs are using the business sector. Computers left over from the corporate sector are periodically donated to charitable causes or academic institutions for reuse, or they are commonly sold at auction (Gupta et al., 2011).

27.6 WHAT IS SITUATION OF THE UNIVERSAL E-WASTE PRODUCTION?

The projected 20–50 Mt of e-waste produced globally each year accounts for 1%–3% of the projected 1636 Mt of annual urban waste generation globally. Separately, TVs, PCs, and cell phones will produce 5.5 Mt in 2010 and 2015, respectively. E-waste will make up about 8% of all urban garbage in developed countries (Idrees et al., 2018). The estimated annual contribution of each digital item to the formation of e-waste, E (kg/year), is determined by multiplying the mass, M (kg), by the amount (number of electronic items sold and consumed, N), and by the normal life cycle, L.

$$E = MN / L$$

The amount produced in the EU between 1990 and 1999 fluctuated between 3.3 and 3.6 kg/inhabitant, however the estimates for the years 2000 to 2010 range between 3.9 and 4.3 kg per resident. The computation shows that the population of Europe produces 14 kg/person/year, while Switzerland produces 9 kg/person/year. It is projected that the EU-15 will produce 5.5 Mt per year, whereas the EU-27 will produce 8.3–9.1 Mt per year. China produced 2.5 Mt in 2005, whereas the United States produced about 2.6 Mt. Although the amount of e-waste produced in less developed nations is uncertain, it was estimated that in 2007 Thailand and India produced 0.1 and 0.3 Mt, respectively. Gaidajis et al., 2010). According to the reports are the annual production of PCs (0.78 billion units), cellphones (3.4 billion units), landlines (1.2 billion units), Televisions (1.4 billion units), and radios (2.5 billion units) totals 11.7 Mt. (Robinson, 2009).

27.7 HOW DOES E-WASTE HARM THE ENVIRONMENT?

Since e-waste can harm individuals in two different ways—direct exposure in disposal regions and food chain contamination—environmental circumstances have a substantial impact on human health. Most ecological and physiological research is conducted in developing nations, particularly in Asia, which includes China, India, and Vietnam. Human health will be impacted by the toxic degradation of the air, land, and water caused by ineffective e-waste treatment. A study carried out in Vietnam established that soil at e-waste processing facilities was contaminated with dioxins. The concentration exceeded what the WHO considers to be the maximum permissible concentration due to open combustion and open storage practices (Suzuki et al., 2016).

Sediment samples from a nearby river and soil samples from areas with open combustion pits, e-waste dumping sites, and rice paddies were included (Gogoi et al., 2024), according to a Chinese study. Soil and sediment samples had notable quantities of seven metals: zinc, copper, lead, nickel, tin, calcium, and molybdenum. According to multivariate research, there is a large concentration of components directly related to the processing of e-waste. According to Uchida et al. (2018), research revealed Cu deposition in the soil from open-burning regions, demonstrating the influence of open-burning operations. In four Indian areas (Rampur, Shahjahanpur, Moradabad, and Bareilly), illicit e-waste breakdown and open-air combustion were the causes of high amounts of cadmium in groundwater (Idrees et al., 2018). Because of the overabundance of cadmium poisoning, the groundwater system is extremely toxic. Aquatic life will be impacted by pollutants in open water, which will result in bio-magnification.

27.8 THE HUMAN HEALTH IS SUFFERING AS A RESULT OF E-WASTE

The identification of hazardous metals and other compounds in e-waste exposure has been the subject of numerous researches. A study carried out in South America examined the manual metal-collection practices used by low-income children there and the relationship between those practices and elevated blood lead levels (BLLs) as a result of lead exposure. The BLLs for children and adolescents were found to be greater than the BLLs expected, according to (Pascale et al. 2016). Two investigations tracked exposure in e-waste locations using a variety of indicators. Few studies evaluated residents' susceptibility to phthalic acid esters (PAEs), which are commonly discovered in e-waste recycling facilities, by collecting urine samples from them. Unlike these samples, the reference group consisted of residents in the non-e-waste area. According to (Wang et al. (2018), the results demonstrated that the mPAE levels in the urine of e-waste site residents were significantly higher than those of reference area people. Exposure to heavy metals can have both short-term and long-term consequences, including inflammation, cardiovascular illness, reproductive and respiratory issues, and urinary disease. Moreover, human DNA may be impacted by exposure to e-waste contaminants (Wang et al., 2018). An analysis looking at possible genomic alterations in e-waste recyclers revealed a substantial correlation between DNA damage and the amount of time spent processing the garbage. An analysis of the health of those living in temporary e-waste recycling facilities supports this finding. Furthermore, spontaneous abortions, early births, shorter and lighter births, stillbirths, and lower birth weights are all brought on by exposure to e-waste (Grant et al., 2013).

27.9 ECONOMIC ANALYSIS OF E-WASTE IN INDIAN SOCIETY

The use of electronic components is expanding as a result of falling costs, increased internet usage, and advancements in information and communications technology. Because labor is so cheap in Asian countries like China and India, more than half of the world's produced e-waste is sent for recycling (Kahhat and others, 2008). The electronics industry's rising use of both quantity and toxicity is the cause of India's e-waste rise and the country's high productivity expansion. Every year, India imports more than 50,000 tons of electronic waste. One of the industries in India that is growing the fastest is the manufacturing of electronics and E-waste, which is a result of increased use in the electronics sector. The prediction states that 52 million tons will be produced in India. According to the forecast, India will produce 52 million tons of e-waste by 2020, accounting for almost 40% of the total amount of e-waste produced worldwide (Vaidya, 2016).

Many pollutants are released into the environment by the antiquated recycling techniques. The Control Pollution Control Board (CPCB) officially registers recyclers once a year. There are

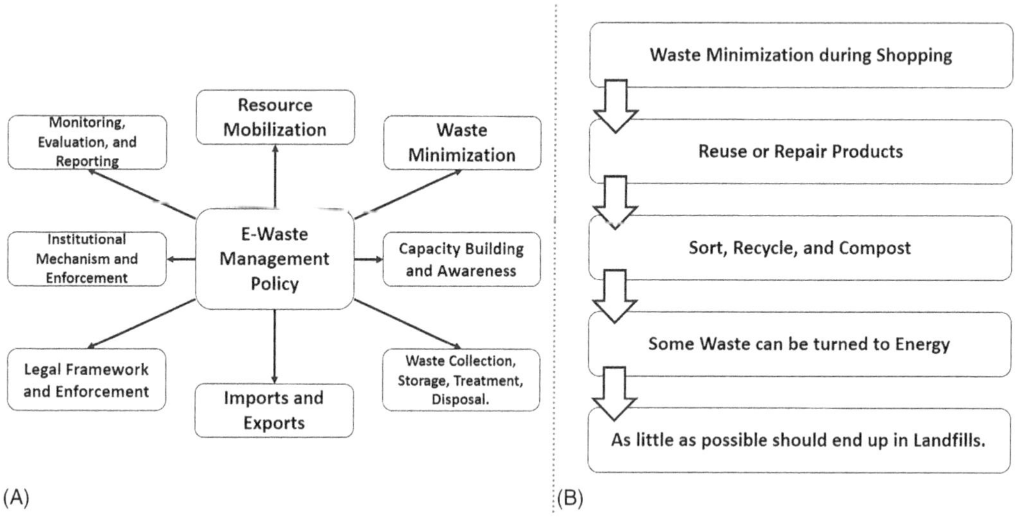

FIGURE 27.1 (A) Strategies necessary to manage E-waste in environment; (B) steps to regulate the production and consumption of E-waste.

currently 23 approved formal recyclers in the nation that handle the produced e-waste through formal procedures. Only a certain quantity of e-waste is provided to these recyclers to manage. India produces the fifth most e-waste globally (Vats and Singh, 2014). E-waste can be drastically decreased by recycling and reusing the products. Reusing and recycling materials helps to reduce energy use and remove dangerous substances from the environment. The different elements of the e-waste management policy are depicted in Figures 27.1 (A) and (B).

27.10 E-WASTE OR A UNIQUE RESOURCE FOR FUTURE NEEDS

When appropriate extraction techniques are utilized, e-waste can produce significant amounts of the valuable fractions that can be reintroduced into production streams. A fraction's resale value is increased by proper processing, which guarantees that the extraction is carried out while maintaining the material's integrity. Processes for recycling and disassembly that are environmentally friendly are crucial for this reason. They can be utilized to stimulate a recycling-based economy, resulting in the creation of jobs and financial gains as well as the promotion of resource preservation.

27.11 START WITH PREVENTION WHEN HANDLING E-WASTES

Profits of size could help India move toward a circular economy, where no resource is wasted. Waste reorganization can benefit the environment even in nations with high rates of recovery and recycling. Unfortunately, there aren't many incentives for limiting and reducing waste in our existing production and consumption patterns. Before using "leftovers" from one process as an input for another, waste prevention must be considered throughout the entire value chain, from material selection to product design and packaging. To advance up the waste hierarchy, all parties involved—consumers, manufacturers, lawmakers, local governments, trash treatment facilities, etc.—must collaborate. When consumers are prepared to separate their waste, recycling will only be possible if the necessary infrastructure is in place for collection. On the other hand, towns can only raise their recycling rates via recycling a smaller proportion of household waste. In the end, how we manage garbage will determine whether it is a resource or a burden.

27.12 CONCLUSION

There are definitely many lessons to be gained from the complexity of India's e-waste situation, as well as from the stark economic differences and diverse regional and cultural environments of the nation. Education and awareness in the area of trash and waste management are becoming increasingly important from the standpoint of global resource management.

REFERENCES

Gaidajis, G., Angelakoglou, K., Aktsoglou, D. (2010). E-Waste: Environmental problems and current management. *Journal of Engineering Science and Technology Review*, 3, 193–199.

Grant, K., Coldizen, F. C., Sly, P. D., Brune, M. B., Neira, M., van den Berg, M., Norman, R. E. (2013). Health consequences of exposure to e-waste: A systematic review. *Lancet Global Health*, 1, 350–361.

Gupta, R., Sangita, Kaur V., (2011). Electronic waste: A case study. *Research Journal of Chemical Sciences*, 1, 49–56.

Idrees, N., Tabassum, B., Abd-Allah, E. F., Hasehm, A., Sarah, R., Hashim, M. (2018). Groundwater contamination with cadmium concentrations in some West U.P. regions, India. *Saudi Journal of Biological Sciences*, 25, 1365–1368.

Kahhat, R., Kim, J., Xu, M., Allenby, B., Williams, E. Zhang, P. (2008). Exploring E-waste management system in the United States. *Resources Conservation and Recycling*, 52, 955–964.

Lakshmi, S., & Raj, A. (2017). A Review study of E-waste management in India. *Asian Journal of Applied Science and Technology (AJAST) Volume*, 1, 33–36.

Pascale, A., Sosa, A., Bares, C., Battocletti, A., Moll, M.J., Pose, D., Laborde, A., Gonzales, H., Feola, G. (2016). E-Waste informal recycling: An emerging source of lead exposure in South America. *Annals of Global Health*, 82, 197–201.

Robinson, B. (2009). E-waste: An assessment of global production and environmental impacts. *Science of the Total Environment*, 408, 183–191.

Gogoi, N., Sarma, A., Choudhury, M., Samanta, P., Faizan, M., & Sadhak, S. (2024). Soil heavy metal pollution and ecological risk assessment in disturbed and undisturbed soil of Morigaon, Assam. *Discover Environment*, 2(1), 13. https://doi.org/10.1007/s44274-024-00039-z

Suzuki, G., Masayuki, S., Matsukami, H., Tue, N.M., Natsuyo, U., Tuyen, L.H., Viet, P.H., Takahashi, S., Tanabe, S., Brouwer, A., Takigami, H. (2016). Comprehensive evaluation of dioxins and dioxin-like compounds in surface soils and river sediments from e-waste processing sites in a village in northern Vietnam: Heading towards the environmentally sound management of e-waste. *Emerging Contaminants*, 2, 98–108.

Uchida, N., Matsukami, H., Someya, M., Tue, N. M., Tuyen, L. H., Viet, P. H., Takahashi, S., Tanabe, S., Suzuki, G. (2018). Hazardous metals emissions from e-waste-processing sites in a village in northern Vietnam. *Emerging Contaminants*, 4, 11–21.

Vaidya, A. (2016). Process and Control management of ever-increasing generation of E-waste. *International Journal of Computer Applications*, 151, 20–22.

Vats, C.M., Singh, S. (2014). Status of e-waste in India- A review. *International Journal of Innovative Research in Science, Engineering and Technology*, 3, 16917–16931.

Wang, Y., Sun, X., Fang, L., Li, K., Yang, P., Du, L., Ji, K., Wang, J. Liu, Q., Xu, C., Li, G., Giesy, J. P., Hecker, M. (2018). Genomic instability in adult men involved in processing electronic waste in Northern China. *Environment International*, 117, 69–81.

28 Electronic Wastes Handling, Impact on Biodiversity and Legal Implications

A Study Related to India

Monruskin M. Calma
Calma Expeditions Inc., Brockton, Massachusetts, United States

Susmita Dhar
Sister Nivedita University, Kolkata, India

Shreya Chakraborty
Independent Researcher, Kolkata, India

28.1 IMPORTANCE OF BIODIVERSITY IN ECOLOGICAL BALANCE

Biodiversity is comprised of genetic diversity, species diversity, and ecosystem diversity which are the main components of life on the Earth. It comes from the word bio that means life and diversity mean variability. It is the variety of all living things including humans, animals, microorganisms, flora, and fauna. Generally, a variety of ecosystem services can be seen as a component of biodiversity. By protecting the total ecosystem, the conservation of biodiversity can be done. Biodiversity is essential for life on Earth and has immense value. For a healthy ecosystem, we need to protect biological diversity so that human health, well-being, economic prosperity, food safety, and security can be protected.

Biodiversity includes the variety of lives of every living organism on this earth, which is the paramount aspect of life and without which there will be no sustenance. Nature's verity and variability of species are included within this term. Biodiversity can be of three types, genetic biodiversity, species biodiversity, and ecological biodiversity. Biodiversity is important to maintain ecological stability. Ecosystem services are maintained to protect and secure the right to life of every living organism. A diverse eco-system with more productivity has a specific role in this planet (Kumar 2016). All species have the right to exist in a beautiful habitat. Preserving biodiversity is therefore very important for maintaining various cultures and spiritual heritage. Biodiversity, if maintained properly through conservation, will balance the ecosystem which also has social cultural, and educational benefits. It is then very important to maintain the right to life, the right to development, food, shelter, and health.

28.2 E-WASTES AND THE DESTRUCTION OF BIODIVERSITY – LEGAL ASPECTS

The Ministry of Environment and Forests (MoEF), Govt. of India is the main department for incorporating, planning, policy, rule, and regulation regarding electronic waste. *The management of e-waste was covered under the Environment and Forest Hazardous Waste (Management and*

DOI: 10.1201/9781003582311-28

Handling) Rules 2008. E-waste (Management and Handling) Rules 2010 under the Environment Protection Act 1986 is also important regarding e-waste addressing and recycling. Recently E-waste Management Rules 2016 and 2011 deal with Extended Producer Responsibility (EPR). The Central Government in exercising the power conferred by sections 6, 8, and 25 of the Environment Protection Act 1986 enacted the e-waste rule 2011 which came into force on 1st May 2012. The rules consist of 18 directives regarding e-waste management, handling, remediation responsibilities for the collection, centres, and recyclers of e-waste. The rules also provide a procedure seeking authorization and registration for handling e-waste. In India, we need a strong legal mechanism to combat the E-waste menace. Green technology can be used for recycling e-waste in India. Technology transfer from developed countries to developing countries is a new way to deal with the problem of waste generation. In India, the Central Pollution Control Board (CPCB) and the State Pollution Control Board (SPCB) are the main authorities in monitoring and ensuring the whole environment by combating pollution under the Environmental Protection Act (EPA) 1986, Water Act 1974, and AIR Act 1981. With the installation of the online effluent monitoring system, the polluting industries are in check by the CPCB and SPCB in a transparent way. Mass awareness programs, training, and advertisement are initiated by the Government of India to strengthen the situation. Action plans, e-office, hazardous waste tracking system, enhancement of IT, surveillance of swages, and various implementation of waste management rules under the EPA 1986 are some major steps to reducing the disposal of hazardous e-waste.

28.3 INTERNATIONAL SCENARIO AND E-WASTE HANDLING

According to the UN, e-waste is any discarded product having toxic and hazardous substances. In 2021 according to United Nations each person on this planet will produce 7.6 kg of e-waste and 57.4 million tons worldwide. United Nations Environment Programme in a report in 2015 estimated world electronic waste worth USD 19 billion is illegally dumped. E-waste is harmful to the environment and the right to life which contains lead, mercury, arsenic, cadmium, etc., which can cause cancer, miscarriage, neurological damage, etc. The US can contribute towards climate change as every device has a carbon footprint which directly causes biodiversity loss. There are so many international treaty conventions among which the Basel convention on hazardous waste is important. E-waste is hazardous waste having transboundary movement. The Basel guidelines are an important set of global policies regarding electronic waste. UN e-waste coalition talks about collaboration, partnership, and supporting the member states to tackle this problem along with International Labor Organization (ILO), International Telecommunication Union(ITU), United Nations Environment Programme(UNEP), United Nations Industrial Development Organization(UNIDO), World Health Organization (WHO), etc. (https://www.downtoearth.org.in/news/waste/india-collected-just-3-e-waste-generated-in-2018-10-in-2019-cpcb-report-75072).

28.4 INDIAN SCENARIO AND E-WASTE HANDLING

In India, actual e-waste data is not properly available. Those that are available are based on models of obsolescence of electric product. The Central Pollution Control Board 2005 estimated 1.347 lakh MT e-waste was generated in the country in 2005. In 2007 total generated e-waste of India were 3,32,979 MT. Mumbai ranked first in generating e-waste followed by Delhi, Bangalore, Chennai, Kolkata, Ahmedabad, Hyderabad, Pune, Surat and Nagpur. In India MOEF, CPCB, SPCB and Director General of Foreign Trade under Water Act 1974 are monitoring and certifying authority of handling e-waste and importing of second hand goods. The Department of Information Technology in the ministry of communication and information technology Government of India is the main ministry for electronic industry. They are striving towards excellence to developed technology solutions for e-waste management withheld of environmentally friendly technology. In India mostly e-waste are unorganized and following primitive measure for recycling and the awareness of public is also an important factor. According to a CPCB report, India only collected just 3% e-waste in 2018 and 10% in 2019 and in 2020 e-waste generation rose 32% to 1,014,961 tons (https://www.downtoearth.org.in/news/waste/india-collected-just-3-e-waste-generated-in-2018-10-in-2019-cpcb-report-75072). India is the 3rd largest producer as per global e-waste generation behind China and USA in 2019 (https://factly.in/india-generates-the-3rd-highest-volume-of-e-waste-even-as-its-per-capita-generation-is-1-3rd-the-global-average/, Arya & Kumar 2020; Awasthi et al., 2018; Gupta & Kumar, 2014).

28.5 E-WASTES AND TERRESTRIAL ECOSYSTEM

This is a land-based community, comprising of biotic and abiotic ingredients in an area. An example of this ecosystem includes tundra, taigas, tropical rain forest, grass land and desert. Forest (an ecosystem dominated by trees and is important to protect and conserve for right to life), desert (largely barren land having less are rainfall), grassland (any area dominated by grasses like prairies, Steppes, Pampas, Savannahs) can be a part of terrestrial ecosystem. The US generation in the environment and the improper handling of e-waste resulting in loss in terrestrial ecosystem worldwide may accelerate climate change. To protect the earth from the bad effects of e-waste, we need clean technology and efficient management to protect the right to life of present and future generations. The Basel Convention was held to tackle hazardous wastes which include e-waste also. The international communities are after e-waste coalition which aims to include commitment by the signatories in building partnership to support the earth and ecosystem from the worse effect from E-waste generation. For the terrestrial ecosystem, the impact of bio-diversity loss from 1970 is an over-burdened issue. Land-use has changed due to bio-diversity loss and extinction of animals, plants and living organisms having great impact on the eco-system services. The unprecedented expansion of urban areas is causing shrinking of forest, wetland, and grassland. Fresh water ecosystems are facing challenges due to pollution, climate change and bio-diversity loss. Anthropogenic activities have a great impact on marine ecosystem and terrestrial waters. The exploitation of sea-based organisms and river-networks are causing harm to the coastal infrastructure and marine eco-system (Babu et al., 2007).

28.6 E-WASTES AND MARINE ECOSYSTEM

An urgent action is needed to protect human beings, flora and fauna and the whole ecosystem including marine ecosystem from the worse effect of E-waste. The amount of E-waste that is generated globally is increasing day by day through research and development. It was revealed that specifically the marine ecosystem is hampered due to E-waste generation. The toxic chemicals that exist in E-waste are fatal for the marine ecosystem. Chemicals like Polychlorinated biphenyls (PCBs), nonylphenol (NP), and triphenyl phosphate (TPPs), etc., from E-waste, contaminate land, ocean, water, and air. E-waste contains a hazardous material and toxic substance that is harmful for public health. When e-waste is not handled properly or is burnt, the following deadly substances (capable of causing a wide range of health disorders and even death) are released: "(*Arsenic, Barium, Beryllium, Brominated flame retardants, Cadmium, Chlorofluorocarbons, Chromium, Dioxins and furans, Lead, Mercury, Polychlorinated biphenyls (PCBs), Polyvinyl chloride (PVC), Selenium*)". According to IDR environmental service already two thirds of aquatic life is considered endangered species. E-waste is not directly dumped in water, but land filled e-waste can affect the marine ecosystem. Sometime lead in the batteries, alloy, etc., can reach marine ecosystem or mercury from cell phone batteries are causing long term threat for marine life. It can reduce the reproductive abilities of fish, altering blood and oxygen exchange. Contaminated plankton, birds, fish might affect humans in the future. Chromium, cadmium, creates hazards to fish, wildlife, and marine areas. Two thirds of the marine species are still known and some of them have yet to be identified. The ocean species are at risk in marine ecosystem and two thirds of them are an endangered species. The toxic e-waste is harming these organisms which can lead to biodiversity loss at a faster rate. An improperly disposed chemical are polluting the marine ecosystem and kills the sea biodiversity flora, fauna, sea bird, fish any other organism. Land filled e-waste can affect the marine wildlife and contaminated water with toxins, hazardous chemicals like lead, mercury, chromium, cadmium, etc., may worsen the problem (https://www.belmont-trading.com/2017/01/examining-the-impact-of-ewaste-on-marine-life/).

28.7 E-WASTES AND CONTAMINATION OF FOOD CHAIN

Electronics waste is solid waste from electronics material. According to the Global E-waste Monitor Report the E-waste in 2019 is around 53.6 million metric tons. E-waste contributed towards climate change, pollution, hampering of environment and food chain and threat towards human health. Toxins from E-waste will affect the environment and greatly impact on human health through the consumption of food from contaminated water. Micro plastic can be seen in several edible fish from the water of Nigeria which is not stomach friendly. The people having no awareness and the biodiversity are lost day by day due to the contamination. When e-waste enters our platter that interferes with our right to life and right to food. According to the UN over fifty million tons of electronics waste are discarded every year. If E-waste is not recycled in a proper way, it will disrupt the food chain. Various studies revealed that the e-waste generation in food stuff is injuries and dangerous. E-waste can be found in rivers, oceans, and land which can easily enter the food chain. The people eating contaminated food are facing health risks. There is incident when arsenic and mercury positioning from diet can lead to death of various people. E-waste generated by civilization every day is polluting the environment and contaminating our food chain. According to the United Nations 50 million tons of electronic waste are discarded yearly. It should be treated properly otherwise it can affect people's health and food chain. The eggs in tuna in the marine ecosystem are sometimes contaminated with heavy metals and dangerous for human consumption. According to the European food safety authority in a slum called Agbogbloshie, if a child a eats a single chicken egg, they will consume 220 times tolerable daily intake of chlorinated dioxins. Fish caught in the waters are also contaminated due to E-waste and the people are forced to consume heavy metals through this food intake. It is a process which can enhance the toxic chemicals within the predators and the entire food chain and the all organisms can be impacted when the animals consume the prey. The toxic chemicals will the reach the animal and bioaccumulate (Mackay & Fraser, 2000). So firstly, the food chain at the bottom like plankton are infected with chemicals eaten by small fish, having the plankton in their system will be eaten by larger fish and then human or animal. It is through this process which the contaminated chemicals and toxins will reach the highest food chain, causing the greatest danger. So, in bioaccumulation the toxic chemicals are accumulated gradually in an organism and in bio magnifications that will reach successively in the higher level of food chain. In recent years it has been revealed that the consumption of seafood is causing cancer. Consumption of toxic metals in sea birds will affect their egg production. Bio magnifications are causing disruption in the natural food chain which will ultimately cause the loss of biodiversity and destruction of coral reefs.

28.8 CONSEQUENCES OF BIODIVERSITY LOSS AND ECOLOGICAL IMBALANCE

Unprecedented E-waste generation can cause harm to ecological diversity and biodiversity loss can be directly attributed to E-waste generation. We need proper handling of electronic waste, to prevent contamination articles in the soil and biodiversity loss. E-waste releases harmful chemicals which are also harmful for human health. If the biodiversity is facing loss due to various manmade occurrences life on the Earth will be at a stake. Biodiversity also has a great impact on the human body, behavior, immunity, nutrition, and resilience of the community (Gogoi et al., 2024). As a result, biodiversity loss can lead to the spreading of various diseases even now a days we are facing Covid 19 which is a direct consequence of biodiversity loss and Wildlife and its impact on Wildlife. Ebola, Zika, NIPAH virus, Dengue, Malaria all infectious diseases are causing and contributing towards health hazard having direct connection to biodiversity. Climate change is the culprit of these days which is causing bio-diversity loss and throwing challenges to humans. The global diversity outlook report of 2020 points out the status of biodiversity in 2020 which stressed on the topic 'humanity is at stake'. Bio-diversity loss is the declining of flora and fauna which causes ecological imbalance (Lucier & Gareau 2019). In 2019, the UN with IPBS published a report on bio-diversity loss. Nature has vital contribution to this planet which comprises of biodiversity and ecosystem services. They are essential for human right to life and also the protection of flora and fauna. Globally, different species are now-a-days due to the loss of biodiversity which is causing threat to ecological balance (https://www.iberdrola.com/sustainability/biodiversity-loss).

28.9 FUTURE CONCERNS AND RECOMMENDATIONS

1. By spreading the awareness about the biodiversity loss and the great impact of it in the ecosystem to common people through environmental education can bring changes.
2. The necessity of sustainable lifestyle to protect right to life we need to support biodiversity as a sensitive issue which can help the life on the Earth to survive.
3. Through proper research and development this menace of E-waste can be tackled through recycling which can help to protect the Mother Earth.
4. Producer responsibility for end of life waste management should be created to protect the ecosystem mitigates the damages.
5. Awareness, stringent regulation for recycling and disposal, decreasing the purchasing demand of the public can help the e-waste generation.
6. To achieve 2050 vision for biodiversity across the globe we need a paradigm shift to shape the nature towards promise, prospect, and progress with sustainability. Proper climate action, sustainable cities, sustainable agriculture, all can contribute towards a better future which is cherished by us in present.
7. Biodiversity is the root cause of sustenance of life in this earth which must be protected by the conservation of nature, its restoration, and sustainability and through fostering structural and transformative change. The international co-operation in local and global arena can really be set as a relevant measure to achieve the goal of 2030. Sustainable development goals are based on these motives to protect the extinction of the species in this beautiful habitat. We need to follow the dictate of the international authorities in the municipal arena with utmost care and caution to protect our bio-diversity.
8. The end of life waste management and producer responsibility should be fixed and regulated with stringent regulation can be a way out to protect biodiversity and ecosystem as well as research and development to mitigate the damage which may produce nontoxic products easy to repair having longer life cycles will help to protect the environment and the world.
9. Though there are hazardous waste rules and e-waste rules under the environment protection Act 1986. India needs a comprehensive exhaustive legislation to handling of e-waste and to save the environment it is need of the hour.
10. Green technology transfer from the developed countries to India by lifting the IPR barrier can be helpful to protect the environment and mitigate climate change and biodiversity loss. Right to life with human dignity in whole some environment is our basic fundamental right which needs to be protected through proper legal mechanism.

28.10 CONCLUSION

Electronic waste has posed a significant challenge to the global north and south. There are multiple aspects of e-waste, and to reduce it, we need recycling to protect Mother Earth in a harmonized way. The consequences of e-waste generation are very vast and affect the right to life of the consumers as well as the global citizens. If not checked now, it can be dangerous for future generations' right to life.

REFERENCES

Arya, S., & Kumar, S. (2020). E-waste in India at a glance: Current trends, regulations, challenges and management strategies. *Journal of Cleaner Production*, Vol. *271*, 122707.

Awasthi, A. K., Wang, M., Wang, Z., Awasthi, M. K., & Li, J. (2018). E-waste management in India: A mini-review. *Waste management & research: the journal of the International Solid Wastes and Public Cleansing Association, ISWA*, Vol. *36*, Issue 5, 408–414. https://doi.org/10.1177/0734242X18767038

Babu, B. R., Parande, A. K., & Basha, C. A. (2007). Electrical and electronic waste: A global environmental problem. *Waste Management & Research*, Vol. *25*, Issue 4, 307–318.

India collected just 3% e-waste generated in 2018, 10% in 2019: CPCB report (2021). https://www.downtoearth.org.in/news/waste/india-collected-just-3-e-waste-generated-in-2018-10-in-2019-cpcb-report-75072

Biodiversity loss: causes, consequences and solutions. (n.d.) https://www.iberdrola.com/sustainability/biodiversity-loss

Examining the impact of eWaste on marine life. (2017). https://www.belmont-trading.com/2017/01/examining-the-impact-of-ewaste-on-marine-life/

Gogoi, N., Sarma, A., Choudhury, M., Samanta, P., Faizan, M., & Sadhak, S. (2024). Soil heavy metal pollution and ecological risk assessment in disturbed and undisturbed soil of Morigaon, Assam. *Discover Environment*, Vol. 2, Article 13. https://doi.org/10.1007/s44274-024-00039-z

Gupta, V., & Kumar, A. (2014). E-waste status and management in India. *Jowurnal of Information Engineering and Applications*, Vol. *4*, Issue 9, 41–48.

Kumar, R. (2016). Current Scenario of e-waste management in India: issues and strategies.

Lucier, C. A., & Gareau, B. J. (2019). Electronic waste recycling and disposal: An overview. *Assessment and Management of Radioactive and Electronic Wastes*, 1–12. https://doi.org/10.5772/intechopen.85983

Mackay, D., & Fraser, A. (2000). Bioaccumulation of persistent organic chemicals: Mechanisms and models. *Environmental pollution*, Vol. *110*, Issue 3, 375–391.

Index

Pages in *italics* refer to figures and pages in **bold** refer to tables.